Muster
Fr. Grössl!

**Artificial Photosynthesis**

*Edited by*
*Anthony F. Collings and Christa Critchley*

*Related Titles from Wiley-VCH*

I. Willner, E. Katz (Eds.)
**Bioelectronics**
From Theory to Applications

2005
ISBN 3-527-30690-9

W. R. Briggs, J. L. Spudich (Eds.)
**Handbook of Photosensory Receptors**
From Theory to Applications

2005
ISBN 3-527-31919-7

C. M. Niemeyer, C. A. Mirkin (Eds.)
**Nanobiotechnology**
Applications and Perspectives

2004
ISBN 3-527-30658-9

V. Balzani et al. (Eds.)
**Electron Transfer in Chemistry**
5 Volumes

2001
ISBN 3-527-29912-6

# Artificial Photosynthesis

From Basic Biology to Industrial Application

*Edited by*
Anthony F. Collings and Christa Critchley

WILEY-VCH Verlag GmbH & Co. KGaA

**Editors:**

**Dr. Anthony F. Collings**
CSIRO Industrial Physics
PO Box 218
Lindfield, NSW 2070
Australia

**Prof. Christa Critchley**
School of Integrative Biology
The University of Queensland
Brisbane, QLD 4072
Australia

■ This book was carefully produced. Nevertheless, editors, authors and publisher do not warrant the information contained therein to be free of errors. Readers are advised to keep in mind that statements, data, illustrations, procedural details or other items may inadvertently be inaccurate.

**Library of Congress Card No.: Applied for**
A catalogue record for this book is available from the British Library

**Bibliographic information published by Die Deutsche Bibliothek**
Die Deutsche Bibliothek lists this publication in the Deutsche Nationalbibliografie; detailed bibliographic data is available in the Internet at http://dnb.ddb.de.

© 2005 WILEY-VCH Verlag GmbH & Co. KGaA, Weinheim

All rights reserved (including those of translation into other languages). No part of this book may be reproduced in any form – by photoprinting, microfilm, or any other means – nor transmitted or translated into machine language without written permission from the publishers. Registered names, trademarks, etc. used in this book, even when not specifically marked as such, are not to be considered unprotected by law.

Printed in the Federal Republic of Germany
Printed on acid-free paper

**Cover Design**  SCHULZ Grafik-Design, Fußgönheim
**Typesetting**  K+V Fotosatz GmbH, Beerfelden
**Printing**  betz-druck GmbH, Darmstadt
**Bookbinding**  Litges & Dopf Buchbinderei GmbH, Heppenheim

**ISBN-13:** 978-3-527-31090-6
**ISBN-10:** 3-527-31090-8

## Foreword

The history of science and technology is full of examples of humans first emulating nature then improving the desired effect far beyond nature's original inspiration. The centenary of human flight in 2003 is a sterling example: the flight of birds and insects inspired a technology that took less than a century to progress from a few seconds of powered flight to commonplace behemoth aircraft that carry hundreds of people at hundreds of kilometers per hour.

In a similar way, science is taking inspiration from one of the most common processes of life on earth to develop a technology to use the same resources and produce some of the same outcomes at rates and scales that far exceed those found in nature. In this field the ubiquitous process is photosynthesis – an ancient process inherent to almost all plants and many prokaryotes on the planet that ultimately enabled the development of earth's animal kingdom.

From a practical perspective, the natural process of photosynthesis is an enviable one. Plants capture the energy from sunlight to combine water with carbon dioxide, producing organic compounds and oxygen. These products enable respiration and growth throughout the biosphere, vital processes in an elegant cycle that has shaped and served life on earth for hundreds of millions of years.

We are at a different starting point in our relationship with plants and photosynthesis than were the pioneers of flight: we already employ more practical knowledge and obtain more practical value from the products of the plant community than did those contemplating powered flight while admiring an eagle soar or a dragonfly hover. But beyond enjoying the fruits of natural photosynthesis comes the prospect of improving upon it. At first glance, the proposition that humans can emulate or excel a process that took hundreds of millions of years for nature to perfect sounds outlandish! But have photosynthetic organisms really perfected the process of capturing solar energy and converting it to other forms of energy? The enzyme combination known as Rubisco – critical to the process of carbon fixation in photosynthesis – is one of the slowest acting enzymes known to science. Where photosynthetic organisms lack efficiency at the molecular scale, they compensate with quantity: Rubisco is also among the most abundant enzymes on the planet.

"Inefficiencies" such as the Rubisco enzyme remind us that research and development inspired by natural processes of photosynthesis will ultimately seek to surpass them. Could better photon capture be developed in the lab? Could we improve upon the plant's conversion of sunlight to energy? Could some algal production of hydrogen from seawater be emulated to scales of practical use? Could we fix carbon dioxide and combine it with water to make organic products of our choice and design? Could we do this in an environmentally friendly, sustainable, and scaleable way?

It is the prospect of improving upon various aspects of photosynthesis that has given birth to the term "artificial photosynthesis." The notions conveyed in this term – clean, readily accessible energy sourced from a clean, daily-replenished source – go straight to the heart of many challenges of the 21st century. From an environmental perspective, replacing the use of carbon-based fossil fuels would obviate the greenhouse effects predicted by climatologists based on current emissions projections. Indeed, if some aspects of artificial photosynthesis actually duplicated the natural formula, $CO_2$ would be actively removed from the atmosphere in the course of energy production. From an entrepreneurial perspective, the new technologies would spawn new industries and thousands of jobs. From a national security perspective, a locally sourced and replenished energy supply would eliminate dependence for oil upon traditional providers whose strategic interests may not always align with those of their customers. Further down this line of thought is the recognition that oil and gas supplies are ultimately finite and that alternative energy sources must be developed in coming decades. Perhaps most importantly from a humanitarian perspective, a local, renewable energy source could bring the basic elements of material comfort – light, heat, energy for appliances – to the billions of people on the planet who now live without the creature comforts our affluent societies take for granted. With widespread improvement in comfort and contentment would come a more stable society overall.

In January 2003, scientists from around the world gathered to discuss the state of the art of technical aspects of artificial photosynthesis in a Boden Research Conference supported by the Australian Department of Education, Science, and Training (DEST). Following these three days of discussions, a U.S.–Australia workshop supported by DEST and the U.S. National Science Foundation explored means of furthering our international collaboration to accelerate progress in the most promising fields.

Technical presentations at these workshops covered the many fields embraced under the "artificial photosynthesis" umbrella. Sessions covered the analysis of natural and model systems for the collection of photons (PS I), natural and artificial systems of conversion of light into energy (PS II), the state of the art of solar cell technology, biological production of hydrogen in the course of photosynthesis, natural and artificial systems of $CO_2$ fixation and sequestration, and the social and political context for developing alternative energy sources. The schedule allowed for ample discussion and networking, and numerous new collaborations germinated in subsequent months.

As the leading nations of the 21st century recognize the need to move beyond dependence upon finite supplies of gas and oil, they are undertaking the first steps toward developing alternative energy sources that could be used on a national scale. The following chapters outline the first steps of research in a field that we offer in service to this energy transformation.

*Miriam Baltuck*
U.S. Embassy Canberra Science Advisor

*Peter McGauran*
Australian Government Minister

# Preface

**Why Artificial Photosynthesis?**

The accepted, but by no means uncontested, scientific view is that the excessive use of fossil fuels has resulted in the greenhouse effect, or global warming, as a consequence of a buildup of $CO_2$ in the earth's atmosphere. These fossil fuels and the cheap energy that they have provided have powered the industrial development of the last two centuries. Regardless of whether or not one believes that global warming is a real effect and is directly linked to increasing atmospheric carbon dioxide, fossil fuel resources are finite. It is now a question of when rather than if these resources will run out, and the gloomier predictions are that it will be sooner rather than later.

For developed and developing nations alike, this represents a confronting need for change in the patterns of energy consumption with concomitant social, political, and economic consequences. The Kyoto Protocol on Global Warming calling for significant reduction in greenhouse gas emissions has hardly had uniform endorsement or acceptance, but one lesson to be learned from Kyoto is that moral rectitude will not be a sufficient incentive to produce the required revolution in energy use. The doubters and the economic rationalists will have to be convinced by economic arguments, and the necessary social and engineering changes will have to be paid for by generating new products. However, these economic arguments and the urgency for remedial action and investment in solutions will change over time.

Artificial or bioengineered photosynthesis represents one possible solution to the problems of global warming. The concept is to mimic the light (photon collection) and dark (energy conversion and $CO_2$ capture) processes of natural photosynthesis that produce energy (electricity and hydrogen) and biopolymers (including food) with high efficiency. This approach may lead to bioanalogue systems for (1) nonpolluting electricity generation, (2) photohydrogen generation, (3) alternative carbon products, and (4) carbon dioxide sequestration, all using solar energy. These artificial photosynthetic technologies could prove especially important because their implementation will reduce our greenhouse gas emissions and water requirements, and reduce or even eliminate our dependence on fossil fuels.

Major social change is inevitable because of the conflict of depleting fossil fuel resources and the loss of industries associated with their use, while increas-

*Artificial Photosynthesis: From Basic Biology to Industrial Application*
Edited by Anthony F. Collings and Christa Critchley
Copyright © 2005 WILEY-VCH Verlag GmbH & Co. KGaA, Weinheim
ISBN: 3-527-31090-8

ing demand for energy per capita remains a measure of economic progress. However, an appropriately staged replacement of fossil fuel with alternative energy sources does not have to threaten the livelihoods of people.

The community must be engaged and their support elicited to ensure that these changes are introduced harmoniously. The public also must be made aware of the challenges to be confronted and, more importantly, must share in the excitement of the scientific discoveries and the possibilities offered by these new technologies, much as they did in the 1960s with Kennedy's challenge of putting a man on the moon. There is plenty of evidence that the community will adopt new technology when it is useful and improves comfort and lifestyle, to wit public adoption/embrace of the home computer, mobile phones, the Internet, and e-mail. On the other hand, public resistance to GM crops provides an example of why the public must be part of the decision-making process. For this reason we begin the book with Ian Lowe's discussion of the social and political issues involved in the implementation of artificial photosynthesis. Many of these issues are discussed in an Australian context, but the findings are likely to be universal. Lowe advocates the tenets of sustainability science for assessing the efficacy of a conventionally based, scientific solution to global warming and the replacement of fossil fuels such as proposed here by artificial photosynthesis.

Artificial photosynthesis is about learning from nature, "stealing nature's secrets" as it were, and not necessarily an attempt to better nature. This is implicit in the comprehensive and integrated model of artificial photosynthesis presented by Ron Pace in Chapter 2. An interesting feature of the model is that it is based on living organisms but that no living organisms as such are utilized. The four bio-analogue systems are described in some detail: organic photovoltaics, hydrogen generation, bioenergetic conversion, and $CO_2$ fixation. Not only does the Pace model present a unique and articulated view of artificial photosynthesis, it also provides a context with respect to which we can examine various pieces of research that come under the umbrella of artificial photosynthesis.

The other chapters in this book describing research in artificial photosynthesis are grouped into four sections that parallel the processes of natural photosynthesis: photon energy capture and conversion, photovoltaic current generation, photoproduction of hydrogen, the fuel of the next century, and carbon dioxide use.

The chapters in Section 2, Capturing the Sun/Sunlight Energy, describe new systems for the capture of the sun's energy and its conversion into charge and current. In keeping with the theme of "stealing nature's secrets," these chapters elucidate the physics underlying natural photosynthesis. The skin pigment melanin evolved to protect the body against the damaging rays of the sun. Paul Meredith and his coworkers exploit its absorptive properties to harvest the photons and trap their energy to generate currents via a photoelectrochemical Grätzel cell using nanocrystalline titania as the semiconductor. Optical spectroscopic studies on the redox-reactive chromophores of photosystem II are described in the chapter by Elmars Krausz and Sindra Årsköld. They have discovered major variability in the electronic structure of P680 from different organisms that may be important for the design of chlorophyll-based artificial photovoltaic systems. Jeff Reimers and Noel Hush present theoretical calculations that describe a four-state model of the primary charge separation in the reaction center of photosystem II. Their model correctly describes the electrical properties of the reaction center chlorophyll radical cation, an important feature also for the design of chlorophyll-based artificial photovoltaic systems. Reza Razeghifard and Tom Wydrzynski have succeeded in making artificial reaction centers with synthetic peptides that incorporate the redox pigments and cofactors, and they foreshadow the construction of membrane assemblies that will mimic natural systems.

Any artificial photovoltaic structure will require matrices and scaffolds/frameworks to which the active chromophores are attached. Ron Warrener's polynorbornane structures are one such scaffold (molrac) that could accommodate the pigments and cofactors required for photon-energy-capturing chromophores. The detailed chemical composition and geometry of the norbornene building blocks are discussed in Chapter 8. Alternative scaffolds are Les Dutton's marquettes – synthetic proteins that can harbor electron transfer components and are easily attached to gold surfaces – and porphyrin arrays, which have been studied by several researchers.

To open the third section, Feeding the Grid from the Sun, Martin Green presents a review of the current state of the silicon photovoltaics industry and discusses his third-generation, high-efficiency silicon photovoltaic devices. The Arizona State University group, the Center for the Study of Early Events in Photosynthesis, has made many important contributions to the physics of photosynthesis. Gust, Moore, and Moore describe in Chapter 10 how artificial molecular systems that mimic bacterial photosynthetic energy conversion can be designed and synthesized.

The fuel of the future is hydrogen, and its production using biological systems and energy from the sun is a major technological goal of the research efforts in developed nations, in particular in the U.S. and Europe. This is the theme of the fourth section of the book. Much progress has been made in the U.S. National Renewable Energy Laboratory with a major program using green algae for the production of hydrogen. The major achievement coming from the Maria Ghirardi–Mike Seibert team is the temporal separation of $O_2$ and hydrogen production because of the $O_2$ sensitivity of the hydrogenase. They use two approaches: sulfur deprivation to reduce the rate of $O_2$ production and molecular engineering of the hydrogenase enzyme to make it more $O_2$ tolerant. Tasso Melis and his coworkers at Berkeley have succeeded in generating green algae with truncated chlorophyll antennae, thereby reducing the effective photon energy absorption and dissipation and improving solar energy conversion efficiency. In Chapter 12, Melis summarizes the state of the art in getting better irradiation of cells deeper in high-density microalgal cultures.

High-energy carbohydrates are the principal product of the natural photosynthesis process, and $CO_2$ is the carbon source for this metabolic/synthetic activity. More than 200 billion tonnes of $CO_2$ annually are consumed in natural photosynthesis; therefore, there is ample raw material available for exploitation via artificial photosynthesis to deliver end products. The final section of the book is devoted to research into more productive or more efficient utilization of atmospheric carbon dioxide. Ribulose bisphosphate carboxylase/oxygenase (Rubisco), the most abundant protein in the world, is the primary carbon-fixing enzyme. John Andrews and Spencer Whitney have used transgenic modification of Rubisco in an attempt to make this enzyme less imperfect: it could support a given photosynthesis rate with 86% less water, 35% less light, and 99% less protein investment in the enzyme. In Chapter 14, Jill Gready addresses the inefficiency of the enzyme, its poor selectivity, and its wasteful fixing of $O_2$ and asks whether this is a failure of evolutionary adaptation. She presents computer simulation studies of the $Mg^{2+}$ binding site that augment experimental studies of the catalytic mechanism and key protein groups. The tantalizing prospect implicit in Chapters 13 and 14 is the reengineering of Rubisco. Tom Sharkey discusses the three major classes of compounds that might be targeted as realistic carbon-based end products of artificial photosynthesis: carbohydrates (sugars), oils, and terpenes. The ability of Cyanobacteria PCC7002 to fix $CO_2$ in a continuous stirred bioreactor has been studied by Dilip Desai and his group. He reports on the development of a mathematical model that agrees well with their experimental observations. Since $CO_2$ emissions are already higher than is generally considered desirable and continue to grow, there is a strong engineering argument that we must "buy time" while awaiting the development of options like artificial photosynthesis and renewable energy. Geological sequestration of $CO_2$ – particularly in the ocean or in oil or natural gas basins, which maintain the $CO_2$ at super-critical pressure – is under evaluation in several countries, including Australia, and is already in place in Norway and North America. Peter Cook reports on what might be only a temporary solution but which does permit breathing room.

This book is comprised of work presented at the Boden Research Conference on Artificial Photosynthesis, which was held 9–12 January 2003 in Sydney. The conference was organized by the Australian Artificial Photosynthesis Network (AAPN), a collective of more than 40 Australian scientists that was formed in September 1999 to promote, encourage, and coordinate research on artificial photosynthesis. The AAPN owes its inception to Dr. John Lowke, former Chief of the CSIRO Division of Applied Physics, and received inspiration and encouragement from the late Sir Rutherford Robertson.

Just a few years back, it would have been appropriate to ask if artificial photosynthesis could be a reality. Now, one might better frame the question as "when" rather than "if." The various chapters of this book may be the initial steps along what might be a long and difficult path. However, the journey has started, even if the difficulties in making artificial photosynthesis a reality are formidable. It remains to be seen whether the approaches presented here will be among the technologies that will ultimately contribute to an economic and practicable reproduction of one of nature's greatest miracles.

## Acknowledgments

This book is dedicated to the memory of Sir Rutherford (Bob) Robertson, for whom artificial photosynthesis was to be the last of his many contributions to Australian and world science. He actively supported the early meetings of the Australian Artificial Photosynthesis Network and, in particular, the vision of "dry" agriculture as a major benefit of artificial photosynthesis. Bob was a mentor to many in the photosynthesis community and an inspiration to all who met him.

The book is comprised of work presented at the Boden Research Conference on Artificial Photosynthesis, which was held 9–12 January 2003 in Sydney under the auspices of the Australian Academy of Science and with contributions from PS2001 Pty. Ltd., the Australian National University, and CSIRO Industrial Physics. The Boden Research Conference was followed by a U.S.–Australia Bilateral Workshop on Artificial Photosynthesis supported by the Department of Education, Science, and Training (Australia) and the National Science Foundation (USA).

We would also like to thank the U.S. Ambassador to Australia, His Excellency, Mr. Thomas Schieffer, for his support and encouragement of this work. Finally, to all of our colleagues in the AAPN, thank you for your whole-hearted support of the project.

July 2005

*Tony Collings*
*Christa Critchley*

# Contents

**Foreword**  V

**Preface**  IX

**List of Contributors**  XXIII

**Part I**  **The Context**  1

**1**  **Artificial Photosynthesis: Social and Political Issues**  3
*Ian Lowe*

1.1  Introduction  3
1.2  The Need for a Transition to Artificial Photosynthesis  4
1.3  Some Associated Social and Political Issues  6
1.4  Using the Available Photons: Towards Sustainability Science  9
1.5  Conclusions  11
    *References*  11

**2**  **An Integrated Artificial Photosynthesis Model**  13
*Ron J. Pace*

2.1  Introduction  13
2.2  Natural Photosynthesis  13
2.3  Artificial Photosynthesis: An Integrated Strategy  17
2.4  A Technological Approach to Photosynthesis  19
2.5  Program 1: Biomimetic Photoelectric Generation  20
2.5.1  Milestones  24
2.6  Program 2: Electrolytic Hydrogen  24
2.6.1  Milestones  28
2.7  Programs 3 and 4: Waterless Agriculture  28
2.7.1  Program 3: Bioenergetic Converters  29
2.7.1.1  Milestones  30
2.7.2  Program 4: The $CO_2$-fixing Enzyme Reactor  31
2.7.2.1  Milestones  32
2.8  Conclusions  33
    *References*  33

| Part II | **Capturing Sunlight** 35 |
|---|---|

| 3 | **Broadband Photon-harvesting Biomolecules for Photovoltaics** 37 |
|---|---|
| | *Paul Meredith, Ben J. Powell, Jenny Riesz, Robert Vogel, David Blake, Indriani Kartini, Geff Will, and Surya Subianto* |
| 3.1 | Introduction 37 |
| 3.2 | The Photoelectrochemical Grätzel Cell (Dye-sensitized Solar Cell) 39 |
| 3.3 | Typical Components and Performance of a DSSC 41 |
| 3.3.1 | Construction and Mode of Operation 41 |
| 3.3.2 | Typical DSSC Performance 45 |
| 3.3.3 | Device Limitations 47 |
| 3.4 | Melanins as Broadband Sensitizers for DSSCs 48 |
| 3.4.1 | Melanin Basics 48 |
| 3.4.2 | Melanin Chemical, Structural, and Spectroscopic Properties 50 |
| 3.4.3 | Melanin Electrical and Photoconductive Properties 58 |
| 3.4.4 | Melanins as Broadband Photon-harvesting Systems 61 |
| 3.4.5 | A DSSC Based Upon Synthetic Eumelanin 62 |
| 3.5 | Conclusions 63 |
| | *References* 64 |

| 4 | **The Design of Natural Photosynthetic Antenna Systems** 67 |
|---|---|
| | *Nancy E. Holt, Harsha M. Vaswani, and Graham R. Fleming* |
| 4.1 | Introduction 67 |
| 4.2 | Confined Geometries: From Weak to Strong Coupling and Everything in Between 68 |
| 4.2.1 | Conventional Förster Theory: B800 to B800 Intra-band Energy Transfer 69 |
| 4.2.2 | Generalized Förster Theory: B800 to B850 Inter-band Energy Transfer 69 |
| 4.2.3 | Generalized Förster Theory with the Transition Density Cube Method: Car to Bchl Inter-pigment Energy Transfer 70 |
| 4.2.4 | Modified Redfield Theory: Intra-band B850 Exciton Dynamics 72 |
| 4.3 | Energetic Disorder Within Light-harvesting Complexes 73 |
| 4.3.1 | From Isolated Complexes to Membranes: Disorder in LH2 73 |
| 4.3.2 | Photosystem I 75 |
| 4.4 | Photochemistry and Photoprotection in the Bacterial Reaction Center 78 |
| 4.5 | The Regulation of Photosynthetic Light Harvesting 79 |
| 4.6 | Concluding Remarks 83 |
| | *References* 83 |

| 5 | Identifying Redox-active Chromophores in Photosystem II by Low-temperature Optical Spectroscopies  87 |
|---|---|
| | Elmars Krausz and Sindra Peterson Årsköld |

| 5.1 | Introduction  87 |
|---|---|
| 5.2 | Experimental Methods  89 |
| 5.2.1 | Sample Preparation  89 |
| 5.2.2 | Illumination  90 |
| 5.2.3 | Spectra  90 |
| 5.3 | Results and Discussion  91 |
| 5.3.1 | Absorption and CD Signatures: Plant PSII Cores and BBYs  91 |
| 5.3.2 | Absorption and CD Signatures: Plant and Cyanobacterial PSII Cores  94 |
| 5.3.3 | Absorption Signatures: The Native and Solubilized Reaction Center  94 |
| 5.3.4 | MCD Signatures: P680 and $Chl_Z$  96 |
| 5.3.5 | Electrochromic Signature: $Pheo_{D1}$ in Active PSII  99 |
| 5.4 | Conclusions  103 |
| 5.4.1 | Low-temperature Precision Polarization Spectroscopies  103 |
| 5.4.2 | Signatures of P680 and $Chl_Z$  103 |
| 5.4.3 | Electrochromism Signature of $Pheo_{D1}$  104 |
| 5.4.4 | Coupling and Robustness in P680 and Biomimetic Systems  104 |
| | References  105 |

| 6 | The Nature of the Special-pair Radical Cation Produced by Primary Charge Separation During Photosynthesis  109 |
|---|---|
| | Jeffrey R. Reimers and Noel S. Hush |

| 6.1 | Introduction  109 |
|---|---|
| 6.2 | The Special Pair  109 |
| 6.3 | The Hole-transfer Band  113 |
| 6.4 | Initial Investigations of the Hole-transfer Band  116 |
| 6.5 | Identification of the SHOMO to HOMO Band  118 |
| 6.6 | Full Spectral Simulations Involving all Bands  119 |
| 6.7 | Predicting Chemical Properties Based on the Spectral Analysis  121 |
| 6.8 | Conclusions  125 |
| | References  125 |

| 7 | Protein-based Artificial Photosynthetic Reaction Centers  127 |
|---|---|
| | Reza Razeghifard and Thomas J. Wydrzynski |

| 7.1 | Introduction  127 |
|---|---|
| 7.2 | Natural Reaction Centers  127 |
| 7.2.1 | Structure and Function  127 |
| 7.2.2 | Creation of a Charge-separated State  129 |
| 7.2.3 | Mutational Studies  129 |

| 7.3 | Synthetic and Semi-synthetic Reaction Centers  *130* |
|---|---|
| 7.3.1 | Multi-layered Films  *131* |
| 7.3.2 | Synthetic Reaction Centers  *132* |
| 7.3.2.1 | Electron Acceptor  *134* |
| 7.3.2.2 | Electron Donor  *136* |
| 7.3.2.3 | Photocatalyst: Photoactive Peptides  *138* |
| 7.4 | Perspective  *140* |
|  | References  *141* |

**8  Novel Geometry Polynorbornane Scaffolds for Chromophore Linkage and Spacing  *147***
*Ronald N. Warrener, Davor Margetic, David A. Mann, Zhi-Long Chen, and Douglas N. Butler*

| 8.1 | Introduction  *147* |
|---|---|
| 8.2 | Results and Discussion  *151* |
| 8.2.1 | Reaction at Carbonyl Groups to Form Unsymmetrical Type III Dyads  *151* |
| 8.2.2 | Extended-frame Dyads  *154* |
| 8.3 | Preliminary Results  *155* |
| 8.3.1 | The Use of Multicarbonyl Reagents for Dyad Formation  *155* |
| 8.4 | Conclusions  *157* |
| 8.5 | Dyad Nomenclature  *158* |
|  | References  *165* |

**Part III  Feeding the Grid from the Sun  *167***

**9  Very High-efficiency in Silico Photovoltaics  *169***
*Martin A. Green*

| 9.1 | Introduction  *169* |
|---|---|
| 9.2 | Silicon Wafer Approach  *171* |
| 9.3 | Thin-film Approaches  *173* |
| 9.4 | Third-generation Technologies  *178* |
| 9.5 | Conclusions  *183* |
|  | References  *184* |

**10  Mimicking Bacterial Photosynthesis  *187***
*Devens Gust, Thomas A. Moore, and Ana L. Moore*

| 10.1 | Introduction  *187* |
|---|---|
| 10.2 | Natural Photosynthesis  *188* |
| 10.3 | Artificial Photosynthesis  *190* |
| 10.3.1 | Artificial Antenna Systems  *190* |
| 10.3.2 | Artificial Reaction Centers  *194* |
| 10.3.3 | Antenna–Reaction Center Complexes  *199* |
| 10.3.4 | Transmembrane Proton Pumping  *201* |

| | | |
|---|---|---|
| 10.3.5 | Synthesis of ATP  204 | |
| 10.3.6 | Transmembrane Calcium Transport  206 | |
| 10.4 | Conclusions  208 | |
| | References  209 | |

**Part IV    Photohydrogen**  *211*

**11    Development of Algal Systems for Hydrogen Photoproduction: Addressing the Hydrogenase Oxygen-sensitivity Problem**  *213*
*Maria L. Ghirardi, Paul King, Sergey Kosourov, Marc Forestier, Liping Zhang, and Michael Seibert*

| | |
|---|---|
| 11.1 | Introduction  213 |
| 11.2 | Sulfur Deprivation and Hydrogen Photoproduction  214 |
| 11.2.1 | Background  214 |
| 11.2.2 | Model of the Interactions Between Different Metabolic Pathways in Sulfur-deprived Cells  215 |
| 11.2.3 | Confirmation of the Model  217 |
| 11.2.4 | Limiting Factors for $H_2$ Photoproduction under Sulfur Deprivation  218 |
| 11.2.5 | Mechanism of Regulation  220 |
| 11.3 | Molecular Engineering of the Algal Hydrogenase  221 |
| 11.3.1 | Algal Hydrogenases and $H_2$ Production  221 |
| 11.3.2 | Cloning and Sequencing of the Two *C. reinhardtii* [FeFe]-Hydrogenases  221 |
| 11.3.3 | Anaerobic Expression of the two *C. reinhardtii* Hydrogenases  223 |
| 11.3.4 | Oxygen Inhibition of Hydrogenase Activity and Molecular Engineering for Increased $O_2$ Tolerance  224 |
| | References  226 |

**12    Bioengineering of Green Algae to Enhance Photosynthesis and Hydrogen Production**  *229*
*Anastasios Melis*

| | |
|---|---|
| 12.1 | Introduction  229 |
| 12.2 | Rationale and Approach  230 |
| 12.3 | Physiological State of the Chl Antenna Size in Green Algae  231 |
| 12.4 | The Genetic Control Mechanism of the Chl Antenna Size in Green Algae  232 |
| 12.5 | Effect of Pigment Mutations on the Chl Antenna Size of Photosynthesis  233 |
| 12.6 | Genes for the Regulation of the Chl Antenna Size of Photosynthesis  235 |
| 12.7 | Conclusions  237 |
| | Acknowledgements  237 |
| | References  237 |

| Part V | The Carbon Connection *241* | |
|---|---|---|
| 13 | **Manipulating Ribulose Bisphosphate Carboxylase/Oxygenase in the Chloroplasts of Higher Plants** *243*<br>*T. John Andrews and Spencer M. Whitney* | |
| 13.1 | Introduction *243* | |
| 13.2 | Why Manipulate Rubisco in Plants? *243* | |
| 13.2.1 | Genetic Manipulation of Higher-plant Rubisco Is Now Feasible *243* | |
| 13.2.2 | The Advantages of "Ecological" Studies of Rubisco "at Home" in Its Physiological Context *244* | |
| 13.2.3 | A Compelling Example of Genome–Phenome Interactions *244* | |
| 13.2.4 | An Improvement in the Resource-use Efficiency of Photosynthesis? *245* | |
| 13.3 | What Constitutes an Efficient Rubisco? *245* | |
| 13.3.1 | Key Kinetic Parameters *245* | |
| 13.3.2 | Physiological Consequences of Rubisco Efficiency *246* | |
| 13.3.3 | Regulatory Properties *247* | |
| 13.3.4 | Evolution of Rubisco Efficiency *248* | |
| 13.4 | How to Find a Better Rubisco? *248* | |
| 13.4.1 | In Nature? *248* | |
| 13.4.2 | By Rational Design? *248* | |
| 13.4.3 | By in Vitro Evolution? *249* | |
| 13.5 | How to Manipulate Rubisco in Plants? *250* | |
| 13.5.1 | Nuclear Transformation *250* | |
| 13.5.2 | Plastid Transformation *252* | |
| 13.6 | What Have We Learned So Far? *252* | |
| 13.6.1 | Both Nuclear and Plastidic Genomes Are Able to Express Both *rbc*L and *Rbc*S Genes *252* | |
| 13.6.2 | Photosynthesis and Growth Can Be Supported by a Foreign Rubisco *254* | |
| 13.6.3 | The Properties of a Mutated or Foreign Rubisco Are Reflected in the Leaf's Gas-exchange Properties *254* | |
| 13.6.4 | The Requirements for Folding and Assembly of the Subunits of Red-type, Form-I Rubisco Are Not Accommodated in Chloroplasts *255* | |
| 13.6.5 | A Better Strategy for Directed Mutagenesis of *rbc*L *256* | |
| 13.6.6 | Subunit Hybrids Can Be Formed *in vivo* *256* | |
| 13.7 | Priorities for Future Manipulation of Rubisco *in vivo* *257* | |
| 13.7.1 | The Structural Foundations of Efficient Properties *257* | |
| 13.7.2 | Regulation of Rubisco Gene Expression *258* | |
| 13.7.3 | Folding and Assembly of Rubisco Subunits *258* | |
| 13.8 | Conclusions *259*<br>References *260* | |

| 14 | **Defining the Inefficiencies in the Chemical Mechanism of the Photosynthetic Enzyme Rubisco by Computational Simulation** 263 |
|---|---|
| | *Jill E. Gready* |
| 14.1 | Introduction 263 |
| 14.1.1 | Catalytic Inefficiencies 263 |
| 14.1.2 | Evolutionary Constraints? 264 |
| 14.1.3 | Experimental Limitations 265 |
| 14.1.4 | Goals of Simulations 265 |
| 14.1.5 | Simulation Options 266 |
| 14.2 | Computational Methods 267 |
| 14.2.1 | Computational Programs 267 |
| 14.2.2 | Enzyme Models 268 |
| 14.2.3 | Active-site Fragment Complexes 268 |
| 14.2.4 | QM/MM Simulations 269 |
| 14.3 | Results and Discussion 271 |
| 14.3.1 | Fragment-complex Calculations 271 |
| 14.3.1.1 | Enolization Step 271 |
| 14.3.1.2 | Carboxylation Step 273 |
| 14.3.1.3 | Hydration Step 275 |
| 14.3.1.4 | Sequential Addition of $CO_2$ and $H_2O$ 275 |
| 14.3.1.5 | Alternative Conformations of the *Gem*-diol 275 |
| 14.3.1.6 | C2-C3 Bond Cleavage: Pathway I 276 |
| 14.3.1.7 | C2-C3 Bond Cleavage: Pathway II 276 |
| 14.3.1.8 | Protonation of C2 276 |
| 14.3.1.9 | Dissociation of Products 277 |
| 14.3.2 | Summary of Main Findings 277 |
| 14.3.3 | QM/MM+MD Calculations 277 |
| 14.3.3.1 | $CO_2$ Addition: Early vs. Late Protonation of the Carboxylate 278 |
| 14.3.3.2 | Hydration of the $\beta$-Keto Acid 280 |
| 14.3.3.3 | His294 Protects Intermediates from Decarboxylation 280 |
| 14.3.3.4 | The Tightly Coupled Active-site Environment 280 |
| 14.4 | Conclusions 281 |
| | References 281 |
| | |
| 15 | **Carbon-based End Products of Artificial Photosynthesis** 283 |
| | *Thomas D. Sharkey* |
| 15.1 | Introduction 283 |
| 15.2 | What Are the End Products of Plant Chloroplast Photosynthesis? 284 |
| 15.3 | Does End-product Synthesis Ever Limit Photosynthesis? 285 |
| 15.4 | What Would Be a Desirable Carbon-based End Product of Photosynthesis? 286 |
| | References 289 |

**16  The Artificial Photosynthesis System: An Engineering Approach**  *291*
*Dilip K. Desai*

16.1 Introduction  *291*
16.2 Engineering Approach to APS  *291*
16.3 Elements of the Engineering Approach  *292*
16.3.1 Economic Value  *292*
16.3.2 Limitations of Natural Photosynthesis Systems (NPS)  *292*
16.3.2.1 Speed of NPS  *292*
16.3.2.2 Energy Efficiency of NPS  *292*
16.3.2.3 Water Requirements of NPS  *293*
16.3.2.4 Land Use for NPS  *293*
16.3.3 Scale of Operation  *293*
16.3.4 Functional Specification  *294*
16.4 Elements of Envisaged System  *294*
16.5 Cyanobacteria  *295*
16.6 Photo-bioreactor  *296*
16.7 Theory  *296*
16.8 Results  *298*
16.9 Conclusions  *299*
*References*  *299*

**17  Greenhouse Gas Technologies: A Pathway to Decreasing Carbon Intensity**  *301*
*Peter J. Cook*

17.1 Introduction  *301*
17.2 $CO_2$ Capture  *301*
17.3 Storing $CO_2$  *303*
17.4 Australian Initiatives: Capture and Storage Technologies  *306*
17.5 Conclusions  *307*
*References*  *308*

**Subject Index**  *309*

# List of Contributors

**T. John Andrews**
Australian National University
Molecular Plant Physiology
Research School of Biological Sciences
PO Box 475
Canberra, ACT 2601
Australia

**Sindra Peterson Årsköld**
Lund University
Department of Biochemistry
PO Box 124
22100 Lund
Sweden

**Miriam Baltuck**
U.S. Embassy
Scientific Advisor
Canberra, ACT 2600
Australia

**David Blake**
The University of Queensland
Department of Physics
Brisbane, QLD 4072
Australia

**Douglas N. Butler**
Central Queensland University
Center for Molecular Architecture
Rockhampton, QLD 4702
Australia

**Zhi-Long Chen**
Central Queensland University
Center for Molecular Architecture
Rockhampton, QLD 4702
Australia

**Anthony F. Collings**
CSIRO Industrial Physics
PO Box 218
Lindfield, NSW 2070
Australia

**Peter J. Cook**
Cooperative Research Center
for Greenhouse Gas Technologies
PO Box 463
Canberra, ACT 2601
Australia

**Christa Critchley**
The University of Queensland
School of Integrative Biology
Brisbane, QLD 4072
Australia

**Dilip K. Desai**
CSIRO Manufacturing
and Infrastructure Technology
Graham Road
Highett, Victoria 3190
Australia

*Artificial Photosynthesis: From Basic Biology to Industrial Application*
Edited by Anthony F. Collings and Christa Critchley
Copyright © 2005 WILEY-VCH Verlag GmbH & Co. KGaA, Weinheim
ISBN: 3-527-31090-8

## List of Contributors

**Graham R. Fleming**
University of California, Berkeley
Department of Chemistry
Berkeley, CA 94720-1460
USA

**Marc Forestier**
University of Zurich
Plant Biology Department
Limnological Station
8802 Kilchberg
Switzerland

**Maria L. Ghirardi**
National Renewable Energy Laboratory
1617 Cole Blvd.
Golden, CO 80401
USA

**Jill E. Gready**
Australian National University
John Curtin School of Medical Research
PO Box 334
Canberra, ACT 2601
Australia

**Martin A. Green**
University of New South Wales
Centre of Excellence for Advanced Silicon Photovoltaics and Photonics
Sydney, NSW 2052
Australia

**Devens Gust**
Arizona State University
Department of Chemistry and Biochemistry
Tempe, AZ 85287-1604
USA

**Nancy E. Holt**
University of California, Berkeley
Department of Chemistry
Berkeley, CA 94720-1460
USA

**Noel S. Hush**
The University of Sydney
School of Chemistry and School of Molecular and Microbial Biosciences
Sydney, NSW 2006
Australia

**Indriani Kartini**
The University of Queensland
Department of Chemical Engineering
Brisbane, QLD 4072
Australia

**Paul King**
National Renewable Energy Laboratory
1617 Cole Blvd.
Golden, CO 80401
USA

**Sergey Kosourov**
Russian Academy of Science, Pushchino
Institute for Basic Biological Problems
Moscow 142290
Russia

**Elmars Krausz**
Australian National University
Research School of Chemistry
Canberra, ACT 0200
Australia

**Ian Lowe**
Griffith University
School of Science
Nathan, QLD 4111
Australia

**David A. Mann**
Central Queensland University
Center for Molecular Architecture
Rockhampton, QLD 4702
Australia

**Davor Margetic**
Ruder Boskovic Institute
Laboratory for Physical Organic Chemistry
Division of Organic Chemistry and Biochemistry
Bijenicka c. 54
10000 Zagreb
Croatia

**Peter McGauran**
Australian Government Minister
Parliament House
Canberra, ACT 2600
Australia

**Anastasios Melis**
University of California, Berkeley
Department of Plant & Microbial Biology
111 Koshland Hall
Berkeley, CA 94720-3102
USA

**Paul Meredith**
The University of Queensland
Department of Physics
Brisbane, QLD 4072
Australia

**Ana L. Moore**
Arizona State University
Department of Chemistry and Biochemistry
Tempe, AZ 85287-1604
USA

**Thomas A. Moore**
Arizona State University
Department of Chemistry and Biochemistry
Tempe, AZ 85287-160
USA

**Ron J. Pace**
Australian National University
Department of Chemistry
Canberra, ACT 0200
Australia

**Ben J. Powell**
The University of Queensland
Department of Physics
Brisbane, QLD 4072
Australia

**Reza Razeghifard**
Australian National University
Research School of Biological Sciences
Canberra, ACT 0200
Australia

**Jeffrey R. Reimers**
The University of Sydney
School of Chemistry
Sydney, NSW 2006
Australia

**Jenny Riesz**
The University of Queensland
Department of Physics
Brisbane, QLD 4072
Australia

**Michael Seibert**
National Renewable Energy Laboratory
1617 Cole Blvd.
Golden, CO 80401
USA

**Thomas D. Sharkey**
University of Wisconsin-Madison
Department of Botany
430 Lincoln Drive
Madison, WI 53706
USA

**Surya Subianto**
Queensland University of Technology
School of Physical and Chemical Sciences
Brisbane, QLD 4072
Australia

**Harsha M. Vaswani**
University of California, Berkeley
Department of Chemistry
Berkeley, CA 94720-1460
USA

**Robert Vogel**
The University of Queensland
Department of Chemistry
Brisbane, QLD 4072
Australia

**Ronald N. Warrener**
Central Queensland University
Center for Molecular Architecture
Rockhampton, QLD 4702
Australia

**Spencer M. Whitney**
Australian National University
Molecular Plant Physiology
Research School of Biological Sciences
PO Box 475
Canberra, ACT 2601
Australia

**Geff Will**
Queensland University of Technology
School of Physical and Chemical Sciences
Brisbane, QLD 4072
Australia

**Thomas J. Wydrzynski**
Australian National University
Research School of Biological Sciences
Canberra, ACT 0200
Australia

**Liping Zhang**
Ceres, Inc.
Malibu, CA 90265
USA

# Part I
# The Context

# 1
# Artificial Photosynthesis: Social and Political Issues

*Ian Lowe*

## 1.1
## Introduction

All human life is dependent, directly or indirectly, on photosynthesis. Its direct effects provide the source of all our food, either as plant material or as the plants that feed the animals, birds, and fish we eat. The plants that harness solar energy are also the source of the oxygen we breathe and an essential component of the water cycle. At a more abstract level, the plants around us also contribute to our cultural identity, as well as being a source of spiritual sustenance to many people.

The feature that distinguishes modern industrial society from all previous epochs is our use of fuel energy. A modern Australian uses energy at a rate of about 6 kW (Lowe 1989). To express that in human terms, our energy use is equivalent to having about 50 slaves working in relays around the clock for each of us. Fuel energy does for us what slaves did for feudal despots: it cooks our food, washes our clothes, heats our water, entertains us, fans us when we are hot, carries us about, and so on. Most of the energy comes from the stored end products of millions of years of photosynthesis – peat, lignite, coal, oil, and natural gas. Earlier human civilizations depended on short-term stocks of stored photosynthesis products, especially wood. Harnessing the resources of coal, oil, and gas has enabled a dramatic expansion in energy use. The level of energy use per head in Australia is about 50 times that of societies that still use short-term photosynthesis products.

There is a clear link between the level of energy use in different societies and their material living standards, though not the simple causal connection that is sometimes assumed. For example, Hoyle (1978) argued that the standard of living in the U.S. was higher than that in the UK because of greater levels of energy use; therefore, the UK would be able to improve its living standard simply by expanding its rate of energy use. This claim is clearly false. The simplest way to increase energy use would be to make the process of converting energy

*Artificial Photosynthesis: From Basic Biology to Industrial Application*
Edited by Anthony F. Collings and Christa Critchley
Copyright © 2005 WILEY-VCH Verlag GmbH & Co. KGaA, Weinheim
ISBN: 3-527-31090-8

into services less efficient, thus expanding energy use without any improvement at all in living standards; indeed, higher energy costs would probably reduce funds available for other purposes and lead to a perceived fall in material prosperity. This example illustrates an important point. Material comfort does not flow directly from the level of energy use, but from the level of services that the energy provides. As Lovins (1977) argued, people don't want energy; they want hot showers and cold beer – and a range of other energy services. A large four-wheel-drive vehicle uses three times as much fuel to travel one kilometer as an efficient small sedan, but the passengers have been transported exactly the same distance in similar comfort. An inefficient refrigerator uses two to three times as much electricity as an efficient one, but the contents are kept at a similar temperature. As modern technology has developed in an age of cheap energy, efficiency has not been a priority. A recent study suggested that it would be quite feasible to reduce energy use in the industrialized world to 25% of the present level without any significant loss of amenity, a goal that has since been adopted by such European countries as Denmark and Norway (Spangenberg 2000).

## 1.2
### The Need for a Transition to Artificial Photosynthesis

In historical terms, the epoch of stored photosynthesis has been comparatively brief. Coal was seen as an inferior substitute when its use became widespread as a result of a shortage of wood in the late eighteenth century (Wilkinson 1974). The era of petroleum fuels began in the late nineteenth century. There are three reasons for believing that we are approaching the end of the present epoch: resource limitations, environmental problems, and social issues.

In resource terms, it could be said until recently that pessimists feared the peak of world oil production might be only five years away, and optimists thought it might be as far away as 2020 (Deffyes 2001). A special series of reports on energy in *New Scientist* recently pointed out that pessimists now believe that the peak of world oil production was actually in the year 2000 and that we are already on the downhill slope (Holmes and Jones 2003). There are still optimists who think that it might be 10 years away or more, but there is no substantial disagreement with the geological fact that the peak of world oil production, if it hasn't already happened, will happen in most of our lifetimes. After that we'll see the real show for which the 1970s was an out-of-town tryout, coming soon to a planet near you. Make sure you are sitting comfortably, because a long run is assured. In that near-future world, oil will become scarcer and more expensive. We will have to change the basis of our energy use for transport, which is implicitly posited on the assumption that there will always be cheap, readily available petroleum fuels. While current expectations are that we will have cheap petroleum fuels for a few decades, this belief is based on two heroic assumptions. The first is that there will be continuing stability in, and willingness to export oil from, the region we call the Middle East, despite the Bush administration's bumbling interventions. The second assumption is that the majority of the world's population will continue to do without the transport options we take for granted while we dissipate the dwindling supplies of petroleum in such selfish indulgences as car races, jet skis, motor boats, and using heavy four-wheel-drive vehicles for suburban trips. If the entire world used oil at the rate Australians do per person, and it could be pumped out fast enough, the global resources would last less than two years! So the first and most basic reason for moving away from the present pattern of fuel energy use is that we are dissipating a limited resource, making change inevitable.

The second reason for change is that the use of fossil fuels is causing serious environmental problems, at all levels from the local to the global. At the local level, fuel use in urban areas is the main cause of air pollution that is bad enough to pose serious health risks in many large cities (UNEP 2002). Technological change has led to improved air quality despite increasing fuel use in many cities of the industrialized world (UNEP 2002), but these gains are likely to be cancelled out by increasing vehicle use (Brisbane City Council 2002). At the regional level, the problem of acid precipitation has caused policy changes to reduce the production of sulfur dioxide as a byproduct of using fossil fuels (UNEP 2002). At the global level, the burning of increasing amounts of fossil fuels is the main cause of human-induced climate change; concern about this problem led to the development of the Framework Convention on Climate Change and its associated Kyoto Protocol, an agreement to curb emissions of carbon dioxide and other "greenhouse gases." While the science of climate change is still developing, the current scientific opinion is that emissions of carbon dioxide are about 2.5 times the capacity of natural systems to absorb the gas (IPCC 2001). In other words, global use of fossil fuels needs to be reduced to about 40% of the present level to bring emissions into balance with the natural carbon cycle. Even then, the long residence time of carbon dioxide molecules in the atmosphere means that concentrations will continue to increase for decades; thus, global temperatures will continue to increase for about a century and sea levels will continue to rise for several centuries. Recent scientific thinking suggests that climate change is accelerating and influencing other global cycles (Steffen et al. 2004), posing a very serious threat to the future of human civilization.

The third reason for change is an associated social issue. As discussed earlier in this chapter, the present pattern of fuel energy use is grossly unequal. At one extreme, average per capita energy use in the U.S. is more than 10 kW, while the figure in poor countries is about 0.1 kW – lower by two orders of magnitude. Both the resource limits and environmental problems discussed above mean that it is impossible for the entire world to use energy as the U.S. now does. On the other hand, the provision of such basic services as clean drinking water, adequate shelter, and reasonable nutrition in the poorer parts of the world will require increasing energy use. The only feasible way of squaring this circle is to move away from stored photosynthetic products to new forms of artificial photosynthesis. The case for moving in this direction recognizes the scale of the available resources. The natural flows of solar energy are four orders of magnitude greater than the present global energy use and therefore are larger than any conceivable future energy demand. In fact, the amount of solar energy that hits Australia in one summer day is of the same order of magnitude as the global annual energy use for all purposes (Lowe 1994). As other chapters in this volume show, there are many ways of capturing and storing enough solar energy to meet human needs.

## 1.3
## Some Associated Social and Political Issues

The traditional approach in the physical sciences and engineering is a positivist model that sees science as objective knowledge, devoid of social or political content, while technology is seen as an unmitigated good. There are some serious problems with this approach to both science and technology. In the case of science, the great creative scientists always put considerable stock in their instinct or the aesthetic appeal of the theoretical models they derived, thus celebrating the emotional dimension of their science. It is now appreciated more generally, though not yet universally, that our perception of the world is inevitably influenced by our values, our culture, and our predisposition, so that our engagement with nature and the production of knowledge are invariably social processes. Some knowledge is neutral and can be used for good or ill. The laser has no political content in itself, but there is great political significance in the choice to use it for entertainment, for healing, or to guide weapons of mass destruction. Other knowledge has embodied political content: the neutron bomb or biological weapons can only be used to kill people.

Deciding how much public money will go to science is a political choice. Political considerations also influence the mechanisms for allocating research funds, the membership of granting bodies, the priorities for research spending, and even the assessment of specific proposals. Though most researchers would like to believe that the peer review process leads to the funding of the most-deserving applications, there are some serious problems with that belief. The first is that the process can never be objective. We all see the world through the lenses of our values and experiences. Economics is probably the extreme case of a discipline in which one ideological position has effectively crowded out all others, but there are only differences of degree between it and other fields. It is chastening to recall that the peer review process failed to fund the two crucial pieces of research that together showed that the ozone layer was being depleted; the work was done only because the researchers had access to discretionary resources that allowed them to proceed against the considered opinion of their academic peers (Lowe 1989). It is thus very important for the peer review process to be supervised by researchers who are open-minded and who together repre-sent the mainstream of thought in their broad field.

That raises the second problem, namely, that Research Council panels have enormous influence. They choose the reviewers for proposals, and they also have the right to vary the rankings that emerge from the review process. But the choice of the funding panels is always political, in the sense that there are always many individuals who are qualified for appointment to the small number of positions. Governments tend to choose what Sir Humphrey Appleby called "sound chaps" – mostly males, in senior positions and meeting the definition of being "sound" because their values are similar to those making the choice. The Howard government has gone further than any in Australian history to make explicitly political choices in its appointments. While ideology may not play an obvious role in ranking proposals in pure mathematics, it certainly does in some other fields of inquiry – including some of the most obvious forms of artificial photosynthesis.

The range of fields covered poses the third and most intractable problem. If we leave aside the issue of objectivity, it would be possible, in principle, for the peer review process to rank research proposals in a narrow area where accurate comparisons can be made: inorganic chemistry, number theory, behavioral psychology, or medieval history. But there is no way, even in principle, of determining the relative merit of the number theory proposals and the projects in behavioral psychology; there are no polymaths who could make such comparisons. Consequently, the granting process effectively decides the share of funds that will go into each area through a process of horse-trading and then allocates within each area by a ranking process. When I studied the ARC grants going into different fields of science several years ago, I found that the success rate in a discipline like chemistry varied widely from year to year, but the share of the funds going to chemistry stayed remarkably constant (Lowe 1987). Therefore the chance of a research proposal in artificial photosynthesis being funded is affected by the fraction of the total research budget available for the broad field in which the project lies, and that fraction is in turn determined by political considerations.

The final problem is that those allocating limited funds will always tend to err on the conservative side, so that a shortage of funds almost guarantees that there will be no support for radical proposals that cross the boundaries of traditional disciplines or question long-established theories. In the modern world, the most important research problems often involve several disciplines. In areas where conventional wisdom is clearly failing us, we desperately need to be supporting new approaches. Some of the most promising ideas in the field of artificial photosynthesis struggle for funding because they cross the traditional disciplinary boundaries that still shape priorities for research funding. As an extreme example of this problem, funding of energy-related R&D in Australia since the 1970s has consistently been dominated by the two vectors least likely to be significant in the next century: coal and nuclear energy (Lowe 1983). As this chapter was being finalized, greater government support was going to investigate the speculative technology of geo-sequestration, capturing and trapping carbon dioxide in geological formations to allow expansion of coal burning, while support for solar energy was actually reduced by a decision not to renew funding for the Cooperative Research Centre in that field. Clearly, values influence the opinion of decision-makers determining which field is more likely to contribute to solving the problem of global warming. This became a political issue in 2004, with Democrat and Green senators questioning whether it was appropriate for the Australian government to be advised in this area by its current Chief Scientist, who is also Chief Technologist for the mining corporation Rio Tinto (Guilliatt 2004). Since the interests of Rio Tinto are clearly served by an approach that provides for increasing use of coal, those who disagree with this emphasis logically see the Chief Scientist's advice as reflecting his values.

Applying science to produce new technology is even more blatantly political. We can never change only one thing in a complex system, so the use of new technology always produces losers as well as winners and costs as well as benefits. Scientists and technologists have a responsibility to be aware of the consequences of their work and to realize that no technology is ever purely and universally beneficial. It will always benefit some people and some groups more than others. In the extreme case of military technology, its entire purpose is to give one group or nation an advantage over another. In the case of Concorde, a huge R&D budget and large operating subsidies were used to produce a technical marvel that for 25 years allowed a small number of rich people to cross the Atlantic faster, with little benefit to the broader public, many of whom suffered a greater level of noise nuisance. The opposition to some technological advances is based on the perception of some people that they will not actually benefit. In many cases, that perception is clearly accurate; as Alan Roberts says, wherever uranium is enriched, the taxpayer is impoverished! Examining the costs and benefits can often lead to a clear conclusion about the winners and losers. In other cases, the jury is still out. It might, for example, turn out to be true in the long term that the benefits of genetic manipulation of food crops will outweigh the risks, as is now claimed by its proponents. It might also turn out that the negative effects will outweigh the benefits, as is now claimed by opponents. In the short term, all we can do is scrutinize the competing claims and attempt to assess their credibility.

The traditional process for assessing technology has concentrated on its technical efficacy and its microeconomics: whether it is a cost-effective way of achieving the stated goal. We now know that a wider canvas should be used. The former Resource Assessment Commission developed a framework for considering complex issues (RAC 1991). The RAC argued that it is possible to assess separately the economic costs and benefits of a proposal, the environmental risks, and the social costs and benefits, with suitable qualifications on the accuracy of these estimates. The three "columns" cannot be weighted and added together in some sort of modern felicific calculus to show whether a proposal provides a net benefit or not, the RAC said. An appropriate process is to document the three areas and have a public debate to decide whether the net economic benefits justify the environmental risks and the net social costs (or benefits). The political problem is that governments are uncomfortable about making such overt value judgments; as I have argued elsewhere, they prefer to give the impression that "expert advice" leads to a logically compelling conclusion (Lowe 2003).

## 1.4
### Using the Available Photons: Towards Sustainability Science

The various types of artificial photosynthesis are effectively competing for the right to harness incoming photons and use them for the community's benefit. Other chapters in this book discuss the technical merits of particular possibilities, dealing with such issues as the efficiency of turning the incoming sunlight into useful energy. That is necessary but is not sufficient to enable wise decisions. It is equally important to assess alternative ways of using photons for their social and environmental impacts. Whether we engage in large-scale production of silicon-based photovoltaic cells, polymeric cells, photosynthetic hydrogen production, or bacterial energy production, scaling up the process to meet a significant fraction of human energy needs would have a range of demands on resources and social organization, as well as causing a range of local and global environmental effects. The emerging field of "sustainability science" provides guidance for the research approach that should be used to avoid making some of the past mistakes (Kates et al. 2001).

Sustainability science recognizes that our understanding of nature-society interactions is still limited. Although there have been substantial advances in recent decades through work in the environmental sciences that factors in human impacts and work in social and development studies that takes account of environmental influences, we still have to accept that modern science can be described as islands of understanding in oceans of ignorance (Ehrenfeld 1999). We are constantly engaged in land reclamation, but there is no chance of filling in the oceans. We need to set some broad priorities for our limited scientific effort. At the top of the list should be the urgent need for a better general understanding of the complex dynamic interactions between society and nature. That will require major advances in our ability to analyze the behavior of complex self-organizing systems, as well as developing better understanding of the irreversible impacts of interacting stresses. We need to work at multiple scales of organization and consider the impacts on natural systems of various social actors with different agendas, ranging from environmentalists standing in front of trees to bulldozer drivers pushing them over and Cabinet ministers telling us it is justified because it promotes economic growth.

Case studies from all the inhabited continents show that many of our serious environmental problems are the direct result of applying narrow, specialized knowledge to complex systems (Kates et al. 2001). Agronomists have advised farmers on fertilizer use to improve pasture, but the changes have put unacceptable nutrient loads on waterways. Expert advice has allowed fishing vessels to catch more seafood, leading to depletion of fisheries. Irrigation systems have made it possible to grow new crops but have also deprived streams of the flows needed to maintain riverine ecologies. Species introduced to control one pest have driven other native biota to extinction. Coal-fired power stations provide cheap electricity, but their carbon dioxide emissions are now changing the global climate. So there can be no doubt about the worrying conclusion that great damage can be done by the application of narrow, specialized science without an appreciation of the complexity of natural systems.

We need to address these issues through integrated scientific efforts that focus on the social and ecological characteristics of particular places or regions. Therefore, sustainability science must differ fundamentally from most science as we now know it. The traditional scientific method is based on essentially sequential phases of scientific inquiry such as conceptualizing the problem, collecting data, developing theories, and applying the results. But these familiar forms of developing and testing hypotheses have run into difficulties as we study complex nonlinear systems with long time lags between actions and their consequences. The problems are complicated by our inability to stand outside the nature-society system; thus, we cannot even in principle be objective observers of the system. Think of the parallel of two people with different allegiances watching a football match. If you talk to them, it can be difficult to believe they are watching the same game. They will differ about matters of fact, such as whether the ball was over the line, about matters of judgment, such as whether a foul was committed, and sometimes even about the visual acuity of the official in charge, or the official's integrity. Our situation is worse than those biased observers, because we are actually out on the muddy field! We can't see the whole game and we have an interest in the outcome that affects the way we see the action around us. As part of the nature-society system, we cannot stand outside it and be an objective observer.

We therefore have to accept that our engagement with complex natural systems cannot be based on the old model of rational objective science. The traditional sequential steps must become parallel functions of social learning, additionally incorporating the elements of action, adaptive management, and policy as experiment. Sustainability science therefore needs to employ new methods, such as semi-quantitative modeling of qualitative data and case studies, or inverse approaches that work backwards from undesirable consequences to identify pathways that avoid those outcomes. Scientists will need to work with other interested parties, such as land users or manufacturers, to produce trustworthy knowledge that combines scientific excellence with social relevance.

Meeting the challenge of sustainability science will also require new styles of institutional organization to foster interdisciplinary research and to support it over the long term, to build capacity for that research, and to integrate it into coherent systems of research planning, assessment, and decision support. Around the world, researchers are working on the core questions of sustainability science: the fundamental character of nature-society interactions, our ability to guide those interactions along more sustainable trajectories, and ways to promote the social learning we will need to navigate a transition to sustainability (Kates et al. 2001). We urgently have to develop mechanisms that will nurture those research activities. This is essential if we hope to change to social practices that will allow us to use natural systems sustainably.

## 1.5
## Conclusions

As other chapters in this volume show, there are many technically promising avenues of artificial photosynthesis. The entire field deserves urgent support as we near the end of the era of dependence on stored photosynthetic products. Past practice shows that development and use of technologies on a large scale will have economic, social, and environmental consequences. A broader assessment of new technologies is demanded by the increasing capacity of human activity to perturb the natural systems on which our survival depends. The emerging field of sustainability science suggests an approach for handling these complex and difficult issues. They are a challenge not only to the scientific community but also to our political institutions. It is no exaggeration to say that the survival of human civilization depends on our ability to respond to this challenge.

## References

Brisbane City Council (2002), *Brisbane Air Quality Strategy*, Brisbane

K. S. Deffyes (2001), *Hubbert's Peak: the Impending World Oil Shortage*, Princeton University Press, Princeton

D. Ehrenfeld (1999), *The Coming Collapse of the Age of Technology*, Tikkun Jan/Feb issue, http://www.tikkun.org/magazine/index.cfm/action/tikkun/issue/tik9901/article/990111a.html

R. Guilliatt (2004), The Man in the Middle, Good Weekend Magazine 20–21 March, John Fairfax, Sydney & Melbourne, pp 26–32

B. Holmes, N. Jones (2003), Brace yourself for the end of cheap oil, *New Scientist 179*, 2406, 9–11

F. Hoyle (1978), *Energy or Extinction*, Heinemann Books, London

(IPCC) Inter-governmental Panel on Climate Change (2001), *Third Assessment Report*, Cambridge University Press, Cambridge

R. W. Kates, W. C. Clark, R. Corell, J. M. Hall, C. C. Jaeger, I. Lowe, J. McCarthy, H. J. Schellnhuber, B. Bolin, N. M. Dickson, S. Faucheux, G. C. Gallopin, A. Grubler, B. Huntley, J. Jager, N. S. Jodha, R. E. Kasperson, A. Mabogunje, P. Matson, H. Mooney, B. Moore III, T. O'Riordan, U. Svedin (2001), Sustainability Science, *Science 292*, 641–642

A. Lovins (1977), *Soft Energy Paths*, Penguin, Harmondsworth

I. Lowe (1983), Who Benefits from Australian Energy Research?, *Science and Public Policy 10*, 272–284

I. Lowe (1987), University research funding: the wheel still is spinnin', *Australian Universities Review 30*, 2–11

I. Lowe (1989), *Living in the Greenhouse*, Scribe Books, Newham

I. Lowe (1994), Global climate change and the politics of long-term issues, in S. Dovers [ed], *Sustainable Energy Systems*, Cambridge University Press, Cambridge, pp 194–216

I. Lowe (2003), Science, Research and Policy, in S. Dovers, S. Wild River [eds], *Managing Australia's Environment*, Federation Press, Sydney, pp 472–484

(RAC) Resource Assessment Commission (1991), *Kakadu Conservation Zone Final Report*, Australian Government Publishing Service, Canberra

J. Spangenberg (2000), Operationalising Sustainability, in D. Yencken, D. Wilkinson, *Resetting the Compass,* CSIRO Publishing, Collingwood, p 321

W. Steffen, J. Jager, P. Matson, B. Moore, F. Oldfield, K. Richardson, A. Sanderson, J. Schnellnhuber, B. L. Turner, P. Tyson, R. Wasson (2004), *Global Change and the Earth System: A Planet Under Pressure,* Springer, Berlin

(UNEP) United Nations Environment Program (2002), *Global Environmental Outlook 3,* Earthscan Publications, London

R. Wilkinson (1974), *Poverty and Progress,* Methuen, London

# 2
# An Integrated Artificial Photosynthesis Model

*Ron J. Pace*

## 2.1
## Introduction

Artificial photosynthesis (AP) is an umbrella term, embracing totally novel approaches to research into and development of technologies for nonpolluting electricity generation, fuel production and carbon sequestration using solar energy. As the name implies, the inspiration is drawn from natural photosynthetic systems, which developed in organisms that were among the earliest known to exist on earth [1]. The natural systems are thus the product of an extremely long (> 2.5 billion years) process of evolutionary refinement.

The "grand vision" of artificial photosynthesis is to technologically reproduce the components of natural photosynthesis on large scale for efficient solar energy conversion. The program offers the prospect of economical photovoltaic electricity generation and food production requiring negligible water usage compared to conventional agriculture. In addition, totally renewable hydrogen generation from convenient water sources, such as seawater, becomes feasible. Bockris, in a seminal analysis of future energy options [2], concluded that production of hydrogen fuel from electrolysis of water would become a practical strategy if a "super catalyst" for the anodic, water-oxidizing reaction could be found. It has now emerged that nature has solved this problem, within oxygenic photosynthesis, almost to the absolute limit of thermodynamic efficiency.

## 2.2
## Natural Photosynthesis

The overall process of photosynthesis consists of two main phases, the so-called "light" and "dark" reactions (Fig. 2.1). In the first, light energy is absorbed by "antenna" chlorophyll molecules in special cell membranes (thylakoids) and transferred to "reaction center" chlorophylls. Here electrochemical reactions commence that generate two vital "energy-rich" biological compounds; adeno-

*Artificial Photosynthesis: From Basic Biology to Industrial Application*
Edited by Anthony F. Collings and Christa Critchley
Copyright © 2005 WILEY-VCH Verlag GmbH & Co. KGaA, Weinheim
ISBN: 3-527-31090-8

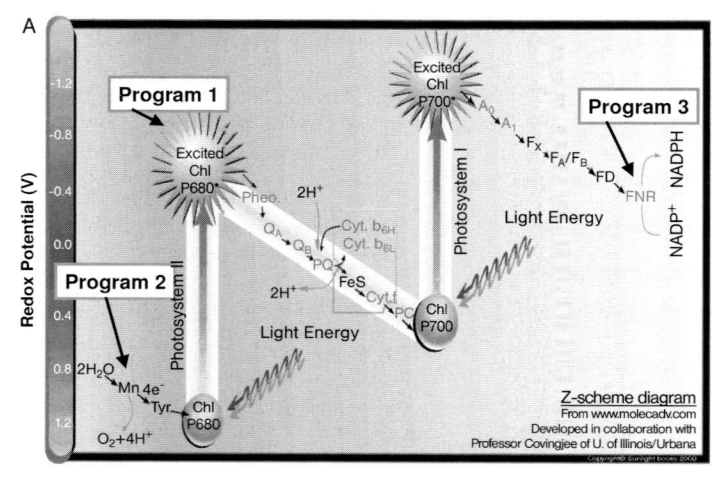

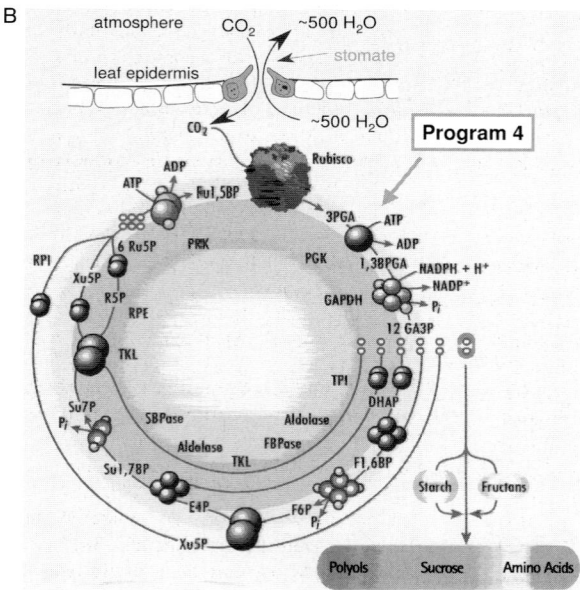

**Fig. 2.1** Diagrammatic summary of the light reactions in photosynthesis (A), known as the 'Z Scheme' (adapted from Govingee, see www. molecadv.com). Electrons originate from water oxidation in photosystem II and are photo-chemically 'boosted' twice on passage through photosystems II and I. The electrons emerge as reducing equivalents in NADPH. The bioelectric current creates a proton gradient across the thylakoid membrane, which powers ATP synthesis by the $F_1/F_0$ proton ATPase.

(B) Diagrammatic summary of the $CO_2$ fixing dark reactions in photosynthesis, known as the Calvin Cycle (adapted from [4]). The overall rate of carbon fixation is limited by the turnover kinetics of the rubisco enzyme. In plants, $CO_2$ uptake is inescapably coupled to large water loss during transpiration through leaves etc. Box numbers refer to the artificial photosynthesis programs described here, which are modelled on or inspired by the indicated steps in the natural photosynthetic processes.

sine triphosphate (ATP) and reduced pyridine nucleotide (NADPH). Oxygen is produced as a byproduct in this process and is released to the atmosphere. The early steps in this chain are actually nature's own photovoltaic energy conversion systems (photosystems), in which the trapped light energy is first converted into electrically stored energy in cell membranes. The light phase requires the cooperation of two different such membrane-bound photochemical assemblies (called photosystems I and II). Each photosystem operates in series, to photochemically "charge" the membrane.

In the dark reactions, the products of the light phase, ATP and NADPH, are used within cells for the formation of carbohydrate (sugars) from carbon dioxide, via a series of biochemical intermediates. The key enzyme in the dark phase is Rubisco, which catalyzes the addition of carbon dioxide to ribulose diphosphate. This process is central to the progressive chemical "assembly" of sugar molecules from carbon dioxide and water. The latter is the only point at which continued water consumption is absolutely required for carbohydrate formation and represents only a tiny fraction of the water applied conventionally in plant growth. The products of the light phase also are essential for the plant's synthesis of lipids and other plant constituents.

The primary photosynthetic energy-transducing processes, involving photosystem II (PSII), the cytochrome $b/f$ complex, and photosystem I (PSI), generate an electron flux and an associated electrochemical gradient across the thylakoid membrane. The electrons come from water, and this oxidation of water to molecular oxygen is the ultimate source of virtually all bioenergetic electrons utilized by living creatures. The process occurs within the oxygen-evolving complex (OEC) located in PSII. All oxygenic photosynthetic organisms share a common set of "core" membrane proteins that contains the D1-D2 reaction center (RC) proteins and inner chlorophyll $a$ (Chla)-binding proteins. The D1-D2 heterodimer (analogous to the type II purple bacteria L/M proteins) contains the photooxidizable special chlorophyll pair "P680" as well as the pigment components associated with primary light-driven charge separation across the thylakoid membrane (Fig. 2.2). The D1 peptide appears to ligate the catalytic Mn cluster responsible for water oxidation.

The water-oxidation process involves three distinct steps, each operating at an efficiency near the theoretical limit and unmatched by any related synthetic system (see Fig. 2.2):

1. Trapping of light energy by chlorophyll pigments and rapid energy transfer to the reaction center (P680), resulting in its oxidation to P680$^+$.
2. Rapid electron donation to P680$^+$ through an oxidizable protein side-chain intermediate (tyrosine 161, $Y_Z$, on the D1 peptide), which stabilizes the charge separation in PS II.
3. Oxidation of water to molecular oxygen within the OEC. This is the most energetically demanding reaction performed by nature, requiring a redox potential of at least 0.8 V. The redox potential developed in P680$^+$ is in excess of 1.2 V.

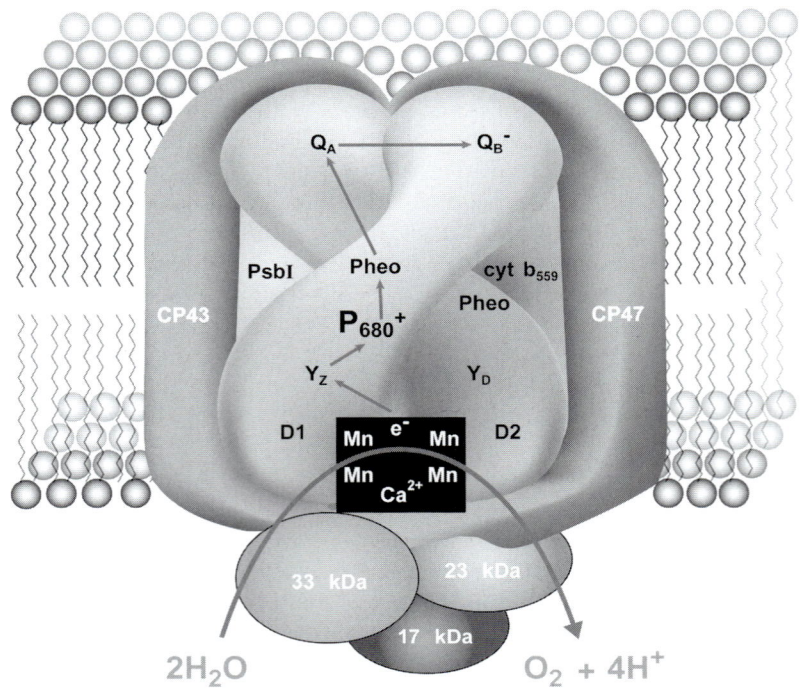

**Fig. 2.2** Schematic representation of photosystem II complex. Light energy is transferred from the chlorophyll antenna regions (CP 43, 47) to the P680 reaction centre special chlorophyll pair, which undergoes photooxidation. The released electron proceeds to the opposite membrane face and reduces a mobile plasto-quinone carrier, $Q_B$. The oxidised reaction centre is re-reduced by electrons, ultimately released from water, through an intermediate electron transfer species $Y_z$ (tyrosine). This stabilises the charge-separated state against back reaction. The Mn/Ca cluster catalyses the water oxidation (4 electron process) and immobilises the reactive intermediates of this reaction. The Mn/Ca cluster operates at a mean redox potential of $\sim 0.8$ V, while the $Y_z$ centre has a redox potential of $\sim 0.95$ V (see Fig. 2.1 A).

The Mn-Ca cluster that comprises the heart of the OEC is the most efficient anodic "electrolysis" system known. It operates under mild conditions of temperature, pH, electrolyte background, etc., with a maximum turnover rate of $\sim 10^3 \text{ s}^{-1}$. Its operation is close to the absolute thermodynamic limit. It catalyzes the reaction $2H_2O \rightarrow 4H^+ + O_2 + 4e^-$. Operation of the primary photosynthetic processes results in formation of ATP and NADPH. The latter is formed as a biochemical output from PSI, using the electron flux originating from PSII, after the electrons receive a second photochemical "boost" (i.e., become highly reducing) on passage through PSI (see Fig. 2.1). The operation of PSII and the cytochrome *b/f* complex results in an electrochemical proton gradient across the thylakoid membrane, which drives production of ATP within the membrane-embedded $F_1/F_0$ proton ATPase [3].

ATP and NADPH provide free-energy input to drive the carbon-fixation reactions (Calvin cycle [4]). This occurs within the cytoplasm of cells or subcellular chloroplast organelles, depending on the photosynthetic species. The process is not light dependent, and the rate-limiting step is the turnover of the Rubisco enzyme, which is in fact the most abundant protein on earth. The carboxylase ($CO_2$-adding) reaction of Rubisco suffers a fundamental, chemically inescapable competition with $O_2$ (oxygenase reaction, see Chapter 13), and extensive biological refinement has occurred in higher plant Rubisco proteins, to operate with much lower $CO_2/O_2$ level balances than occurred in the early atmosphere in which photosynthesis first emerged. $O_2$ also plays the primary role in degrading PSII performance in nature; the photosystem suffers fatal photo-oxidative damage after ca. 20 hours of operation in high light, requiring complete re-synthesis of the D1 protein.

## 2.3
### Artificial Photosynthesis: An Integrated Strategy

There is an emerging recognition that power generation must ultimately come from renewable, nonpolluting sources. It is also apparent that clean water, in many parts of the world, will become an increasingly limited resource and thus agriculture must change to drastically reduce water usage. Therefore, technological strategies are required that recognize these goals and that identify feasible pathways through intermediate and precursor technologies that will achieve them.

Here we propose a strategy of artificial photosynthesis as a means of achieving these aims of clean power generation and dry food production (see Fig. 2.3). Key steps in the natural photosynthesis processes of plants and bacteria provide the models and inspiration for a totally biomimetic, industrial-scale technological approach to achieving the following specific goals:

- Photovoltaic electricity generation using novel, low-cost, synthetic systems with the inherently high photon-capturing and charge-separation efficiency of natural photosystems. These power stations will directly supply the national grid.
- "Dry agriculture," employing enzyme bed reactor systems to fix carbon dioxide from the air or other convenient sources, powered by hydrogen and bioelectric transducers drawing power from the national grid. These will produce carbohydrates (food), liquid fuels, chemical feedstocks, and polymers for fiber production. Water usage will be at or near the absolute chemical minimum and thousands of times lower than in conventional agriculture.
- Hydrogen production from seawater or other suitable water sources. Electrode systems employing catalytic surfaces modeled on the relevant high-efficiency active sites in photosynthetic organisms will achieve the electrolytic decomposition of water (into hydrogen and oxygen).

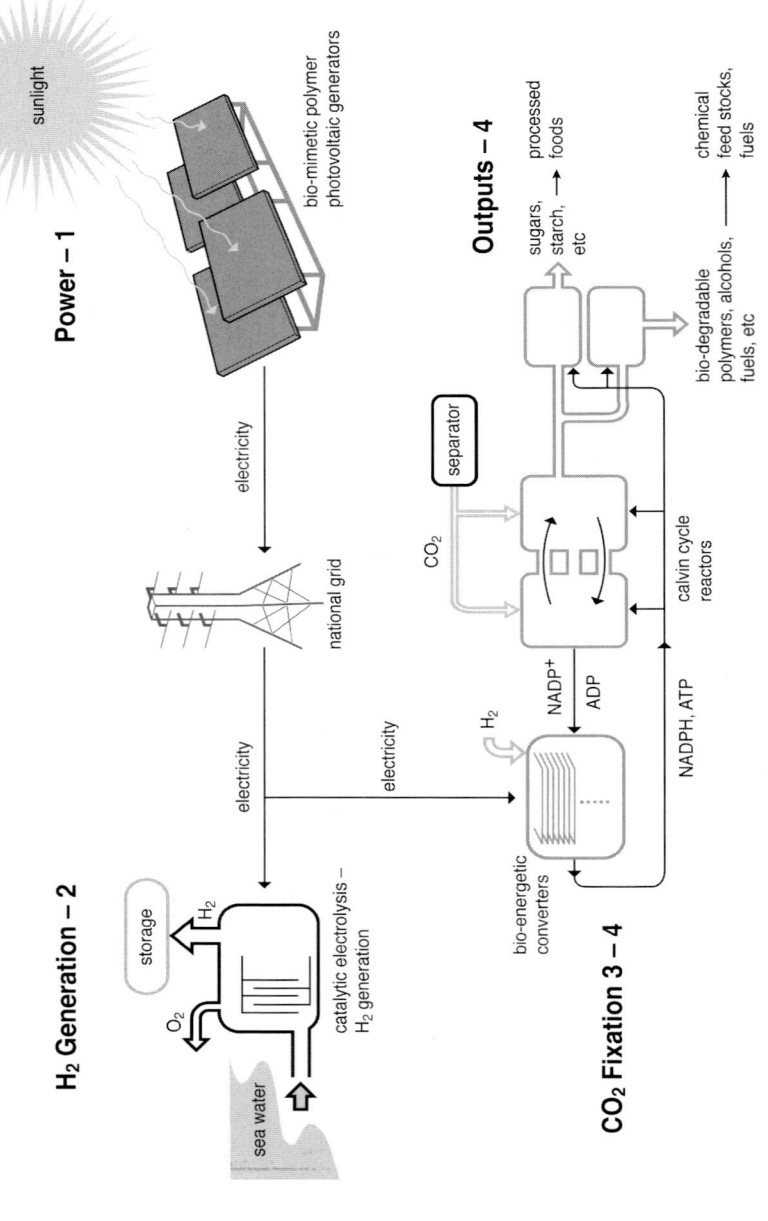

**Fig. 2.3** Artificial photosynthesis concept. This is composed of four programs, whose components are interconnected through energy transfer systems. These are electricity (the national grid) and bulk hydrogen. The energy and mass flows then need not be stoichiometrically coupled. It is expected that the 'dry agriculture' carbon fixation process in programs 3 and 4 would use only a fraction of the outputs from programs 1 and 2, which generate electricity and hydrogen respectively.

## 2.4
## A Technological Approach to Photosynthesis

Natural photosynthesis carries out the following overall reaction in the carbon fixation process:

$CO_2$ + $H_2O$ + [light energy] → $O_2$ + carbohydrate

However, in nature the complex sequence of reactions summarized above is characterized by three fundamental, limiting factors:

1. The "bioelectron flow" generated by the primary photochemical processes must be fully utilized locally within the organism, principally through carbohydrate production in plant cells, etc.
2. The $CO_2$ extracted from the environment is at low (0.03% in air) concentration, which inevitably requires that vast water loss, through transpiration, accompanies the $CO_2$-uptake process. This is the principal reason plants need water.
3. All of the biochemical-biophysical reactions must occur in the presence of oxygen, essentially at atmospheric concentrations. Natural systems expend a major metabolic effort in dealing with the toxic effects of this reactive species. In $H_2$-generating organisms (hydrogenase bacteria, etc.), the $H_2$ production occurs only in subcellular regions from which background $O_2$ has been actively excluded.

Industrial-scale artificial photosynthesis offers the prospect to circumvent these limitations. This is because the light and dark reactions may be totally "uncoupled" in a biomimetic system, both in terms of the actual energy-material flow balance and even the requirement to be physically co-located in space. The interconnection then becomes the existing electricity grid and transport of high-energy fuel intermediates (hydrogen). This modular organization allows most system components to operate totally anaerobically, which will be crucial to the practical viability of the total approach.

Fig. 2.3 illustrates the principal technology subsystems of the total AP concept. A unique feature is that all are linked, through the national grid, by their generation or consumption of electricity. This provides several key elements of flexibility:

- Spatial location of the components, eventually quite large, industrial-scale installations, can be chosen to optimize factors such as environmental siting, access to inputs (seawater, $CO_2$ sources, waste heat sources, high solar irradiance locations), and proximity to end use facilities.
- The different elements are not critically interdependent. We expect the proposed technologies to develop at different rates, and that approaches not included here (e.g., in renewable power generation or hydrogen production) may coexist, either transitionally or in the long term. This will not significantly impact on remaining components of the program and allows implementation in the most technically and economically efficient manner.

- A strict physical-chemical isolation of processes is possible, unlike in natural photosynthetic $H_2$-producing systems. Elimination or drastic reduction of molecular oxygen in the photoelectric and $CO_2$-fixing modules will have major benefits in chemical robustness and longevity of the molecular components.
- Biological or bio-inspired components can be chosen and combined from whatever sources are most appropriate, without any limitations relating to whole-organism viability.

At present none of technologies indicated in Fig. 2.3 exist in integrated form, although some components are available or are under development as part of other research programs. These operate in the emerging AP area itself and in allied fields such as supported membrane technology [5]. Nor is the above program a unique approach to artificial photosynthesis in general. Currently, two groups are pursuing established AP programs, in Europe [6] and North America [7]. In both cases, the focus is more directly aimed at reproducing aspects of photosynthetic function, in vitro, using biomimetic chemical constructs. To date these have demonstrated photogenerated transmembrane charge separation and functional electron transfer from Mn clusters inspired by the OEC site in PSII. The results are highly impressive and give confidence in the biomimetic approach, although the work is essentially laboratory scale at present. Most importantly, it has yet to demonstrate the levels of chemical robustness and turnover efficiency that practical systems would require. These two factors are key to making AP a realistic approach for economic, industrial-scale energy-fuel (including food) production. Recent advances in the detailed, molecular-level understanding of photosynthesis, particularly the primary photochemical processes, have been dramatic [8–11]. The challenge is to exploit these developments, together with other emerging technologies (membrane, nanotechnology), to evolve solutions that may be bio-inspired but utilize strengths that established manufacturing, fabrication, and materials processing provide.

In the following sections, research and development approaches to realize the technologies needed for each of the four system components in Fig. 2.3 will be briefly discussed. These outlines are not intended to be exhaustive or prescriptive but rather to represent plausible programs based on the author's experience with natural photosystems and practical, commercialized, biomimetic membrane devices. Significant milestones that the various programs would be required to achieve are indicated in each case.

## 2.5
### Program 1: Biomimetic Photoelectric Generation

*A novel organic, photovoltaic technology with high conversion efficiency and voltage output to supply the national grid.*

Natural photosystem reaction centers execute light trapping and photoelectric charge separation with extremely high efficiency [12]. This results in the local crea-

Table 2.1 Performance comparisons for photovoltaic technologies.

| | Open circuit voltage ($V_{OC}$) | Max. theoretical efficiency (sunlight) | Current at max. efficiency (mA/cm$^2$) | Present efficiency | Charge carrier recombination time (s) | Quantum efficiency of charge separation |
|---|---|---|---|---|---|---|
| $S_i$ | ~0.6 V | <25% | ~40 | 15–20% | $\leq 10^{-3}$ | ~55%[a] (effective) |
| Present organic photovoltaics | 1–2 V | ? | ≤3 (present operation) | 2–3% | $<10^{-6}$ | 10–30% |
| "PSII" Stacked membrane converter | >20 V | >40% | <4 | – | $>10^{-1}$ | ~100% |

a) $V_{OC}/E_{bandgap}$ (voltage factor).

tion of an ~1-V electrochemical potential difference across a 40-Å low dielectric membrane region. The charge recombination rate (i.e., the back reaction rate) is extremely low (see Table 2.1), as charge stabilization occurs through metastable oxidation and reduction of internal donor and acceptor groups (quinones, etc.) that are separated by ~40 Å. No more than six chlorophyll-type pigments are necessary for this process, and the structure and energy coupling within such natural systems is under active study by numerous groups (e.g., see [13–15]). These systems appear sufficiently tractable that construction of robust, model chromophore assemblies executing the process should be feasible. Purpose-built single-chromophore/acceptor systems have been demonstrated [6, 7].

Several factors will be important in the choice of components and design of such primary photoelectric assemblies:

1. The structures should be convenient to produce, either chemically or biosynthetically, allowing a range of chromophoric species to be readily incorporated. This is to maximize response across the solar spectrum.
2. Charge recombination rates must be much lower than those in current organic photovoltaics to obtain efficiencies even comparable to existing semiconductor systems (e.g., Si solar cell).
3. Photoelectron transfer should be over as large a distance as possible (tens of angstroms) to maximize current density. This also contributes to satisfying requirement 2.
4. Facile electron charge transfer into and out of the complex, at the donor and acceptor regions, must be possible.

Artificial peptide- (see Chapter 7) or specific organic-based structures offer promise as the molecular frameworks within which the chromophore and charge separating moieties can be accommodated. Relevant examples of the latter include highly efficient porphyrin-based assembly chemistries [16] and molecular scaffolds (molracs) [17]. These "made-to-order" molecular frames may be hydrophilic or hydrophobic and can be adorned with a range of chromophores [18]. Efficient photoconversion under high incident light flux is the expected operating condition for a large-scale photovoltaic system. In principle this can be achieved with either (1) large molecular units that contain a high number density of light-capturing chromophores, funneling energy to a single rapid-turnover charge-separating center per unit (analogous to natural light-harvesting/reaction center complexes) or (2) larger numbers of small units, each with few chromophores per unit ($\sim 4$). It seems likely that for biomimetic photoelectric systems, as for natural photosystems, maximal turnover rates will be limited by the kinetics of intramolecular charge transfer into and out of the unit, at the donor and acceptor regions (see Fig. 2.4). This suggests that designs involving small photoreaction units, arranged for maximum density and light-capture/intramolecular charge transfer efficiency, are favored.

An important limitation for any organic-based photovoltaic system, which at present seems fundamental, is that the maximum photocurrent densities achievable are at least an order of magnitude less than those for inorganic semiconductor devices (see Table 2.1). Since the solar spectrum contains light quanta mainly in the 1–3 eV range, full solar radiance ($\sim 1$ kW m$^{-2}$) corresponds to $\sim 50$ mA cm$^{-2}$ for total direct conversion. One way to address this restriction is to operate at low (a few mA) current densities but to multiply the output voltage by series operation. This means "stacking" the photoreaction units internally, and membrane structures provide an obvious means to this end. Recent advances in robust, solid-supported, synthetic bilayer technologies have occurred, which could be applicable here [5]. Although developed for quite different purposes (biosensors), these synthetic membranes are composed of covalently linked hydrophilic and hydrophobic layered structures, constructed from highly stable organic species. They have excellent ionic insulation properties. This system provides a natural framework, when extended, to incorporate orientationally aligned chromophore-acceptor units, which could function in parallel and in series as a photovoltaic conversion assembly. Fig. 2.4 shows the conceptual design.

Photoactivated charge-separation groups are vectorially incorporated into the low-dielectric (lipid-like) regions, which are separated by hydrophilic (polyether) spacer groups containing mobile charge carriers. A series assembly of 20–30 such layered units, which on present experience is feasible, could produce voltages up to $\sim 20$ V (i.e., 10–20 times per unit area performance of conventional semiconductor systems). Moreover, different chromophore species can be combined to absorb across the whole visible to near-infrared region. Suitably engineered chlorophyll-porphyrin-type pigments with a combined, effective surface density of $>10^{16}$ per cm$^2$ would achieve near total solar radiation trapping, with back reflection. Within a layer, the density of charge-separation units would be $\sim 5\times10^{13}$ cm$^{-2}$, and under full load about 10% of these would be carrying charge at steady state. The mobile carriers (redox-active small molecules, ionic

## 2.5 Program 1: Biomimetic Photoelectric Generation

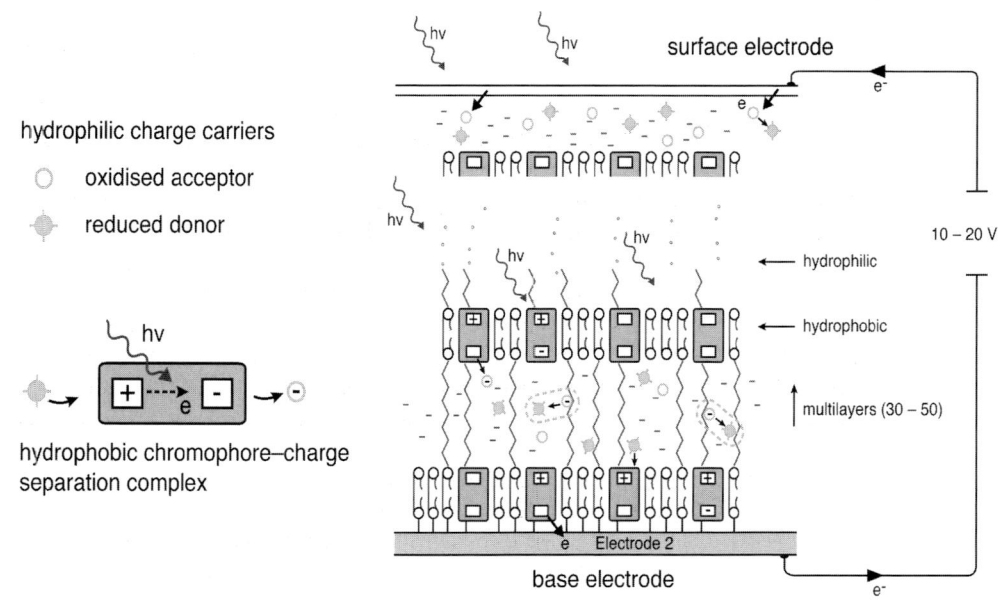

**Fig. 2.4** Bio-mimetic photoelectric generator conceptual design. The basic charge separation units would contain ~ 4 pigment molecules (chlorophyll/porphyrin etc.) to execute the primary photo-oxidation charge separation step. This charge separation would be stabilised, as in natural photosystems, by appropriate donor acceptor moieties (quinone, phenoxides etc.). The system is constructed with alternating hydrophobic-hydrophilic layers, each of several tens of Å thickness, so as to operate at low current density but (relatively) large output voltage. This voltage is developed progressively across the low dielectric layers in which the charge separation units are embedded. Mobile redox carriers operate in the high dielectric regions to transfer charge by diffusion (roughly analogously to mobile carriers in the non junction regions of semi-conductors).

complexes, etc.) would need to have sufficient density and mobility in the high-dielectric layer regions to efficiently transport charge (electrons), essentially by diffusion, across the ~ 20-Å, low-potential drop gap. Edge insulation strategies already demonstrated (for the biosensor system) seem feasible and the main electrical leakage losses will be through the hydrophobic regions directly. The synthetic membranes have far higher breakdown voltages (>1 V) than do natural lipid systems. The overall efficiency of the system should be several-fold higher than current solar cell technologies. The base electrodes would probably be conducting polymer [19] on metal, with surface texturing to optimize overall trapping effectiveness. Exclusion of molecular $O_2$ will be vital to enhance component lifetimes. At present there are few data that bear quantitatively on this, but some experience with conventional organic photovoltaics suggests that the lifetime improvement can be dramatic and that $O_2$-induced damage is the principal source of performance degradation in these systems [20].

The organic, photoactive region would be ~0.1 µ thick, with a mass of ~100 kg km$^{-2}$ of electrode area. With ~10% overall efficiency, output would be 1 GW per ton of photovoltaic layer, from material that is ultimately just carbon-based chemicals. In principle the electrodes could be fabricated as large sheets (tens to hundreds of square meters), and the huge resulting aspect ratio means that edge effects are minimized and $O_2$ permeability of the base and transparent surface electrodes becomes the critical factor. These would probably be evaporated metal and $SnO_2$ on glass.

### 2.5.1
### Milestones

1. Identify and synthesize potential chromophore scaffolds (molrac, peptide, or other constructs) and basic bilayer-hydrophilic repeating units.
2. Develop and test strategies to incorporate scaffolds in bilayer units. Spectroscopically test chromophores and isolated constructs for absorption and charge-separation characteristics.
3. Identify and test appropriate electrode materials (conducting polymer, transparent types). Select and test appropriate mobile charge carriers. Test model photovoltaic systems with single bilayer/chromophore assemblies on metal electrodes.
4. Construct and test photovoltaic assemblies with several repeating bilayer units on metal electrodes. Modify on basis of system performance experience.
5. Optimize long-term photostability and conversion efficiency.

## 2.6
### Program 2: Electrolytic Hydrogen

*New strategies for efficient hydrogen generation from electricity, using electrode systems modeled on natural enzyme centers.*

The chemically demanding reaction in the electrolytic decomposition of water into $H_2$ and $O_2$ is the anodic oxidation of water (or OH$^-$) to molecular oxygen and protons. Current industrial-scale electrolysis systems operate under extreme alkaline conditions (~40% Na/KOH), often at elevated temperatures. The Mn-containing, water-splitting catalytic site in photosystem II (PSII) performs this reaction at close to thermodynamically limiting efficiency (<0.2 V over-voltage), at a high turnover rate (~$10^3$ s$^{-1}$), under mild external pH and in the presence of significant concentrations of environmentally common anions, such as Cl$^-$. A biomimetic electrolysis system based on the natural PSII catalytic site would have substantial thermodynamic and kinetic advantage.

Hydrogen generation by electrolysis is an established industrial process but operates on a limited scale with current technology. A detailed analysis of the prospects for electrolytically generated hydrogen as a renewable fuel, for a "hydrogen economy," was presented by Bockris [2]. Two important practical difficulties limiting the feasibility of such a program were identified:

1. Although the thermodynamic, reversible cell voltage for water electrolysis is $\sim 1.2$ V under moderate conditions (50–100 °C, a few atmospheres of pressure), the anodic reaction is overwhelmingly rate limiting, resulting in substantial electrode over-voltages ($\sim 1$ V) when working at current densities required for practical operation (see Fig. 2.5 A). This impacts directly on conversion efficiency and point 2 below.
2. At cell voltages above $\sim 2$ V, $Cl_2$ will be formed at the anode if $Cl^-$ is present in the feed water (Fig. 2.5 B). This would be environmentally unacceptable for large-scale $H_2$ production, requiring the absence of such salts from the water feed.

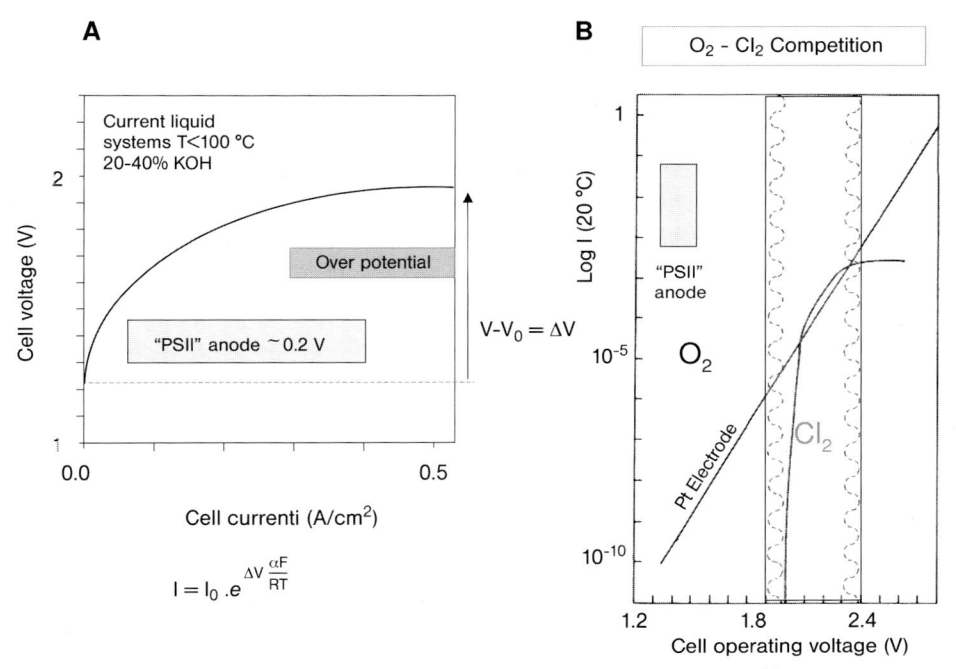

**Fig. 2.5** Summary of basic performance characteristics of conventional anodic water oxidation electrode systems and the catalytic water oxidase within photosystem II. (A) Comparison of electrode over-voltage versus current density for existing alkaline liquid systems and a hypothetical PS II anode, as described in Fig. 2.6. In the electrode potential equation, $I_o$ is the reversible current, F the Faraday constant and $\alpha$ a dimensionless factor ($\sim 0.5$) [2]. Because the reversible current term, $I_0$, is $< 10^{-10}$ amp/cm$^2$ even for the best conventional metal anode surfaces, practical current densities ($\sim 0.2$–1 amp/cm$^2$) require total cell voltages $\sim 2.0$ V. (B) shows how cell over-voltage affects the competition between water oxidation and oxidation of a typically abundant soluble anion species, like $Cl^-$. Cell operating voltage must be maintained below $\sim 1.5$ V, under mild operating conditions, if $Cl_2$ formation is to be excluded.

With present and projected electrolytic technologies (e.g., see [21, 22]), the above issues are addressed either by operating at extreme alkaline conditions (to limit anode over-voltage) or with pure water feed for vapor-phase systems functioning at very high temperatures and pressures. However, Bockris concluded that "conventional" electrolysis, using readily available water sources (seawater, etc.) would be competitive and clean if a "super catalyst" could be found for the anodic reaction. The cathodic ($H_2$-forming) reaction is not limiting with existing electrode materials (e.g., Ni).

The water-splitting site in PSII has, fundamentally, the properties required for the "super catalyst." Its operational performance, in equivalent terms, is indicated in Fig. 2.6 and far exceeds conventional electrode capability, particularly with regard to the factors above. It contains up to four Mn and one Ca in a compact, exchange-coupled cluster [23, 24]. The structure of the site in cyanobacterial PSII is emerging from recent studies [8–10]. Although the full structural detail of the site is yet to be resolved, it is apparent that most of the protein ligands, which define the cluster geometry, are located in a very small region near the C-terminus of the D1 polypeptide of the PSII reaction center [25]. This suggests the possibility that functioning catalytic site analogues might be assembled from small, model peptides. One proposal for such a structure has been made [26]. In addition, some existing Mn-complex OEC models show

| Mn Site Turnover: | $10^3$ s$^{-1}$ (PSII) |
|---|---|
| Site Density: | $10^{15}$ cm$^{-2}$ |
| Current Density: | ~ 0.7 A /cm$^2$ |
| Cell Voltage: | ~ 1.3 V |
| Operating Temp: | 50-70 °C |
| Energy Consumption: (with waste heat input) | ~ **3- 3.3 kWh/ m$^3$ H$_2$** |

ANODE: POLYMER ELECTRODE
Mn catalytic complexes bound to surface
~ $10^{15}$ / cm$^2$

ANODE

CATHODE
Ni Alloy etc.

CONDUCTING IONIC POLYMER MEMBRANE (Sulfonic acid type etc.)

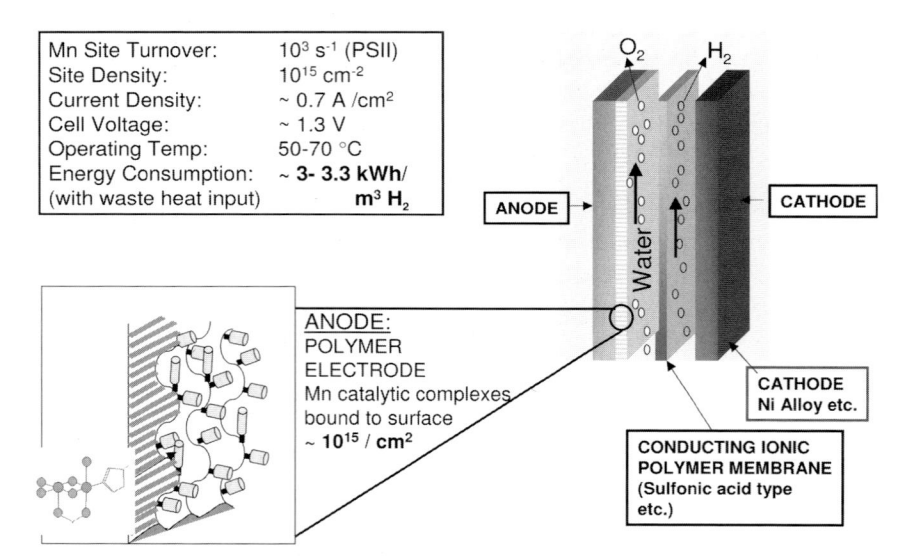

**Fig. 2.6** Possible design for a $H_2$ generating electrolysis system, using anode surfaces incorporating catalytic Mn/Ca clusters modelled on the water oxidase centre in photosystem II. Assuming comparable catalytic behavior of the synthetic clusters to the natural centre, overall performance of the system operating with saline water at ~ neutral pH is as indicated. The cathodic ($H_2$) releasing reaction is not in fact limiting and could be implemented with existing materials. The sulfonic acid electrolyte membrane would be based on existing technologies (Nafion etc. e.g., see [21, 22]).

promise of useful catalytic function [27]. Eventually, totally "synthetic" structures based on molrac scaffolds or minimal peptides from rationalized four-helix bundle proteins (e.g., as in [28]) may be developed for application to large-scale water splitting.

The artificial catalytic clusters would be coupled to an (anodic) electrode surface (e.g., as in Projects 1 and 3). A system with a realistic density of cluster sites ($\sim 10^{15}$ cm$^{-2}$ on a textured conducting polymer surface), operating with natural turnover efficiency, would support $H_2$ generation at $\sim 10$ kA m$^{-2}$ current density and $\sim 3$ kWh m$^{-3}$ $H_2$ power consumption (utilizing waste heat as thermal input). It could function with seawater (as the substrate/electrolyte combination) at moderate temperatures (<100 °C). A conceptual design is shown in Fig. 2.6. This uses an ionically conducting polymer membrane as the internal bridge. Its performance would be at or beyond the projected efficiencies of modern high-temperature, solid-electrolyte systems ("Hot Elly"), which place extreme demands on materials engineering (high-pressure operation at $\sim 1000$ °C [21]).

The system could function with conventional metal cathodes (e.g., Ni) to produce $H_2$, but optimum efficiency and suppression of side reactions could be achieved if active-site models based on natural hydrogenases could be incorporated (e.g., [29]). A critical consideration will be the operational lifetime of the coated catalyst surfaces, particularly the anode. There is little direct data to draw on here, as the turnover inhibition rate of PSII in vivo (ca. once per day) is determined by complex factors involving unavoidable exposure to both light and oxygen (and interaction of these two). The electrolytic system, while running at significant $O_2$ levels, would operate totally dark, and the catalytic molecular cluster would be a much simpler and more robust entity than the D1 peptide in PSII.

Although current expectation is that nonrenewable sources for $H_2$ generation will be most economical in the short term (e.g., [22]), electrolysis offers one of very few long-term renewable strategies for molecular $H_2$ generation in a hydrogen economy. A fraction of the hydrogen so generated would be consumed in the $CO_2$-fixation systems described below. The net effect of this is to eliminate totally the need for freshwater input into carbohydrate production. In effect, the ultimate source for such water becomes the sea (see Scheme 2.1).

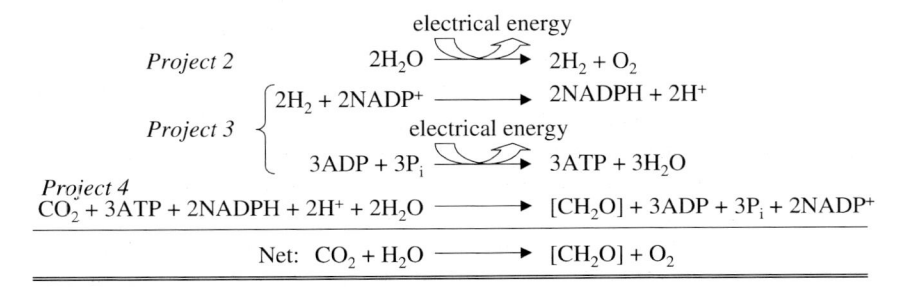

*Project 2*: $2H_2O \xrightarrow{\text{electrical energy}} 2H_2 + O_2$

*Project 3*:
$2H_2 + 2NADP^+ \longrightarrow 2NADPH + 2H^+$
$3ADP + 3P_i \xrightarrow{\text{electrical energy}} 3ATP + 3H_2O$

*Project 4*:
$CO_2 + 3ATP + 2NADPH + 2H^+ + 2H_2O \longrightarrow [CH_2O] + 3ADP + 3P_i + 2NADP^+$

Net: $CO_2 + H_2O \longrightarrow [CH_2O] + O_2$

**Scheme 2.1** Stoichiometry of the combined processes described in Projects 2–4 when producing carbohydrate.

### 2.6.1
**Milestones**

1. Identify potential small-peptide scaffolds (e.g., C-terminus of D1 polypeptide, cytochrome $b_{562}$, rationalized peptides) and catalytic metal clusters. Design metal chelation sites for the various peptides. Clone and express these peptides.
2. Perform biochemical-biophysical characterizations of metal-ligated systems. Identify those giving evidence (spectroscopic, etc.) of similarity to natural catalytic sites.
3. Test most promising candidates for redox activity up to and including water-splitting activity. Modify and optimize.
4. Identify suitable electrode surface types (conducting polymer, membrane functionalized, etc.). Electrically couple catalytically active peptides to electrode surfaces.
5. Optimize long-term reactive stability, ionic background insensitivity, and conversion efficiency.

## 2.7
**Programs 3 and 4: Waterless Agriculture**

These two programs are components of a proposed system of "waterless agriculture." The outputs would include a broad range of "enzymatically fixed" organic products containing carbon, hydrogen, and oxygen (see Fig. 2.3). The carbon and oxygen would be provided by $CO_2$ from atmospheric sources, industrial sources, etc., and the hydrogen from bulk $H_2$ (e.g., from Program 2 above). The basic enzymatic chemistry is that of the $CO_2$-fixing Calvin cycle, found in all oxygenic photosynthetic organisms. Energy input to drive the process comes from $H_2$ and electric power.

An enzyme-based system offers substantial flexibility, as the natural Calvin cycle may be conveniently manipulated near its output stages, using known enzyme reactions, to produce a wide variety of products. These range from basic carbohydrates (food, etc.) to fuels and chemical feedstocks. Although several tens of components are involved in the overall biochemistry, only three are required in substantial amount. Two, NADPH and ATP, must be circulated and continuously regenerated, bioenergetically, from $NADP^+$ and ATP. The third is the Rubisco enzyme, which effects the $CO_2$ addition and would be immobilized in an enzyme bed reactor.

## 2.7.1
**Program 3: Bioenergetic Converters**

These systems combine electrical energy (from the grid) and hydrogen (Program 2) to generate the biochemicals NADPH and ATP required for enzymatic carbon fixation (see Program 4).

In nature, NADPH and ATP are formed in intimate association with the photon-capturing photosystems, which also generate oxygen. $O_2$ is a chemically inescapable inhibitor of key enzyme reactions in the natural carbon-fixation process, as well as a source of damaging, reactive side products. Operating under strictly "dark," anaerobic conditions offers substantial scope for simplifying the biochemistry and greatly enhancing the robustness of the biologically derived components, and this is the approach proposed here.

The technology to generate NADPH and ATP on a large scale does not yet exist. We propose to use oxidation-reduction (redox) enzyme electrode systems, which already function in certain limited commercial applications (as chemical sensors). In natural photosynthesis, the iron-sulfur protein ferredoxin (on photosystem I) reduces $NADP^+$ to NADPH. ATP is formed from ADP by proton flow through the membrane-incorporated ATP synthase. The protons are generated electrochemically by the bioelectron current between photosystems I and II.

It appears plausible that both processes could be implemented biomimetically, using appropriate enzymes incorporated into synthetic membrane-electrode systems. Externally supplied electricity and hydrogen would provide the free energy and reducing equivalents necessary. Again, solid substrate-supported synthetic membrane technology would be the most promising route to this implementation. Functional incorporation of and coupling to the relevant enzymes (ferredoxin, ATP synthase, and specific hydrogenases) would draw on well-established expertise with these systems from natural photosynthesis (e.g. [29, 30]). Through molecular biology, one can "mix and match" protein systems from various organism sources and evolve them artificially to maximize the overall cell performance and stability. Preliminary studies show that the limiting factor will be the turnover efficiency of the ATP synthase [3], whose close packing density ($\sim 10^{12}$ cm$^{-2}$) will determine electrode area. However, with currently available catalytic activities ($\sim 10^3$ s$^{-1}$), realistic electrode current densities ($\sim 1$ mA cm$^{-2}$), and efficient electrode surface stacking ($\sim 10$ per cm, 50-μm channels), the system appears practical. Fig. 2.7 indicates one possible arrangement for the various components necessary. In effect, hydrogen (as reduced quinone) is oxidized at the anode and $NADP^+$ is reduced at the cathode. The resulting proton flux through the tethered membrane at the anode drives ATP synthesis. The plant area required by the bioenergetic converters will in fact be ca. 10 times that for the Calvin ($CO_2$) reactors, which will not be limiting even though the turnover rate for Rubisco is only $\sim 10$ s$^{-1}$.

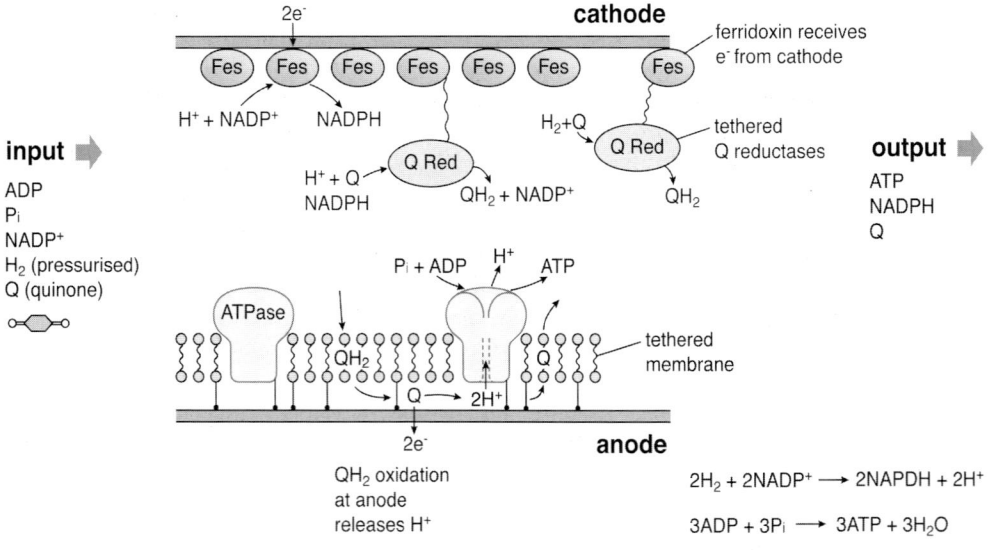

**Fig. 2.7** Schematic picture of the key components of a bio-energetic electrode system, which would accept electrical power and injected $H_2$ as energy inputs to regenerate NADPH from $NADP^+$ and ATP from ADP and phosphate. These components are contained in a recirculating stream, used to provide the metabolic energy necessary to operate Calvin cycle enzymes in an immobilised bed carbon fixation reactor (see Fig. 2.8). The electrode area is determined essentially by the closest packing density of the tethered ATPase ($\sim 1-2 \times 10^{12}/cm^2$). The channel between the planar electrode surfaces would be $\sim 50\ \mu m$ in height and current between the electrodes is conveyed by reduction and oxidation of mobile quinone species. Continuous $H_2$ injection is necessary to provide the correct stoichiometric balance (see Scheme 2.1), and the system operates totally anaerobically.

#### 2.7.1.1 Milestones

1. Identify suitable NADPH reductases (ferredoxin-$NADP^+$ reductase, transition metal complexes, etc.). Attach to an electrode surface and test for $NADP^+$ reduction with redox mediators.
2. Identify and test suitable $H_2$ quinone reductases and mobile quinone carriers for electrolytic proton current generation.
3. Select suitable ATP synthase (bacterially derived) and determine optimal membrane insertion strategy from established reconstitution techniques for hydrophobic proteins. Incorporate into supported membrane electrode with functional proton mobile reservoir.
4. Test and optimize function of ATP synthase system and improve functional stability.
5. Assemble and test final electrode-membrane system for NADPH-ATP generation efficiency.

## 2.7.2
### Program 4: The $CO_2$-fixing Enzyme Reactor

*Artificial enzyme reactors to fix $CO_2$, using biochemical inputs from Program 3.*

The artificial carbon-fixation modules would be fixed/fluidized bed reactors containing the relevant Calvin cycle and related enzymes (recombinant, artificially evolved, or synthetic versions, matrix immobilized) in appropriate stoichiometry (e.g., see [31, 32]). The reactors take $CO_2$ and produce carbohydrates (e.g., sugars and polysaccharides), fuels (e.g., ethanol and isoprene), chemical feedstocks, and polymers for fiber production (e.g., polybetahydroxyalkanoates). Water usage would be near the absolute theoretical minimum and thousands of times lower than required in conventional agriculture. In principle, operation of the bioenergetic converter and enzyme reactor yields net water production (see Scheme 2.1). Products would be separated using macro- and microporous membrane technologies (particularly easy when the product is a gas, e.g., isoprene).

The reactors would operate anaerobically, coupled to or integral with the bioenergetic converter subsystems (Program 3) that provide the biochemical inputs. Carbon dioxide injection would come from high-$CO_2$ (>5%), low-$O_2$ (<5%) streams, probably from industrial combustion sources (including power generation from fossil carbon) in the first instances. Eventually, concentrators (chemical-physical methods now available, e.g., see [33]) will be required for atmospheric-sourced $CO_2$. The enzyme reactors would most likely be located near $CO_2$ sources.

Conversion of carbon dioxide and water into sugars occurs within a series of $\sim 20$ linked dark-phase, enzymatic reactions driven by ATP and NADPH. The latter provides both biochemical energy and "reducing equivalents" (effectively atomic hydrogen), which are necessary for the cycle function. In the bioreactor, as in vivo, virtually all of the enzyme mass is devoted to Rubisco, due to the low turnover rate of this enzyme ($\sim 10 \text{ s}^{-1}$). Specific (currently available) $CO_2$-permeable polymer membranes would be used to introduce this substrate from the gas stream, minimizing water vapor loss from the reactors. Operating at low-$O_2$ and high-$CO_2$ internal concentrations has two advantages. First, it allows the simplest and most biochemically robust forms of Rubisco and other enzymes from thermophilic bacteria to be used as starting points for artificial evolution of the proteins to suit conditions in the reactor beds. Second, it enables the use of high temperatures, which promote high rates of conversion. The only large-volume inputs are $CO_2$, $H_2$, and electricity. No net inputs of "fertilizers" or other nutrient sources are required. An industrial system that would produce 25,000 tons of carbohydrate per year is indicated conceptually in Fig. 2.8. The electrical power requirement is $\sim 30$ MW (continuously averaged), assuming $\sim 50\%$ overall efficiency, and the $H_2$ consumption is $\sim 3,000$ tons per year.

The normal biochemical outputs from the $CO_2$-fixation cycle are sugars (trioses, pentoses, and hexoses), which may be used directly, or converted to polysaccharides, as high-volume feedstocks for food processing. Straightforward

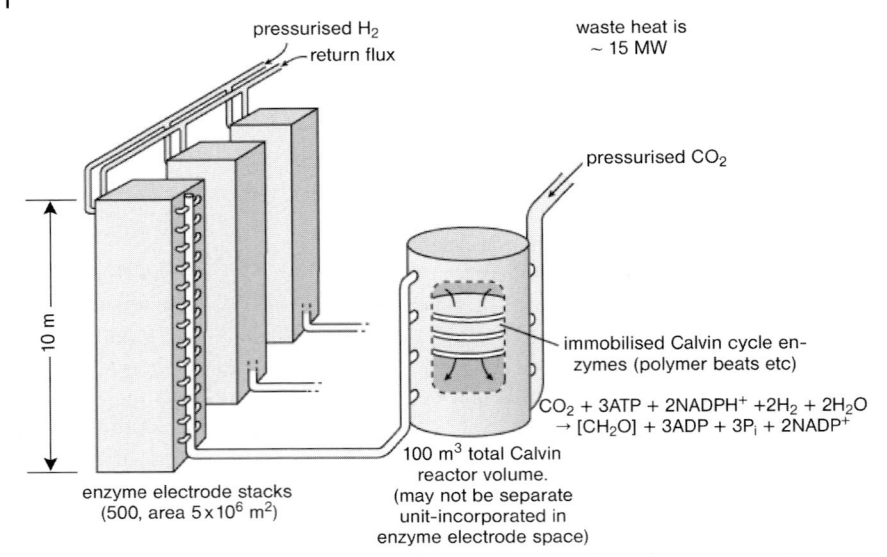

**Fig. 2.8** Conceptual arrangement of components for a carbon fixation plant, using bioenergetic converters and enzyme bed reactors. This would produce ~ 25,000 tons pa of carbohydrate, requiring an electrical power input of 30 MW (continuously averaged) and $H_2$ consumption of ~ 3,000 tons pa. $CO_2$ would be from convenient sources, with low $O_2$ content. System is intended to run totally dark and anaerobic.

Labels in figure: pressurised $H_2$ return flux; waste heat is ~ 15 MW; pressurised $CO_2$; immobilised Calvin cycle enzymes (polymer beats etc); $CO_2 + 3ATP + 2NADPH^+ + 2H_2 + 2H_2O \rightarrow [CH_2O] + 3ADP + 3P_i + 2NADP^+$; 100 m³ total Calvin reactor volume. (may not be separate unit-incorporated in enzyme electrode space); enzyme electrode stacks (500, area $5 \times 10^6$ m²); 10 m.

enzymatic pathways exist from $CO_2$-fixation cycle intermediates to industrially important compounds, such as betahydroxybutyrate (feedstocks for biodegradable plastics manufacture). Practical liquid fuel alternatives, such as isoprene and alcohols, are easily obtained from $CO_2$-fixation cycle products by a small number of additional enzymatic steps. These could provide useful fuel options even after implementation of a full hydrogen economy.

#### 2.7.2.1 Milestones

1. Clone the appropriate genes for ~ 20 $CO_2$-fixation cycle enzymes. Modify for expression of affinity-tagged variants. Preliminary test of ability to fix $CO_2$ and cycle.
2. Optimize *E. coli* expression systems. Develop optimal affinity purification protocols. Artificially evolve limiting components (e.g., Rubisco) to suit reactor conditions.
3. Develop protocols for assembling all 20 proteins appropriately tethered in suitable stoichiometries. Further modification of some proteins to suit reactor if necessary.
4. Test strategies for delivering ATP and $NADPH_2$ to reactor. Test strategies for delivering $CO_2$ and removing products.

5. Test for ability to fix $CO_2$ continuously into products. Modify proteins as necessary. Assess possibilities for combining $CO_2$ reactor with $ATP/NADPH_2$ generator. Add additional enzymes for alternative products.

## 2.8
## Conclusions

The research and development program necessary to realize this vision of artificial photosynthesis is nothing if not ambitious. It embraces a broad range of technologies and scientific disciplines, from chemical and electrical engineering, to chemistry, advanced molecular biology and photon physics. However, it is modular and seeks always to gain inspiration from nature ("steal its secrets") rather than simply "artificially" reproduce plants. The intention is to implement solutions in a manner that is most practical and plays to the individual strengths of both conventional and biomimetic technology. For instance, if sufficient operational lifetime could be achieved, the organic photovoltaic approach described here could offer the first realistic prospect for direct large-scale (national grid) electricity generation from sunlight. Around 10 tons of photoactive material would power a country like Australia. This is something that semiconductor-based systems could probably never match because of materials and fabrication costs, as well as the environmental cost of their manufacture.

Developed nations, in a number of instances, face a second challenge – land degradation through excessive water usage in crop food production. Only a tiny fraction of the water employed in conventional cultivation is incorporated into the final agricultural product. In a sense, the bulk of the agriculturally applied water is wasted, and it contributes to salinity damage of the land. Our current practices in energy and agriculture are highly efficient economically in terms of the prevailing cost structures. They involve huge investments in infrastructure and social capital. Fossil fuels are cheap and large-scale agriculture is highly competitive and productive. However, we should begin now to map our future without cheap fossil fuels and with serious water limitations/shortages.

## References

1 R. E. Blankenship, H. Hartman, *Trends in Biochemical Sciences* **1998**, *23*: 94–97
2 J. O. M. Bockris, Energy – The Solar-Hydrogen Alternative, Chap. 9. Wiley & Sons, New York 1977
3 W. Junge, H. Lill, S. Engelbrecht, *Trends in Biochemical Sciences* **1997**, *22*, 420–423
4 R. C. Leegood, T. D. Sharkey, S. von Caemmerer (Eds.) *Photosynthesis: Physiology and Metabolism*, Advances in Photosynthesis, Vol. 9, Chap. 2 and 3. Kluwer, Dordrecht 2000
5 B. Cornell, G. Khrishna, P. Osman, R. Pace, L. Wieczorek, *Biochemical Soc. Trans.* **2001**, *29*, 613–617
6 A. Magnuson, H. Berglund, P. Korall, L. Hammarström, B. Åkermark, S. Styring, L. Sun, *J. Am. Chem. Soc.* **1997**, *119*, 10720–10725

7 G. Steinberg-Yfrach, P. A. Liddell, S.-C. Hung, A. L. Moore, D. A. Gust, T. A. Moore, *Nature* **1997**, *385*, 239–241
8 A. Zouni, H. T. Witt, J. Kern, P. Fromme, N. Krauss, W. Saenger, P. Orth, *Nature* **2001**, *409*, 739–743
9 N. Kamiya, J.-R. Shen, *Proc. Natl. Acad. Sci. USA* **2003**, *100*, 98–103
10 K. N. Ferreira, T. M. Iverson, K. Maghlaoui, J. Barber, S. Iwata, *Science* **2004**, *303*, 1831–1838
11 P. Jordan, P. Fromme, H. T. Witt, O. Klukas, W. Saenger, N. Krauss, *Nature* **2001**, *411*, 909–917
12 L. M. C. Barter, M. Bianchietti, C. Jeans, M. J. Schilstra, B. Hankamer, B. A. Diner, J. Barber, J. A. Durant, D. R. Klug, *Biochemistry* **2002**, *40*, 4026–4034
13 P. J. Smith, S. Peterson, V. Masters, T. Wydrzynski, S. Styring, E. Krausz, R. J. Pace, *Biochemistry* **2002**, *41*, 1981–1989
14 T. Reinot, V. Zazubovich, J. M. Hayes, G. T. Small, *J. Phys. Chem. B* **2001**, *105*, 5083–5098
15 J. R. Durant, D. R. Klug, S. L. S. Kwa, R. van Grondelle, G. Porter, J. P. Dekker, *Proc. Natl. Acad. Sci. USA* **1995**, *92*, 4798–4802
16 P. Thoradson, A. Marquis, M. J. Crossley, *Org. Biomol. Chem.* **2003**, *1*, 1216–1225
17 R. N. Warrener, D. Margetic, A. S. Amarasekara, D. N. Butler, I. B. Mahadevan, R. A. Russell, *Organic Letters* **1999**, *1(2)*, 199–202
18 M. R. Johnston, M. J. Latter, R. N. Warrener *Australian Journal of Chemistry* **2001**, *54*, 633–636
19 A. K. Burrell, D. L. Officer, P. G. Plieger, D. C. W. Reid. *Chemical Reviews*, **2001**, *101*, 2751–2796
20 G. W. Wallace, P. C. Dastoor, D. L. Officer, C. O. Too, *Chemical Innovation*, April **2000**, 15–22
21 *International Atomic Energy Agency* TECDOC – 1085, **2002**, Chap. 5, Vienna
22 *Report to Congress on Status and Progress of the DOE Hydrogen Program*, US Department of Energy, Feb 4, 1999
23 T. G. Carrell, A. M. Tyryshkin, G. C. Dismukes, *J. Biol. Inorg. Chem.* **2002**, *7*, 2–22
24 J. S. Vrettos, J. Limburg, G. W. Brudvig, *Biochim. Biophys. Acta* **2001**, *1503*, 229–245
25 R. J. Debus, *Biochim. Biophys. Acta* **2001**, *1503*, 164–186
26 R. J. Pace, K. Åhrling, *Biochim. Biophys. Acta*, **2004**, *1655*, 172–178
27 M. Yagi, M. Kaneko, *Chem. Rev.* **2001**, *101*, 21-35
28 R. E. Sharp, C. C. Moser, F. Rabanal, P. L. Dutton, *Proc. Natl. Acad. Sci. USA* **1998**, *95*, 10465–10470
29 B. Hankamer, E. P. Morris, J. Barber, *Nature Struct. Biol.* **1999**, *6*, 560–564.
30 W. S. Chow, A. B. Hope, In: Handbook of Plant Growth: pH as the Master Variable (Ed. Z. Rengel), pp. 149–171. Marcel Dekker, New York 2002
31 S. M. Whitney, T. J. Andrews, *Proc. Natl. Acad. Sci. USA* **2001**, *98*, 14738–14743
32 H. Mauser, W. A. King, J. E. Gready, T. J. Andrews, *J. Am. Chem. Soc.* **2001**, *123*, 10821–10829
33 Capture of $CO_2$. Multiple authors. In: Greenhouse Gas Control Technologies – GHGT – 5 (Eds. D. Williams, R. Durie, P. McMullan, P. Caulson, A. Smith), pp. 85–240. CSIRO Publishing, Melbourne 2001

# Part II
# Capturing Sunlight

# 3
# Broadband Photon-harvesting Biomolecules for Photovoltaics

*Paul Meredith, Ben J. Powell, Jenny Riesz, Robert Vogel, David Blake, Indriani Kartini, Geff Will, and Surya Subianto*

## 3.1
### Introduction

In Chapter 9, we saw how and why silicon-based solid-state photovoltaics remain dominant in the solar cell market. We also saw that even the most advanced "third-generation" silicon devices are based upon the conventional *pn* semiconductor junction. As such, the photon-absorption and charge-separation events occur at the same point in space within the depletion region. The electric field that drives the spatial separation of the electron-hole pair is essentially derived from the differences in electron affinities between the *n*- and *p*-type materials. This correspondence between the photon-absorption and charge-separation events is a key point of differentiation between the photovoltaic effect in a semiconductor junction and the photon-induced generation of a chemical potential in natural systems, i.e., photosynthesis. In the latter, and this is a very simplistic but highly relevant interpretation in the context of artificial photosynthetic systems, the point in space at which the primary light absorption event occurs can be very different from the point at which the "effect" is ultimately felt through the generation of spatially separated charge carriers. Whether or not photon absorption leads to the production of electricity or the initiation of some useful chemical reaction, removal of the "spatial correspondence" constraint has a profound effect upon device design, namely, we are not forced into providing a local electric field at the point of initial photoexcitation.

If we take as a classic example the PSI and PSII photosystems of plants, we see that dozens of chlorophyll molecules within the light-harvesting complex are capable of the primary absorption event. This energy is subsequently transferred with near-unity quantum efficiency to a reaction center chlorophyll in the photosystem complex, where it can be released as a pair of spatially separated charge carriers. Clearly, the separation event requires some local electric field (as in the case of the *pn* junction), but that field need not exist at the absorption

*Artificial Photosynthesis: From Basic Biology to Industrial Application*
Edited by Anthony F. Collings and Christa Critchley
Copyright © 2005 WILEY-VCH Verlag GmbH & Co. KGaA, Weinheim
ISBN: 3-527-31090-8

site. The key advantage of this strategy is clear: the tasks of photon absorption and charge separation can be decoupled, and therefore different molecules can be engineered and optimized for each function. It also means that a very large number of molecules can be used for photon harvesting (presenting a substantial total surface area for light absorption), while maintaining a relatively small number of specialist reaction centers to facilitate charge separation.

In this chapter, we will explore this decoupling concept from the point of view of biomimetic solar cell design. We will focus on a non-silicon device platform that utilizes large surface area collection through nanoscale engineering, namely photoelectrochemical Grätzel cells. This platform has the potential to become a serious low-cost, medium-efficiency alternative to silicon. Furthermore, we will explore the possibilities of using bioorganic pigments with broadband light-absorbing properties as the primary photon-harvesting component. In this context, we will concentrate on a class of biomacromolecules called the melanins, and we will also highlight several other potential materials. What will hopefully become clear during this discussion is the need to understand the detailed photophysical and photochemical properties of these molecules and how these properties relate to their chemical and electronic structure. It is also important to note that we will not deal in this chapter with the concept of creating artificial photosynthetic membranes by immobilizing and assembling photofunctional molecules on various substrates – although this approach has the potential to create direct conversion photovoltaic devices. The interested reader is directed towards a growing body of literature on this subject dealing with, for example, self-assembled monolayers, Langmuir-Blodgett membranes, and cast films [1–3].

Finally, throughout this chapter we will adopt the specific definition that an artificial photosynthetic system for photovoltaic energy conversion is characterized by the following key features:

1. A large surface area for photon absorption created through molecular-scale or nanoscale engineering.
2. Spatial decoupling of the primary photon-absorption event from the charge-separation event.
3. A photoinduced, excited state sufficiently long-lived to allow charge separation to occur without significant quenching from wasteful back reactions and recombination events.
4. Strong and rapid coupling between the absorber (or the charge-separation vehicle in any harvesting chain system) and the electronic sink used to connect the photoactive component to the external circuit.
5. Ultimate generation of a potential difference across a nanostructured thin membrane, which requires charge separation to be directional.
6. Photochemically stable components capable of multiple oxidation-reduction cycles (or multiple exciton generation and separation events in a solid-state semiconductor system).

These are similar criteria to the key elements of any efficient light-driven system that separates charge and stores or converts the resultant chemical energy [4].

## 3.2
**The Photoelectrochemical Grätzel Cell (Dye-sensitized Solar Cell)**

The direct conversion of solar energy into electrical energy (photovoltaics) is a persuasive concept. Ever since the discovery in 1839 of the photoelectric effect by Edmond Becquerel [5], scientists and engineers have been obsessed with the idea of tapping this seemingly "free" and plentiful energy source. Photovoltaics relies upon the fact that solar photons (UV, visible, or infrared) falling upon a semiconductor with a suitable bandgap can create electron-hole pairs. This effect can ultimately lead to the creation of a potential difference at an interface between two regions within the semiconductor possessing different electron affinities (for example, in a *pn* junction) – or indeed at an interface between two different materials, as in a heterojunction device. To date, photovoltaics has been dominated by devices in which the junctions are made from crystalline or amorphous semiconductors. Doped silicon devices remain the lead technology, but compound semiconductor devices made from III–V compounds have made a limited impact in areas such as high-efficiency aerospace applications.

More recently, from the early 1990s onwards, advances in the nanostructuring of crystalline semiconductors have opened up the possibility of creating very large internal surface areas in relatively thin films ($< 10\,\mu m$) [6]. Titanium dioxide (titania), a wide bandgap semiconductor ($E_g \sim 3.2\,eV$), is a particularly interesting material since it is cheap, readily available, and has a commendable environmental profile. Total surface areas in titania nanostructured films can exceed a massive $150\,m^2\,g^{-1}$. There is significant activity in the materials physics and chemistry community focused on increasing porosity (surface area) while controlling pore size, geometry, and orientation [7]. Additionally, as we shall see later in this chapter, from a device perspective (whether it be photovoltaic or some other optoelectronic platform), the electrical conductivity of the nanocrystalline semiconductor network is a key consideration. In the case of titania, for example, the anatase crystalline phase has the highest electrical conductivity and thus is most desirable. Titania can also form two other crystal types, rutile and brookite, both of which suppress electrical conductivity of the network. Clearly, if one could create an interpenetrating mesoporous network of *n*- and *p*-type nanocrystalline titania (or of nanocrystalline titania and a different semiconductor/alternative hole transport medium), this would essentially correspond to a large number of nanojunctions for the absorption of light- and interface-driven exciton separation, i.e., a very high surface area and potentially very efficient solar cells.

However, there are significant difficulties associated with this utopian view of a nanostructured photovoltaic system. Firstly, it is extremely difficult to create a nanostructured, interpenetrating network of two materials. Secondly, titania and many of the other nanocrystalline semiconductors are wide bandgap materials, i.e., they only absorb UV photons. Narrower bandgap semiconductors tend to be susceptible to photocorrosion. This is a serious drawback since the majority of the sun's energy is delivered as visible light. Thirdly, both components have to form a percolated network throughout the film. Holes have to migrate through one component to a suitable cathode, and electrons though the other component to a suitable anode. Mobilities for each charge carrier type must be high, and opportunities for recombination events must be minimized. This is clearly a very difficult task considering that the film consists of two interpenetrating, nanostructured networks with many interfaces and tortuous percolation pathways.

A solution to these problems was suggested by Michael Grätzel and his group in 1991 [8]. Inspired by natural photosynthetic systems, Grätzel demonstrated a photoelectrochemical device based upon decoupling the primary light-absorption and charge-separation events. Retaining the nanocrystalline concept, the design used a visible light-absorbing pigment chemically bonded in a monolayer to the surface of the semiconductor. This device became known as a nanocrystalline dye-sensitized solar cell (DSSC). The "sensitizing component" was chosen such that, in the photoexcited state, it was capable of injecting an electron into the conduction band of its semiconductor host. Additionally, Grätzel solved the interpenetration problem by using a liquid electrolyte containing a redox couple as the hole transport medium. The theory of semiconductor-electrolyte interfaces is well established: a clear and elegant summary of the energetics is given in Ref. [5]. Conventional thought would suggest that a potential difference is established at such an interface. The nature of this interface potential depends very much upon the details of the electrolyte solution (specifically its redox potential) and the Fermi energy of the electrons in the solid. The surface of the semiconductor can be *p*-type or *n*-type depending upon whether a depletion or accumulation layer is formed upon contact, i.e., whether the electrolyte preferentially accepts or donates electrons. Both situations can be manufactured by including electron- or hole-scavenging components in the electrolyte. The creation of a suitable contact potential would certainly aid the separation of a bound electron-hole pair at the "molecular" junction formed by the semiconductor and the photoexcited sensitizer. However, this simplistic picture is complicated by the fact that nanostructuring of the semiconductor has a profound effect upon its photoelectrochemical properties. Critically, a depletion (or indeed accumulation) layer cannot be formed in the solid since the nanocrystallites are tiny (on the order of tens of nanometers). The voltage drop within the particles remains very small and a significant local electric field cannot be maintained. Hence, it would seem that the photoresponse of such a system is dependent upon the relative rates of reactions of the positive and negative charge carriers with the electrolyte couple. Confirmation of this theory was provided by the demonstration of a switch between anodic and cathodic photoresponse simply by changing the hole/electron-scavenging nature of the electrolyte [5].

Irrespective of whether a local electric field aids the separation of charge, several facts are very clear. Firstly, the primary light-absorption and charge-separation events are decoupled in a DSSC. Secondly, the sensitizer provides a means of harvesting visible as well as UV solar photons. Thirdly, the liquid electrolyte is capable of fully percolating the nanocrystallite network and provides an effective medium for hole transport to the cathode. Hence, it would seem that the Grätzel cell design is an elegant solution to the problems associated with creating a large surface area for photon absorption and charge separation. Additionally, according to the criteria outlined in the introduction, the nanocrystallite dye-sensitized solar cell can clearly be thought of as an artificial photosynthetic system.

## 3.3
## Typical Components and Performance of a DSSC

### 3.3.1
### Construction and Mode of Operation

The schematic in Fig. 3.1 shows a typical DSSC design based upon nanocrystalline $TiO_2$. The support substrate is normally glass [5], although it is perfectly possible to use a flexible plastic substrate. Light is coupled into the cell through this side, and thus the support substrate must be transparent in the visible and near UV. A thin film of a transparent, conducting material is vacuum deposited on the inside of the support substrate and acts as the anode electrode. Semiconductors such as indium tin oxide (ITO) or fluorine-doped tin oxide ($SnO_2$:F) are normally used for this purpose. A porous film of the nanocrystalline semiconductor forms the actual photoanode. These films are usually between 1 and 10 microns thick and can be fabricated using a number of different techniques. The most popular method is to cast a slurry of the nanocrystals using spray, dip, spin, or drag coating and then to calcine the film at 400–450 °C to confer structural stability and create a percolated network. P25 is a nanocrystalline

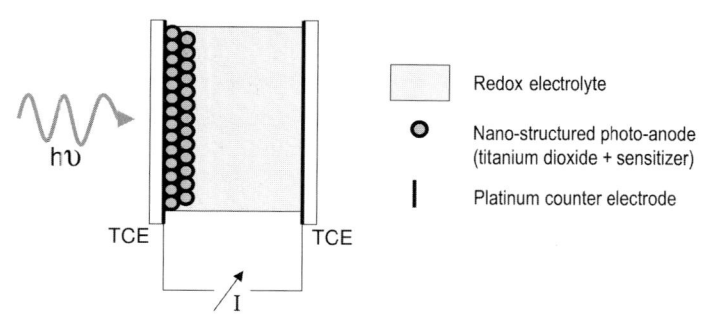

**Fig. 3.1** Schematic representation of a typical nanocrystalline titania dye-sensitized solar cell. Light is incident upon the device through the front transparent substrate and electrode (TCE = transparent conducting electrode). The sensitizer is bound to the surface of the nanocrystalline titania in a monolayer, and the whole photoanode is percolated with an electrolyte containing a redox couple. The circuit is complete by a counter electrode (cathode) made from a thin film of a low work-function material such as platinum.

product sold by BASF. It has a high anatase fraction (>80%) and, depending upon preparation of the colloid, contains 10–25 nm nanocrystallites. We have found that P25 films made using a standard method [9] have total surface areas of $\sim 50 \text{ m}^2 \text{ g}^{-1}$ and average pore diameters of 50–100 nm [10]. Several other hydrothermal methods can be used to produce $TiO_2$ of varying particle size and crystallinity [11]. Of particular interest is the microwave method of Wilson et al. [12], which produces particularly small crystallites (4–5 nm) with a high anatase fraction.

An alternative approach to producing highly porous nanostructured titania photoanodes is via surfactant templating of a titania precursor, e.g., an alkoxide such as titanium isopropoxide or titanium chloride templated with the pluronic surfactant P123 [7]. This sol-gel technique relies upon controlled hydrolysis and condensation of the precursor to create a structured network of titania, followed by removal of the template and calcination at temperatures exceeding 300 °C. Using this method, highly ordered mesophases can be created with well-controlled pore dimensions and geometries [7]. The films produced via this method have a completely different nanomorphology from those created using slurries of nanocrystals. In the former case, the titania forms the pore walls, while in the latter, the pores are actually the nanocrystal interstices. Fig. 3.2 shows a set of transmission electron micrographs highlighting the morphological differences. Although the nanocrystal slurry method is rather "quick and dirty," it

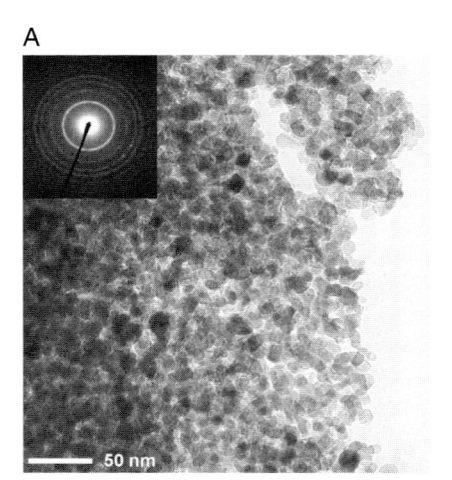

**Fig. 3.2** Bright field transmission electron microscopy (TEM) images showing the different nanomorphologies in nanocrystalline films made by hydrothermal (A) and mesoporous templating (B) techniques. The hydrothermal film contains individual nanocrystals sintered together to form a percolated network by calcination. The templated film (created from the hydrolysis and condensation of titanium isopropoxide and templated by the triblock copolymer P123) contains an ordered mesophase of pores whose walls are composed of nanocrystalline titania. The inset in Fig. 3.2A contains an electron diffraction pattern with circular diffraction rings characteristic of a polycrystalline material.

does seem to produce higher-efficiency photoanodes and is the technique of choice in most laboratories. This is likely due to better uptake of the sensitizer because of the larger pores and the higher degree of crystallinity. The templated mesoporous materials tend to contain some amorphous titania that hinders conduction. A relatively new and rather interesting approach is to combine the two methods in a two-step process: first, create extremely small (<5 nm) nanocrystals of anatase, and second, self-assemble these nanocrystals around a suitable surfactant template to create an ordered mesophase. This approach may ultimately lead to better control over morphology and crystallinity, and hence to improved engineering of the photoanode. We have recently adopted this approach with encouraging early results [13].

Irrespective of the exact morphology and crystallinity of the photoanode, the sensitizer must be adsorbed to its surface in a monolayer [5]. A significant amount of attention has been directed towards optimizing the sensitizer over the last decade or so [4]. The three key design parameters are:

1. The spectral absorption of the sensitizer: ideally, the material should be "black" in the UVA, visible, and near IR, with an extremely high, broadband absorption.
2. The energetics and dynamics of the coupling between the photoexcited state of the sensitizer and the conduction band of the photoanode: the transfer of an electron between the photoexcited sensitizer and the photoanode must be rapid (when compared to the lifetime of any other relaxation or quenching processes) and energetically favorable.
3. The photostability of the sensitizer: the material must be capable of many reduction-oxidation cycles without decomposition.

The best photovoltaic performances (power conversion efficiency and stability) have been achieved with polypyridyl complexes of ruthenium or osmium. The N3 ruthenium complex $cis$-$RuL_2(NCS)_2$, where L=2,2'-bipyridyl-4-4'-dicarboxylic acid as shown in Fig. 3.3, is the standard sensitizer, surpassed in performance only by "ruthenium black" (tri(cyanato)-2,2',2''-terpyridyl-4,4',4''-tricarboxylate RuII), which holds the world record power conversion efficiency for a DSSC (10.4% under AM1.5 global sunlight) [14]. The fully protonated N3 dye has absorption maxima at 380 nm and 518 nm, with respective extinction coefficients of $1.33 \times 10^4$ $M^{-1}$ $cm^{-1}$ and $1.3 \times 10^4$ $M^{-1}$ $cm^{-1}$. The dye is anchored to the titania surface through carboxylate ligands, and the main optical transitions have a metal-to-ligand charge transfer character (MLCT): an excited electron is transferred from the metal center to the $\pi^*$ system of the carboxylate ligand. This electron is subsequently injected into the titania conduction band within femto- to picoseconds, producing charge with near-unity quantum yield. If allowed to relax radiatively, the complex emits at 750 nm with a characteristic lifetime of 60 ns. Hence, the conduction band injection process is by far the most rapid process.

In order for the injection process to also be energetically favorable, the potential of the excited-state electron in the sensitizer must be higher than the conduction band edge of the titania [15]. Fig. 3.4 shows these parameters for N3 adsorbed onto

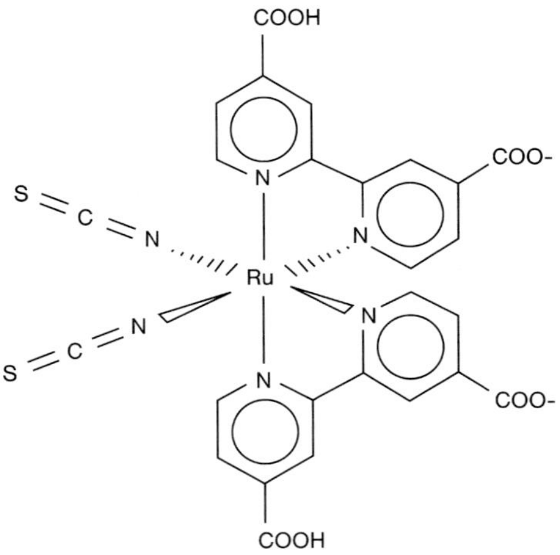

**Fig. 3.3** Structure of the ruthenium complex N3 (adapted from [4]).

undoped titania with a bandgap of 3.2 eV. Also shown in Fig. 3.4 is the redox potential of the electrolyte containing a typical iodine-iodide couple. The role of the electrolyte is to "transport" the hole that is generated by charge separation to the cathode. Stated another way, the electrolyte must re-reduce (donate an electron from the redox couple) the oxidized sensitizer molecule in order to regenerate the system for another cycle. The electrolyte itself is reduced at the counter electrode, usually a low work-function metal such as platinum, in order to complete the circuit. Critically, the redox potential of the couple must be higher than the ground-state potential of the sensitizer. The difference between the quasi-Fermi level potential of the semiconductor and the redox potential of the electrolyte defines the maximum open-circuit voltage that can be generated by the cell. It is also worth noting that this type of cell is a majority carrier device, since the hole (although we describe it as being "transported" to the cathode) remains localized on the oxidized dye molecule. The overall electric field that will tend to draw electrons to the anode, and holes to the cathode, is generated by the difference in work functions between the transparent electrode (ITO or $SnO_2$:F) and the metallic counter electrode. This is typically on the order of $\sim 0.5$–$0.6$ volts.

Summarizing the mode of operation of a typical DSSC such as the one shown in Fig. 3.1:

1. A photon is absorbed by the sensitizer; the photoanode (nanostructured titania plus sensitizer) presents a very large surface area for photon absorption.
2. The photoexcited sensitizer injects an electron into the conduction band of the titania within femto- to picoseconds of the primary absorption event.
3. The injected electron will diffuse through the nanostructured titania and be delivered into the external circuit by the transparent anode electrode.
4. The oxidized dye is reduced by donation of an electron from the redox couple, and the electrolyte is in turn reduced at the counter electrode as the circuit is completed.

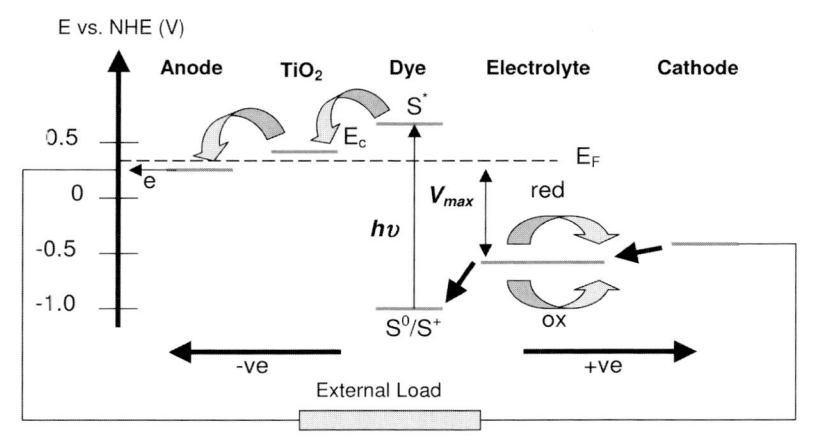

**Fig. 3.4** A simplified energy level scheme for a DSSC based upon nanocrystalline titania and a ruthenium charge transfer complex as sensitizer. NHE refers to the standard hydrogen electrode reference.

For a more comprehensive discussion of the energetics and thermodynamics of light-induced redox reactions in such systems, the interested reader is directed towards Ref. [15].

### 3.3.2
### Typical DSSC Performance

The performance of a solar cell is defined by a number of key parameters:

1. The short-circuit current ($I_{sc}$) and open-circuit voltage ($V_{oc}$) generated under standard illumination conditions (global AM1.5).

2. The fill factor (FF) of the cell under AM1.5 illumination. This quantity is obtained by a full current-voltage characterization and is essentially a measure of the diode behavior of the cell. The FF is given by

$$\text{FF} = \frac{P_{\max}}{V_{oc}I_{sc}} = \frac{(VI)_{\max}}{V_{oc}I_{sc}} \tag{1}$$

3. The incident photon conversion efficiency (IPCE). This is an incident energy-dependent quantity and is a measure of the useful range of the cell. The IPCE at a given illumination wavelength $\lambda$ under incident optical power $P_{in}$ is given by

$$\text{IPCE} = \frac{P_{in}}{I_{sc}} \cdot \frac{e\lambda}{hc} \tag{2}$$

where $e$ is the fundamental electronic charge, $h$ is Planck's constant, and $c$ is the speed of light in vacuum.

4. The global power conversion efficiency ($\eta$) of a cell producing $P_{out}$ electrical power under standard AM1.5 illumination conditions given by

$$\eta = \frac{P_{out}}{P_{in}} = I_{sc} V_{oc} \frac{FF}{P_{in}} \quad (3)$$

Under full sunlight conditions (simulated by the AM1.5 standard illumination), with an incident radiant flux of 100 mW cm$^{-2}$, the best nanocrystalline DSSCs based upon N3 or a similar ruthenium-based dye, produce short-circuit current densities of 16–22 mA cm$^{-2}$ and open-circuit voltages of 0.65–0.75 V. With fill factors of 0.65–0.75, these figures correspond to a maximum global power conversion efficiency of ~10.4%. Many laboratories report respectable efficiencies of 5–8%. Although these figures are fully competitive with the better amorphous silicon devices, the DSSCs are limited by dye lifetimes. In order to be a credible commercial alternative, the dye molecule must sustain at least 10$^8$ redox cycles to give a device a working life of 20 years. With the current ruthenium-based systems, this does not appear to be possible, although the use of purified solvents such as $\gamma$-butyrolactone has produced devices capable of passing standard quality assurance tests for outdoor use [16].

Fig. 3.5 shows a typical normalized IPCE profile obtained in our laboratory using the N3 dye in combination with a templated hydrothermal seed nanocrystalline TiO$_2$ photoanode. The electrolyte for this cell contained 0.5 M LiI and 0.04 M I$_2$ as the redox couple, in 0.05 M 4-*ter*-butylpyridine and acetonitrile. Clearly, the N3 dye is capable of producing a photoresponse from ~400 nm to ~650 nm, i.e., spanning a good proportion of the visible spectrum. The I–V characteristics of this cell are shown in Fig. 3.6, and we see an open-circuit voltage of 0.717 V, a short-circuit current density of 12.12 mA cm$^{-2}$, and a fill factor of 0.58, yielding a respectable power conversion efficiency of ~5% under AM1.5 illumination of

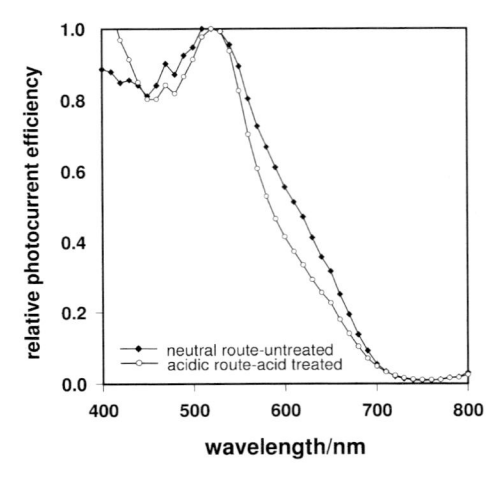

**Fig. 3.5** Incident photon conversion efficiency (IPCE) for a DSSC made from a templated hydrothermal seed nanocrystalline titania (Fig. 3.2A) sensitized with N3. The IPCE has been normalized at the peak response for clarity, and the graph shows two nanocrystal preparation methods [13].

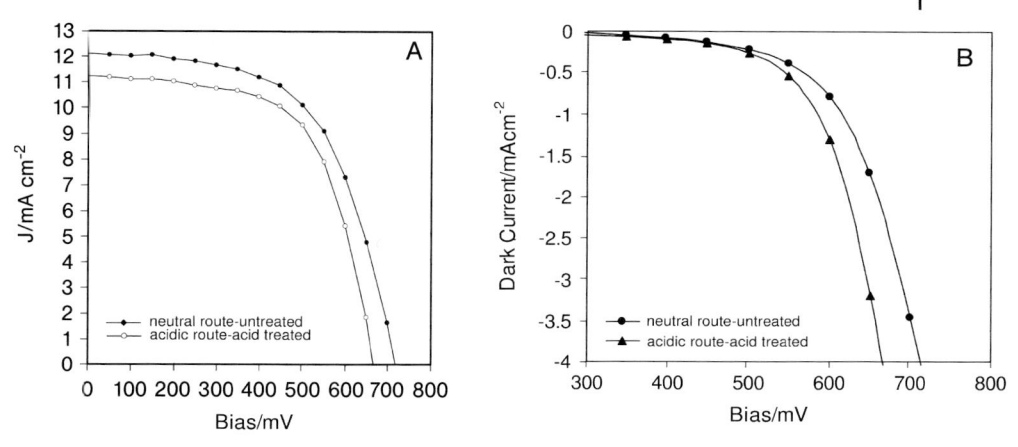

**Fig. 3.6** I–V response of the N3 cell from Fig. 3.5 under white light AM1.5 illumination (A) and in the dark (B).

$\sim 150$ mW cm$^{-2}$. No corrections to this data were made for factors such as back and front surface reflection losses or back surface leakage.

### 3.3.3
### Device Limitations

Currently DSSCs are limited by the following factors:

- the lifetime (stability) of the sensitizer,
- the low electrical conductivity of the nanocrystalline photoanode,
- the stability of the electrolyte system, and
- the overall stability of the cell, given that it contains a liquid component capable of leaking and evaporating if not completely sealed.

Numerous approaches have been suggested to address each of these limitations. Notably, replacement of the liquid electrolyte with a solid hole transport or ionic medium would solve the stability- and leakage-related issues. For example, cells based upon a spiro bis-fluorene-connected arylamine hole transmitter have produced a full sunlight efficiency of 2.56% [17]. Likewise, systems incorporating inorganic p-type semiconductors such as CuI or CuSCN have yielded 1% power conversion efficiencies [18]. In the inorganic case, difficulties associated with semiconductor photostability, and in creating good electrical contact with the nanocrystalline TiO$_2$, will probably limit the use of such materials. Currently, organic hole conductors, which can be applied as liquids and then cured into solids, would seem the most likely candidates to replace the liquid electrolyte [19].

The way forward with respect to increasing photoanode conductivity probably lies in better control of the film nanomorphology. The optimum system is likely to be one in which 100% anatase nanocrystals are assembled into pseudo 1D nanostructures orientated perpendicular to the support substrate. This would favor electron transport to the anode by minimizing lateral percolation pathways. In principle, it is already possible to build TiO$_2$ "micro-pillars" [20], and if this approach could be extended into the nano-regime in order to maintain high internal surface areas, it may be possible to build an orientated photoanode for DSSC applications. An alternative approach, and one that is more realistic in the short term, is to create morphology control with templating. It appears as though the type of mesoporous networks that can be routinely produced using precursors such as titanium isopropoxide and templates such as block copolymer surfactants are not suitable for DSSCs [7], even though they have surface areas exceeding 150 m$^2$ g$^{-1}$. This is likely due to the reduced pore size (relative to the random nanocrystal films) and the "closed" nature of the mesophase, i.e., the sensitizer molecules cannot penetrate the structure. A possible solution to these issues would be to manufacture a mesostructured film with lamellar pores orientated perpendicular to the substrate and pore walls of pure nanocrystalline anatase. To the authors' knowledge, such a film has yet to be fabricated, but with the rapid advances being made in molecular self-assembly, it may be possible in the near future. Finally, a two-stage anatase nanocrystal templating process as discussed in Section 3.3.2 may be a realistic near-term solution. If tiny "seed" crystals can be assembled around a block copolymer template, and the system subsequently calcined to remove the surfactant and form the percolated nanocrystal network, then better morphology control would undoubtedly lead to improved photoanode conductivity and hence device performance.

The ruthenium N3 and ruthenium black dyes are the best sensitizers (in terms of efficiency) demonstrated so far. They have relatively broad absorption profiles, display very rapid injection dynamics, have long-lived excited states, and can be engineered to adsorb strongly to titania. On the downside, they are expensive to manufacture, do not display the desired long-term stability, and do not have the ability to harvest near-infrared photons. Several alternatives to the ruthenium or osmium MLCT complexes have been suggested [21–23]. The next section will deal with one material in particular – a family of broadband-absorbing and ultra-stable biomacromolecular pigments called the melanins.

## 3.4
## Melanins as Broadband Sensitizers for DSSCs

### 3.4.1
### Melanin Basics

The melanins are an important class of pigmentary macromolecules found throughout nature [24]. Eumelanin is the predominant form in humans and acts as the primary photoprotectant in our skin and eyes. Unique among biomacromolecules, the melanins are broadband UV and visible light absorbers (Fig. 3.7). Their spectra display no chromophoric bands and follow a smooth single exponential dependence on wavelength increasing monotonically towards

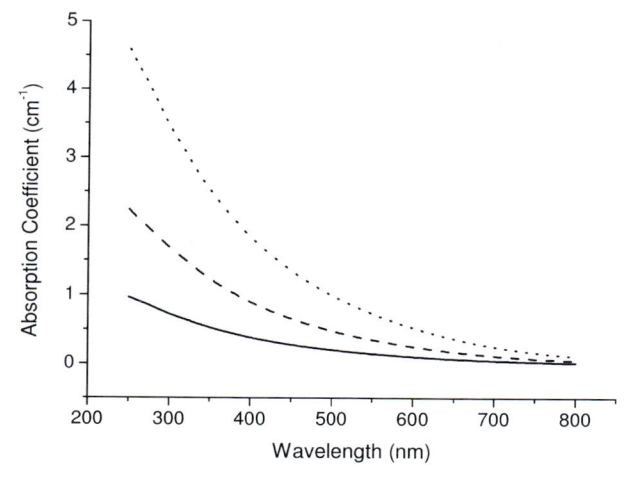

**Fig. 3.7** Absorption coefficient vs. wavelength in the near UV, visible, and near IR of a synthetic eumelanin (aqueous solution) at three different concentrations (0.005% by weight, dotted; 0.0025% by weight, dashed; 0.001% by weight, solid).

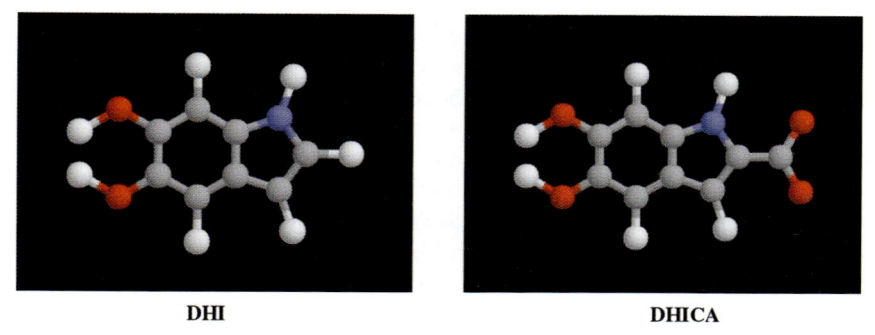

**DHI**  **DHICA**

**Fig. 3.8** Structures of 5,6-dihydroxyindole (DHI) and 5,6-dihydroxyindole, 2-carboxylic acid (DHICA), the monomeric precursors of eumelanin.

the UV [25]. Most melanins are also chemically and photochemically very stable and are potent free radical scavengers and antioxidants [24]. In direct contradiction with its photoprotective properties, eumelanin (along with pheomelanin, the less prevalent red-brown pigment also found in humans) is implicated in the cytotoxic chain of events that ultimately lead to melanoma skin cancer [26]. For this reason, the photophysics, photochemistry, and photobiology of melanins are subjects of intense scientific interest.

In a more general sense, the broader structure-property-function relationships that dictate the behavior of these important biological macromolecules are still poorly understood [27]. In particular, major questions still remain concerning the basic structural unit [28]. It is fairly well accepted that eumelanins are macromolecules of 5,6-dihydroxyindole (DHI) and 5,6-dihydroxyindole-2-carboxylic acid (DHICA) (Fig. 3.8), and that pheomelanins are cysteinyl-dopa derivatives [24, 29]. However, it is still a matter of debate as to whether eumelanin (in particular) is actually a highly cross-linked extended heteropolymer or is composed of DHI/DHICA oligomers condensed into four or five oligomer nanoaggregates [30]. This is an absolutely fundamental issue and is the starting point for the construction of consistent structure-property-function relationships. The answer to this question also has profound implications for our understanding of the condensed-phase properties of melanins. In 1974, McGinness, Corry, and Proctor showed that a pellet of dopa melanin could be made to behave as an amorphous electrical switch [31]. They postulated that these materials might be disordered organic semiconductors. Several studies since have claimed to show that melanins in the condensed solid state are indeed semiconductors [32, 33]. However, it is by no means certain that the conductivity reported is electronic in nature. A clear idea of the basic structural unit is fundamental to developing a consistent model for condensed-phase charge transport in such disordered organic systems.

From the preceding discussion it is clear that melanins may have some potential as photon-harvesting systems. Their broadband absorbance and photostability are key attributes, as are their biocompatibility and bioavailability. However, as we saw in Section 3.3, sensitizer excited-state energetics and compatibility with a suitable nanostuctured semiconducting host are also key considerations in DSSC applications. Work in our laboratory over the past few years has concentrated on understanding these important properties with respect to natural melanins and tailored synthetic analogues. This application-focused work is part of a broader program of condensed matter physics, synthetic organic chemistry, quantum chemistry, and molecular biophysics aimed at gaining a more complete understanding of melanin structure-property-function relationships. In the following sections, we outline some recent results pertinent to the photon-harvesting concept and, ultimately, demonstrate that it is possible to couple these materials to a nanocrystalline $TiO_2$ host in order to produce visible light sensitization and hence a regenerative DSSC system.

3.4.2
**Melanin Chemical, Structural, and Spectroscopic Properties**

As highlighted above, melanins are broadband absorbers with a monotonic exponential dependence of extinction upon wavelength. Fig. 3.7 shows typical absorption spectra for a synthetic dihydroxyindole eumelanin in aqueous solution. Fig. 3.9 summarizes the absorption coefficient values at 380 nm for the three concentrations of Fig. 3.7. It is instructive to use this data to calculate molar extinction coefficients and to compare with a typical ruthenium sensitizer such as N3. Unfortunately, due to the difficulties associated with determining the molecular weight of the eumelanin sample in solution [34], a molar-based compari-

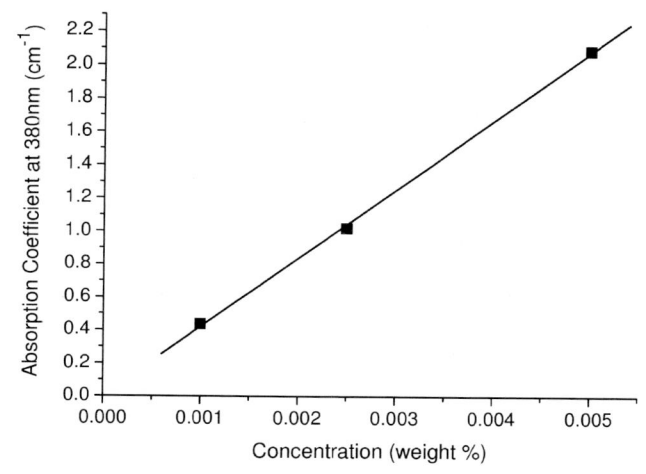

**Fig. 3.9** Absorption coefficient at 380 nm vs. concentration for the three synthetic eumelanin solutions in Fig. 3.7. Concentration errors were estimated to be $1\times10^{-4}\%$ by weight.

son cannot be performed. However, it is possible to calculate a weight-based extinction coefficient, and this comparison (Table 3.1) shows that the synthetic eumelanin is a more effective absorber gram for gram than N3 at both UV and visible wavelengths.

Melanins have also been shown to exhibit some rather exotic emission and excitation as well as absorption behavior. Several steady-state photoluminescence studies have demonstrated that melanins possess an extremely low radiative quantum yield (i.e., the fraction of excitation photons that ultimately lead to a spontaneously emitted photon) [35]. We have recently reported the first accurate measurement of yield for a synthetic eumelanin and found it to be $<6\times10^{-4}$, i.e., 99.94% of absorbed photon energy is dissipated non-radiatively [34]. This measurement is particularly difficult for a number of reasons: firstly, the emission is extremely weak (as evidenced by the very low quantum yield), and secondly, melanins have strong broadband absorption – a fact that leads to heavy

**Table 3.1** Comparison of extinction coefficients at 380 nm and 518 nm (weight-based) for the ruthenium N3 complex and synthetic eumelanin.

| Compound | Extinction coefficient at 380 nm (g cm$^{-1}$) | Extinction coefficient at 518 nm (g cm$^{-1}$) |
| --- | --- | --- |
| N3 | 17.9 | 42.0 |
| Synthetic eumelanin | 17.5 | 24.0 |

emission reabsorption and probe beam attenuation. We have developed a method for accounting for these factors and hence have been able to properly quantify the photoluminescence emission spectra. Fig. 3.10 A, B shows a series of these spectra for three aqueous solutions of eumelanin (0.001%, 0.0025%, and 0.005% by weight). Fig. 3.10 A contains the raw data – heavily affected by probe beam attenuation and emission reabsorption (as evidenced by the inconsistent peak position and lack of correspondence between peak height and concentration). Fig. 3.10 B, C shows how the expected linear relationship between emission and concentration is recovered after re-correction, as is the consistency in peak position. In general, the photoluminescence is characterized by a broad, single peak with a red shift of $\sim 100$ nm relative to the excitation wavelength.

Eumelanin photoluminescence shows a number of other unusual features. Fig. 3.11 demonstrates how the emission is strongly dependent upon excitation energy (wavelength). Peak position and height change as a function of pump energy, with the red edge appearing to collapse towards a fixed value. Surprisingly, the radiative quantum yield (Fig. 3.12) also shows strong excitation dependence – an extremely unusual property among organic chromophores. We believe that our observations are consistent with an ensemble model: many chemically distinct species existing within the system, each with a different HOMO-LUMO gap (difference in energy between the highest occupied and lowest unoccupied molecular orbitals), emission characteristics, and radiative quantum yield. The broad emission is a result of the overlap of individual emission peaks, and the excitation energy dependence arises due to selective pumping of a particular subset of the ensemble. However, these chemically distinct species could only produce different optical signatures if they were small (i.e., non-polymeric molecules). Hence, our data lend further credence to the argument that the basic structural unit of melanins (eumelanin in particular) is oligomeric rather than heteropolymeric in nature. This view is also supported by time-resolved photoluminescence emission studies on eumelanin solutions, which consistently show non-exponential behavior and can only be reconciled by fitting the data to a distribution of lifetimes (Fig. 3.13). The ensemble model could also explain the broadband absorption of melanins. The distribution of HOMO-LUMO gaps (each manifesting itself as an inhomogeneously broadened Gaussian feature in the absorption spectra) could overlap to produce a single monotonic profile.

The above argument clearly relies upon the fact that small macromolecules can form and that, furthermore, these distinct macromolecules have different fundamental absorptions (HOMO-LUMO gaps). Hence, we are once again forced to return to the vexing question of melanin fundamental structure. In the case of eumelanins, the macromolecular entities are based upon DHI and DHICA. It is well known that these two monomers exist in various redox forms, namely the indole quinone (IQ), semiquinone (SQ), and hydroquinone (HQ) [36]. These structures are shown in Fig. 3.14. Quantum chemical methods have proven somewhat useful in predicting the electronic and vibronic properties of these monomeric redox forms and the macromolecules that can be as-

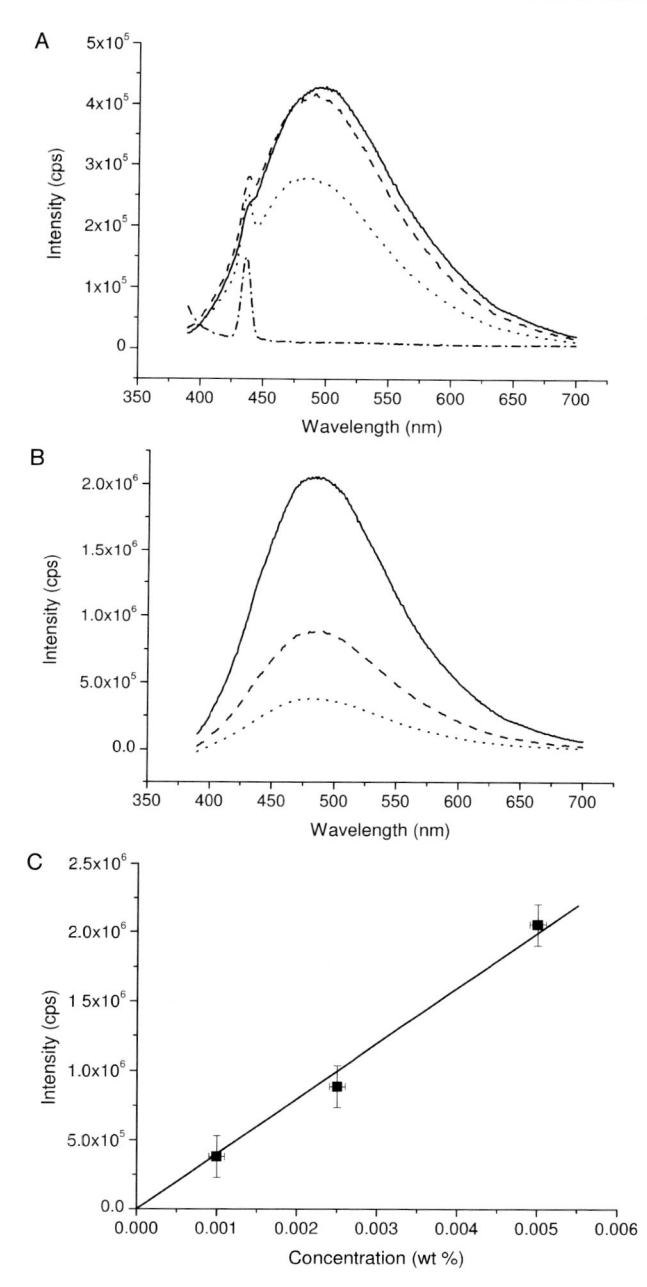

**Fig. 3.10** Raw (A) and re-corrected (B) photoluminescence (PL) emission spectra (pumped at 380 nm) for three synthetic eumelanin solutions: 0.005% (dotted line); 0.0025% (dashed line); and 0.001% (solid line) by weight concentration, and solvent background (dot-dash line). (C) Corrected PL emission peak intensity vs. concentration for the three synthetic eumelanin solutions.

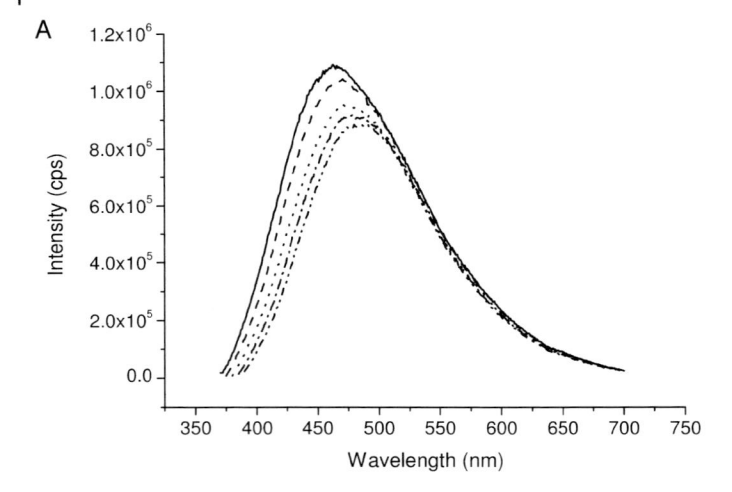

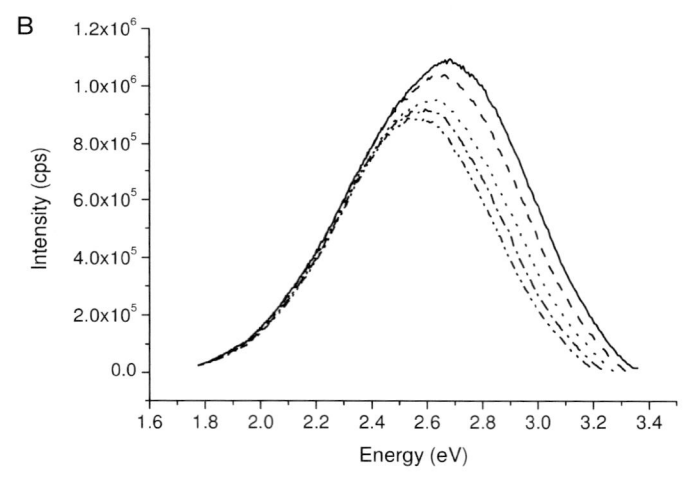

**Fig. 3.11** Re-corrected PL emission spectra for a 0.0025% by weight synthetic eumelanin solution: (A) plotted vs. wavelength and (B) plotted vs. energy for five pump wavelengths: 360 nm (solid line) to 380 nm (inner dashed line) in 5-nm increments.

sembled from them. The first major attempt to study melanins in this way was due to Longuet-Higgins [37], who proposed that the optical (and indeed electronic) properties could be explained if the macromolecule was an infinite homopolymer – an assumption that we now know to be incorrect. Later work used semi-empirical methods [38] such as Hückel theory [39–42] and the intermediate neglect of differential overlap (INDO) [43, 44]. These techniques were applied up to the small oligomer level, but, crucially, the simulated spectra did not show the expected broadband absorption.

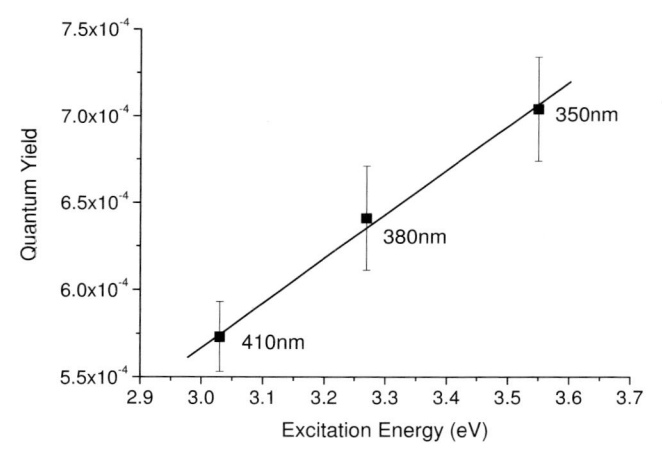

**Fig. 3.12** Radiative relaxation quantum yield for synthetic eumelanin pumped at three wavelengths (350 nm, 380 nm, and 410 nm). The solid line is a linear fit that is meant only as a guide to the eye.

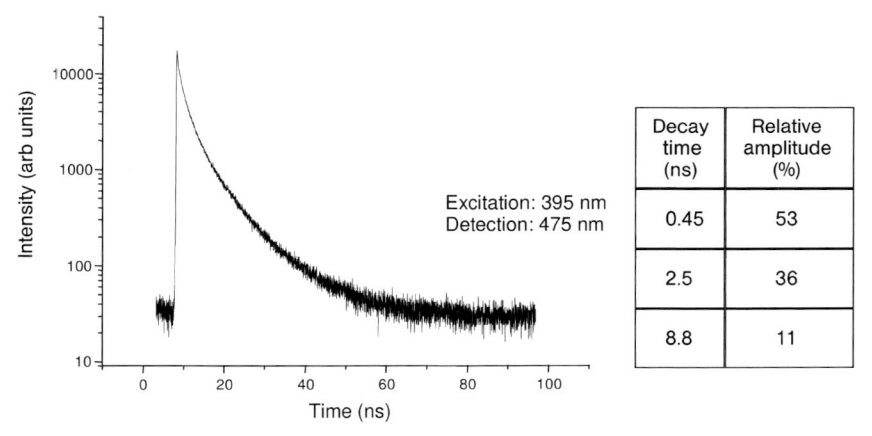

| Decay time (ns) | Relative amplitude (%) |
|---|---|
| 0.45 | 53 |
| 2.5 | 36 |
| 8.8 | 11 |

**Fig. 3.13** Time-resolved PL emission spectrum of a 0.001% by weight synthetic eumelanin solution pumped at 395 nm and monitored at 475 nm. The response is clearly not single exponential, and the table shows the major lifetime components (with their relative amplitudes).

More recently we (and others) have applied ab initio methods, particularly density functional theory (DFT) [45], to both DHI [27, 36, 46] and DHICA. In DFT one maps the Schrödinger equation onto a non-interacting equation for the electronic density, known as the Kohn-Sham equation. While it can be shown that this mapping gives (at least in principle) the exact ground-state energy, in practice one important term called the exchange correlation function ($E_{xc}$) must be approximated. There are basically two approaches to approximating $E_{xc}$: either the fact that DFT is in principle exact is used to derive $E_{xc}$ (the

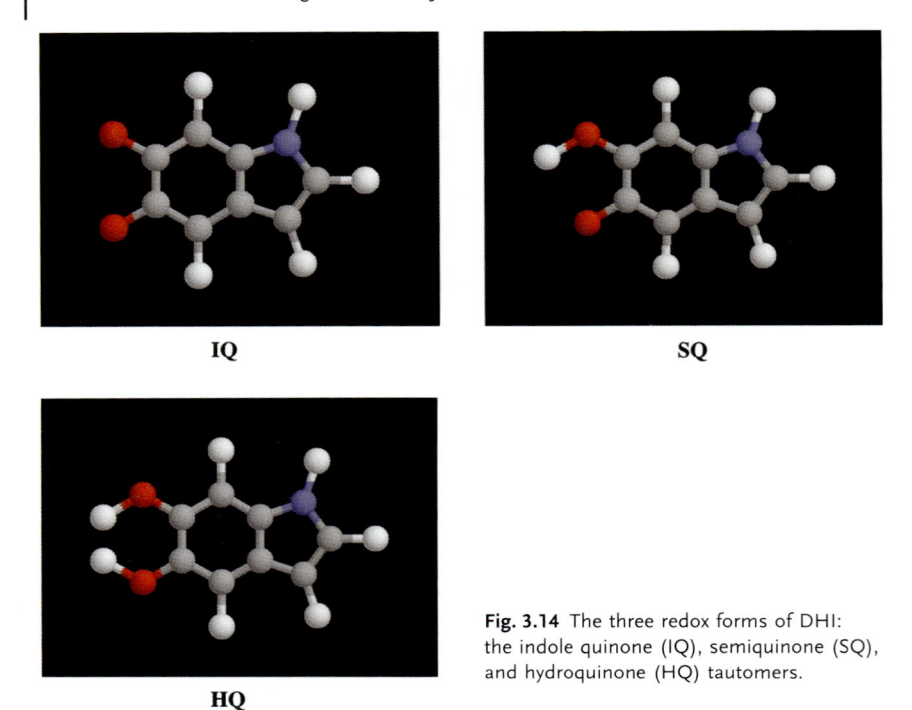

**Fig. 3.14** The three redox forms of DHI: the indole quinone (IQ), semiquinone (SQ), and hydroquinone (HQ) tautomers.

current state-of-the-art functional of this type is due to Perdew, Burke, and Ernzerhof (PBE) [47]), or else $E_{xc}$ is extracted empirically from a large number of experimental results (for example, the hybrid of Becke's three-parameter exchange functional and the gradient-corrected functional of Lee, Yang, and Paar (B3LYP) [48]). Both types of functionals have been successfully applied to study the building blocks of eumelanin. One major drawback with DFT is that, as it is a theory of the ground state, excited-state properties are not given accurately by the simple interpretation of the Kohn-Sham eigenvalues. This is known as the bandgap problem [45]. The optical absorption of a molecule is essentially an excited-state property. Both time-dependent DFT (TDDFT) calculations and the difference of self-consistent fields ($\Delta$SCF) method have been applied to DHI and DHICA to avoid the bandgap problem. However, TDDFT is a rather immature method; the functionals are not as well developed as those for standard DFT. In the $\Delta$SCF method, DFT is recast so that instead of calculating the true ground state, one calculates the ground state subject to the constraint of a given set of excited-state quantum numbers. This adaptation appears to predict excited-state properties rather well [45].

In Table 3.2, we compare the results from the various published calculations. It can be seen that the trends found by all of the methods are in broad agreement. More importantly, the quantitative results of the ab initio calculations (TDDFT and $\Delta$SCF) are in excellent agreement. Critically, these results indicate that the HOMO-LUMO gaps of the DHI redox forms (IQ, SQ, and HQ) are appreciably different. In addition, it has been recently demonstrated that several redox forms and charge states of DHI [27, 30, 36, 44] and DHICA are thermodynamically stable. Hence, given that the basic monomer can cross-link in a number of possible positions (2, 3, 4, and 7), we are led to conclude that a wide variety of "chemically distinct" oligomers, each with a different HOMO-LUMO gap, can form. As the HOMO-LUMO gap is closely related to the fundamental absorption, it is reasonable to expect that absorption spectra derived from such an ensemble of monomers and oligomers would contain a large number of overlapping optical absorption peaks. In the limiting case, if the system contained enough of this "chemical disorder," then the individual peaks would become smeared into a monotonic absorption, or a single broad emission feature in the case of photoluminescence.

Presently, this disordered ensemble model is no more than a hypothesis. More advanced DFT calculations are underway to predict the stability and HOMO-LUMO gaps of larger eumelanin oligomers that result from the cross-linking of DHI and DHICA redox forms and charged states. We are also extending this study to pheomelanin (a cysteinyl-dopa-derived melanin), and other synthetic eumelanins derived from protected forms of DHI and DHICA.

Once validated by direct comparison with experiment, this new structural information will be useful from a number of perspectives. Firstly, it will form the basis of mesoscopic models to explain phenomena such as electrical conductivity and photoconductivity; secondly, it will guide molecular engineering strategies as we seek to optimize functionality for a particular application (e.g., photovoltaics); and thirdly, it will form an integral part of broader attempts to explain melanin structure-property-function relationships in biology. With respect to the

**Table 3.2** The calculated HOMO-LUMO gaps (in eV) for some key eumelanin monomers. The reduced forms of DHI and DHICA are indicated by numbers in parentheses after the name. These numbers correspond to the location from which the hydrogen atom is removed. The functional used is listed in parentheses after the name of the method.

| Molecule | TDDFT [46] (PBE/B3LYP) | TDDFT [46] (B3LYP) | $\Delta$SCF (PBE) [36] | DFT (PBE) [36] | Hückel [36, 40] |
|---|---|---|---|---|---|
| DHI | 4.53 | 4.30 | 3.61 | 3.48 | 3.40 |
| DHI (1,5) | 1.50 | 1.43 | 1.12 | 0.80 | 0.84 |
| DHI (5,6) | 1.82 | 1.79 | 2.02 | 1.07 | 1.30 |
| DHICA | – | – | 3.04 | 2.85 | – |
| DHICA (5) | – | – | 2.67 | 2.24 | – |
| DHICA (6) | – | – | 2.64 | 2.36 | – |
| DHICA (5,6) | – | – | 1.96 | 0.87 | – |
| DHICA (5,1) | – | – | 1.10 | 0.78 | – |
| DHICA (6,1) | – | – | 1.25 | 0.89 | – |

last point, it is an interesting fact (especially in light of the subject matter of this book – artificial photosynthesis) that biomimetic strategies often concentrate on imitating the exquisite structural order that pervades the natural world. In a general sense, many biological macromolecules derive their functionality from precise and directed ordering at several structural levels. Structural biologists and molecular biophysicists are preprogrammed into looking for this directed structure to explain properties and function. If the disordered ensemble model of melanins is correct, the structure-property-function relationships of these materials could represent a significant paradigm shift – functional utility, flexibility, and robustness are derived from chemical and structural *disorder* rather than *order*. It is a common mantra in melanin biophysics that no two melanin molecules are ever made the same. In the context of the disorder model, there may well be a very good reason for this fact.

### 3.4.3
### Melanin Electrical and Photoconductive Properties

In 1974, McGinness et al. [31] showed that a solid pellet of eumelanin (the material is normally a powder when extracted from pigment-containing tissues or when synthesized chemically) could conduct electricity. At applied electric fields of $\sim 350\,\mathrm{V\,cm^{-1}}$, the material was observed to switch between "low" ($\sim 10^{-5}\,\mathrm{S\,cm^{-1}}$) and "high" ($10^{-3}\,\mathrm{S\,cm^{-1}}$) conducting states. This behavior led the authors to postulate that eumelanin was acting as an amorphous semiconductor, a hypothesis originally advanced in a landmark theoretical paper by Longuet-Higgins in 1960 [37]. The amorphous semiconductor model of eumelanin is now relatively widely accepted. A number of experimental studies have claimed not only to confirm this viewpoint but also to derive bandgaps, activation energies, carrier types, and densities, etc., using DC and AC conductivity measurements, thermopower, photoconductivity, and optical absorption in combination with standard Mott-Davies theory [32, 49]. What is worrying about many of these studies is the lack of any serious consideration of the effect of adsorbed water: as demonstrated by Jastrzebska et al., eumelanin conductivity is highly dependent upon atmospheric relative humidity [50]. The latter authors also claim to show some temperature dependence of the conductivity ($\sigma$) but make measurements only over a very limited range (290–340 K in vacuum). They go on to fit the data using the standard thermally activated semiconductor model (Eq. 4) and extract an activation energy ($E_\mathrm{a}$). This is clearly inadequate: normally, one requires at least two to three orders of magnitude in temperature before drawing such conclusions.

$$\sigma = \sigma_0 \exp(-\Delta E_\mathrm{a}/kT) \tag{4}$$

A general model for electrical conductivity in condensed-phase eumelanin remains elusive. Decoupling the effect of adsorbed water requires high vacuum, and to extract any meaningful information from DC conductivity vs. temperature data requires measurements over several orders of magnitude. Fig. 3.15 shows the conductivity of a pressed pellet of synthetic eumelanin (derived from DC current-voltage measurements) as a function of relative humidity at room temperature. It shows an extremely strong dependence – greater than five orders of magnitude in 80%. Apart from suggesting that these pellets may be very sensitive relative humidity sensors, it confirms the dominating nature of adsorbed water [51]. Measurements under vacuum as a function of temperature produce dramatically different results. At $10^{-6}$ torr, even below 10 K, we observed the samples to be too insulating for a conventional I–V system to make a meaningful measurement of current at applied voltages of < 20 V. Consequently, we can say that the resistance of the sample was > 1 G$\Omega$, with corresponding electrical conductivities of < $10^{-9}$ S cm$^{-1}$. Clearly, any electronic contribution to the conductivity, which would have been evident under these conditions, is extremely small.

Considering Fig. 3.15 and the vacuum conductivity measurements in isolation, one might be led to conclude that eumelanin is essentially an insulator in the condensed solid state and that all the conductivity was derived from ionic sources (adsorbed water). However, there are several reports [50] of solid films or pellets of eumelanin producing a significant photocurrent under white light illumination. Fig. 3.16 shows the photocurrent produced by a synthetic eumelanin film produced by the electropolymerization of dopa [52]. These films are structurally more continuous than the pressed powder pellets and display higher room temperature electrical conductivities (although from the preceding discussion, this is likely due to water absorption capacity). The photocurrent was

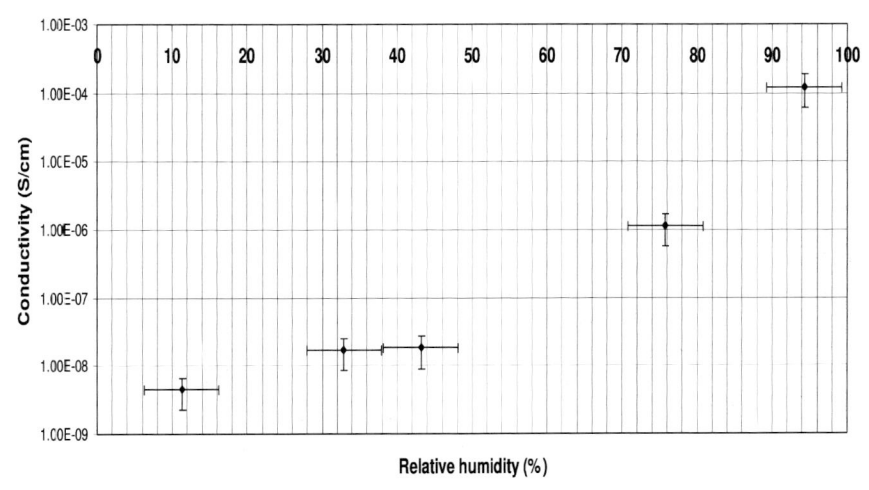

**Fig. 3.15** Conductivity (log) vs. relative humidity at room temperature for a solid pellet of synthetic eumelanin. Different relative humidity values were achieved using a range of saturated salt solutions, and the electrical measurements were made in a van der Pauw configuration.

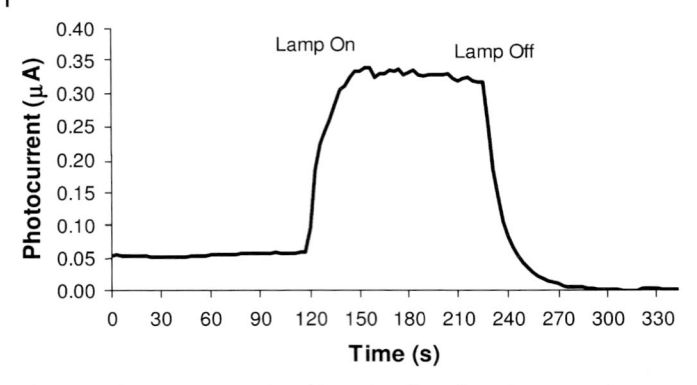

**Fig. 3.16** Photocurrent produced by a thin film of synthetic eumelanin produced by electropolymerization. The photocurrent was generated by illuminating the film with white light from a 150-W Hg vapor lamp, and an electric field of $\sim 65\,V\,cm^{-1}$ was used in a two-electrode configuration.

produced by dropping a 20-V potential across electrodes ($\sim 3$ mm apart) deposited on the film surface. Illumination was provided by an Hg vapor lamp (150 W). Under these conditions a photocurrent of $\sim 0.3\,\mu A$ was generated. The effect was stable and repeatable. The film showed capacitive recovery after the lamp was switched off, consistent with a semi-insulating material. Additionally, illumination caused a heating of the material and subsequent loss of bound water. This was shown by the fact that the instantaneous resistance of the film after illumination was higher than before illumination, but recovered to the original equilibrium value after several minutes.

In general, the production of a photocurrent under white light illumination is indicative of some "semiconductor-like" behavior. Ultraviolet and visible light stimulates a significant number of charge carriers, an observation that can only be reconciled with a bandgap (at least of one component within the system) of less than $\sim 3.5$ eV. Hence, given the conductivity and photoconductivity data, we are forced to consider a hybrid model for charge transport in eumelanins (and likely all other melanins in the condensed solid state): small semiconducting regions (grains) containing delocalized electrons, coupled together by an essentially ionic (and dominating) percolated medium of hydrated amorphous material. It does appear as though the amorphous semiconductor theory first proposed by Longuet-Higgins [37], and supported by McGinness et al. [31] and many others since [49], may be an oversimplification of a much more complicated hybrid situation. More detailed electrical studies and a predictive mesoscopic model based upon an accurate description of the melanin basic structural unit are required in order to advance this hypothesis.

### 3.4.4
### Melanins as Broadband Photon-harvesting Systems

Returning to the key motivation of this chapter, i.e., biomaterials as broadband photon-harvesting systems, in way of summary, we now ask the two key questions: have the melanins got what it takes to be photovoltaic materials, and can they be integrated into a suitable device platform?

Addressing the first question, clearly melanins (eumelanin in particular) have strong, broadband absorbance, comparable with a typical sensitizer such as N3. At least some components within the "eumelanin" ensemble have the potential to absorb solar radiation (UV and visible) and generate photoexcited charge carriers. These are both positives from the photovoltaic perspective, as are the aforementioned attributes of biocompatibility, bioavailability, ease of synthesis, chemical and photochemical stability (capability to cycle through many redox cycles), and general "green credentials." On the downside, the extremely low radiative quantum yield is indicative of strong excited-state phonon coupling. Thermal relaxation processes are characteristically very rapid (picoseconds). Ideally (referring back to the Introduction), a long excited-state lifetime in the photon-harvesting molecule is desirable in order to promote charge separation. However, the origins of the non-radiative relaxation modes in melanins are by no means clear. Paradoxically, any radiative decay is characterized by nanosecond lifetimes – an encouraging fact from the photovoltaic perspective. Clearly, we need to understand more about the energy dissipation pathways in these systems before attempting to engineer a long excited-state lifetime. Finally, with reference to the charge transport in eumelanins, we have seen that electrical conductivity is dominated by adsorbed water and not by electronic processes. Hence, one could conclude that, in a solid thin film of these materials, the mean free path of any delocalized electron or hole would be very limited indeed. If one takes the oligomeric nanoaggregate model as a realistic description, the mean free path would likely correspond to a single oligomer unit, i.e., a few nanometers. It should be noted that no attempts have yet been made to dope these systems. So-called "inherently conducting polymers" display similarly low electrical conductivities, poor mobilities, and mean free paths in their undoped states.

Addressing the second question, on the positive side it is possible to produce tailored synthetic analogues of eumelanins (and other melanins) in order to maximize adsorption to and coupling with a substrate surface. For example, DHICA macromolecules can be synthesized with a high degree of carboxylation in the 2 position. This is a favored ligand for adsorbing and binding ruthenium sensitizers to titania in the DSSC platform. It is also possible to electropolymerize synthetic eumelanins into the pores of nanocrystalline titania. However, sufficient control has not yet been achieved to produce a monolayer of the pigment. Chemisorption of the macromolecule onto nanoporous titania tends to produce "pore blocking," which inhibits electrolyte ingression and efficient coupling of the photoexcited molecule to the titania host.

### 3.4.5
### A DSSC Based Upon Synthetic Eumelanin

Given what we now know about the physics and chemistry of melanins, and given the criteria for sensitization of a narrow bandgap nanocrystalline photoanode (e.g., titania), is it possible to use eumelanin as the light-harvesting component in a DSSC? Critically, for such a system to function as a regenerative photovoltaic device, the photoexcited state of the eumelanin molecule (whatever that is) must be strongly coupled to the semiconductor conduction band.

We have recently fabricated a DSSC based upon nanocrystalline P25 titania sensitized with a synthetic eumelanin. The eumelanin was electropolymerized into the $TiO_2$ pores from a buffered aqueous solution of dopa [52]. To avoid any chance of over-deposition and hence pore blocking, the titania substrate was only partially sensitized, and the resultant photoanode was a light-brown color. Clearly this is non-optimum in terms of total light capture, but the object of the exercise was to determine whether or not the eumelanin was capable of injecting an excited electron into the conduction band of its host. As per Fig. 3.1, $SnO_2$:F on glass was used as the transparent anode electrode and support substrate, an iodine-triiodide couple in acetonitrile was used as the redox electrolyte, and a thin film of platinum was used as the counter electrode (deposited on $F:SnO_2$ on glass). The cell was sealed with a silicone-based polymer to prevent electrolyte leakage and was tested versus a plain titania photoanode cell under standard Air Mass 1.5 global illumination using an Oriel Solar Simulator. Current-voltage scans (I-V) were obtained using a Keithley SMU 2400 Source Meter Unit controlled using in-house computer code. Lamp output was measured using a NIST standard photodiode in order to establish the incident optical power density. Equations (1) and (3) were used to determine fill factors and power conversion efficiencies from the I–V data.

The I–V performance of the bare titania photoanode cell and the eumelanin-sensitized cell are shown in Fig. 3.17. Under AM1.5 conditions, a relatively small amount of the incident light is ultraviolet (as per the natural solar spectrum). The titania photoanode cell shows the expected response. Titania itself produces a photocurrent if illuminated with light above its band edge ($\sim 3.2$ eV). The eumelanin-based cell produces a significantly higher short-circuit current (photocurrent). This clearly indicates conversion of visible light into electrical energy, i.e., the eumelanin is sensitizing the titania and electron injection is taking place.

This is an important result since it proves that coupling between the photoexcited state of eumelanin and the titania conduction band is possible. The fill factor of the cell was calculated to be $\sim 0.4$, and given a short-circuit current of 0.81 µA and an open-circuit voltage of 0.42 V, this yields a global power conversion efficiency of $\sim 0.1\%$. Clearly, this performance is a long way short of the 10–15% achieved with commercial silicon cells and the 5% with a typical N3 DSSC. However, the eumelanin cell was far from optimum – in particular, the photoanode was only partially sensitized. These early prototype results are en-

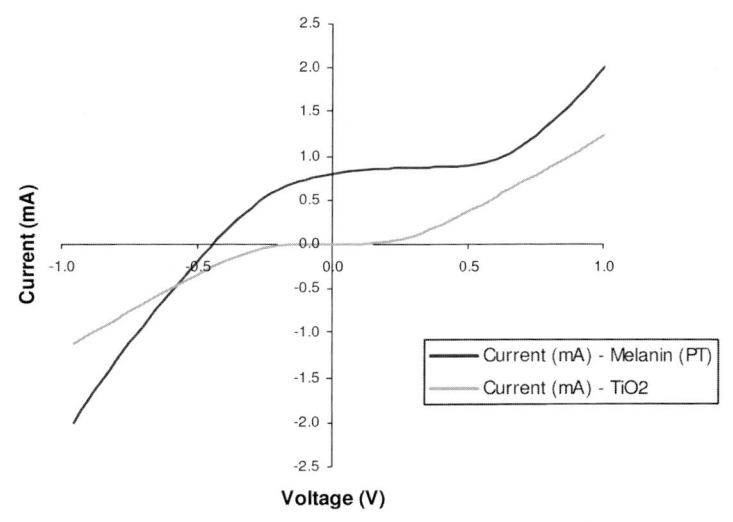

**Fig. 3.17** I–V response under AM1.5 illumination (150 mW cm$^{-2}$) of a nanocrystalline titania DSSC with and without sensitization with a synthetic eumelanin. The sensitization was achieved using direct electropolymerization of eumelanin onto the photoanode.

couraging, and armed with the knowledge that eumelanin sensitization of titania is possible, we are attempting to optimize the cell for greater power conversion efficiency.

## 3.5 Conclusions

In this chapter we have discussed the criteria for classifying a photovoltaic system as artificially photosynthetic (AP). A key difference between semiconductor-junction-based cells and AP systems is the removal of the need for spatial correspondence between the photon-absorption and charge-separation events. The photoelectrochemical Grätzel cell (or dye-sensitized solar cell) is a typical example of a biomimetic solar energy conversion system. These devices make use of the principles of artificial photosynthesis and have competitive power conversion efficiencies relative to current silicon technology. Sensitizers based upon ruthenium charge transfer complexes (the N3 dye is a good example) produce a broadband UV and visible response. However, these complexes have long-term stability problems and are expensive to manufacture.

We have suggested that broadband-absorbing pigments based upon the melanins – a class of biological macromolecules – may be suitable replacements for the ruthenium complexes as sensitizers. Eumelanin in particular is chemically and photochemically very stable, has strong broadband absorption, can be synthesized and engineered to maximize coupling and adsorption to a semiconductor host, and is biocompatible and even bioavailable. However, before we can truly realize the potential of these molecules as photon-harvesting systems, key questions regarding fundamental structure and mesoscopic physics need to be addressed. Despite these knowledge gaps, we have recently demonstrated a regenerative DSSC based upon an electropolymerized eumelanin photoanode. Although the power conversion efficiency in this non-optimized device was modest, we have proven that it is possible to couple the photoexcited state of the eumelanin to the nanocrystalline titania conduction band. Hence, we have shown that these materials have great potential as broadband light-harvesting components in biomimetic photovoltaics.

## Acknowledgments

This work was funded by the Australian Research Council, under Discovery Grant DP0345309, and through the University of Queensland's Research Infrastructure Fund. We acknowledge the contribution of the Centre for Microscopy and Microanalysis at the University of Queensland for assistance in electron microscopy. We also acknowledge David Menzies from Monash University in Melbourne for advice concerning the N3 cell construction and Dr. Adam Micolich at the University of New South Wales for low-temperature electrical measurements. Finally, we would like to thank Prof. Ross McKenzie (Condensed Matter Theory Group, University of Queensland Physics Department), Prof. John Simon (Duke University, USA), Prof. Tad Sarna (Jagiellonian University, Poland), and Dr. John McGinness for fruitful discussion and guidance.

## References

1 H. Imahori, Y. Mori and Y. Matano, *J. Photochem. Photobiol. C: Photochem. Reviews*, **4**, 51–83 (2003).
2 Y. Saga and H. Tamiaki, *J. Photochem. Photobiol. B: Biology*, **73**, 29–34 (2004).
3 Y. Kureishi, H. Tamiaki, H. Shiraishi and K. Maruyama, *Bioelectrochem. Bioenerg.*, **48**, 95–100 (1999).
4 M. Grätzel, *J. Photochem. Photobiol. C: Photochem. Reviews*, **4**, 145–153 (2003).
5 M. Grätzel, *Nature*, **414**, 338–344 (2001).
6 B.J. Scott, G. Wirnsberger, M.D. Mc Gehee, B.F. Chmelka and G.D. Stucky, *Adv. Mater.*, **13**, 1231–1234 (2001).
7 R. Vogel, P. Meredith, I. Kartini, M. Harvey, J.D. Riches, A. Bishop, N. Heckenberg, M. Trau and H. Rubinsztein-Dunlop, *ChemPhysChem*, **4**, 595–603 (2003).
8 B. O'Regan and M. Grätzel, *Nature*, **353**, 737–740 (1991).
9 M. Grätzel, *J. Am. Chem. Soc.*, **123**, 1613–1624 (1993).
10 I. Kartini, P. Meredith, J.C. da Costa, J.D. Riches and G.Q. Lu, *Curr. Appl. Phys.*, **4**, 160–162 (2004).
11 P.A. Venz, J.T. Kloprogge and R.L. Frost, *Langmuir*, **161(11)**, 4962–4968 (2000).
12 G.J. Wilson, G.D. Will, R.L. Frost and S.A. Montgomery, *J. Mat. Chem.*, **12**, 1787–1791 (2002).
13 I. Kartini, D. Menzies, D. Blake, J.C.D. da Costa, P. Meredith, J. Riches and G.Q. Lu, *submitted J. Mater. Chem.* (2004).

14 M. K. Nazeeruddin, P. Pechy, T. Renouard, S. M. Zakeeruddin, R. Humphrey-Baker, P. Comte, P. Liska, L. Cevey, E. Costa, V. Shklover, L. Spiccia, G. B. Deacon, C. A. Bignozzi and M. Grätzel, *J. Am. Chem. Soc.*, **123**, 1613–1624 (2001).

15 A. Hagsfeldt and M. Grätzel, *Chem. Rev.*, **95**, 49–68 (1995).

16 A. Hinsch, *Proc. 16th Eur. PV Solar Energy Conf.*, Glasgow, 32 (2000).

17 U. Bach, D. Lupo, P. Comte, J. E. Moser, F. Weissortel, J. Salbeck, H. Spreitzer and M. Grätzel, *Nature*, **395**, 583–585 (1998).

18 K. Tennakone, G. R. R. A. Kumara, A. R. Kumarasinghe, K. G. U. Wijayantha and P. M. A. Sirimanne, *Semicond. Sci. Technol.*, **10**, 1689–1693 (1995).

19 J. Kruger, U. Bach and M. Grätzel, *Appl. Phys. Lett.*, **79**, 2085–2087 (2001).

20 S. Z. Chu, K. Wada, S. Inoue and S. Todoroki, *Chem. Mater.*, **14**, 266–272 (2002).

21 P. Meredith, PCT/AU02/01327: "*Components based on melanin and melanin-like biomolecules, and process for their production*", Filed September (2002).

22 F. Odobel, E. Blart, M. Lagree, M. Villieras, H. Boujtitia, N. El Murr, S. Caromori and C. A. Bignozzi, *J. Mater. Chem.*, **13**, 502–510 (2003).

24 G. Prota (1992) *Melanins and Melanogenesis.* Academic Press, San Diego, CA.

23 G. Ramakrishna and H. N. Ghosh, *J. Phys. Chem. B*, **105**, 7000 (2001).

25 M. L. Wolbarsht, A. W. Walsh and G. George, *Appl. Opt.* **20**, 2184–2186 (1981).

26 H. Z. Hill, *Melanin: Its role in human photoprotection* (edited by L. Zeise, M. Chedekel and T. Fitzpatrick), 81–91. Valdenmar Press, Overland Park, KS (1995).

27 K. B. Stark, J. M. Gallas, G. W. Zajac, M. Eisner and J. T. Golab, *J. Phys. Chem. B* **107**, 3061-3067 (2003).

28 Z. W. Zajac, J. M. Gallas, J. Cheng, M. Eisner, S. C. Moss and A. E. Alvarado-Swaisgood, *Biochim. Biophys. Acta*, **1199**, 271–278 (1994).

29 S. Ito, *Biochim. Biophys. Acta*, **83**, 155–161 (1986).

30 C. M. R. Clancy, J. B. Nofsinger, R. K. Hanks and J. D. Simon, *J. Phys. Chem. B* **104**, 7871–7873 (2000).

31 J. McGinness, P. Corry and P. Proctor, *Science* **183**, 853–855 (1974).

32 P. R. Crippa, V. Cristofoletti and N. Romeo, *Biochim. Biophys. Acta*, **538**, 164–170 (1978).

33 M. M. Jastrzebska, H. Isotalo, J. Paloheimo, H. Stubb and B. Pilawa (1996), *J. Biomater. Sci. Polymer Edn.*, **7**, 781–793 (1996).

34 P. Meredith and J. Riesz, *Photochem. Photobiol.*, **79(2)**, 211–216 (2004).

35 J. B. Nofsinger and J. D. Simon (2001), *Photochem. Photobiol.*, **74**, 31–37 (2001).

36 B. Powell, T. Baruah, N. Bernstein, K. Brake, R. H. McKenzie, P. Meredith and M. R. Pederson, *J. Chem. Phys.* **120(18)**, 8608–8615 (2004).

37 H. C. Longuet-Higgins, *Arch. Biochim. Biophys.*, **88**, 231–232 (1960).

38 J. P. Lowe, *Quantum Chemistry*, Academic Press, London (1978).

39 A. Pullman and B. Pullman, *Biochim. Biophys. Acta*, **54**, 384 (1961).

40 D. S. Galvao and M. J. Caldas, *J. Chem. Phys.*, **88**, 4088 (1988).

41 D. S. Galvao and M. J. Caldas, *J. Chem. Phys.*, **92**, 2630 (1990).

42 D. S. Galvao and M. J. Caldas, *J. Chem. Phys.*, **93**, 2848 (1990).

43 L. E. Bolivar-Marinez, D. S. Galvao, and M. J. Caldas, *J. Phys. Chem. B*, **103**, 2993 (1999).

44 K. Bochenek and E. Gudowska-Nowak, *Chem. Phys. Lett.*, **373**, 523 (2003).

45 R. O. Jones and O. Gunnarsson, *Rev. Mod. Phys.*, **61**, 689 (1989).

46 Y. V. Il'ichev and J. D. Simon, *J. Phys. Chem.*, **107**, 7162 (2003).

47 J. P. Perdew, K. Burke, and M. Ernzerhof, *Phys. Rev. Lett.*, **77**, 3865 (1996).

48 A. D. Becke, *J. Chem. Phys.*, **98**, 5648 (1993).

49 T. Strzelecka, *Physiol. Chem. Phys.*, **14**, 219–222 (1982).

50 M. Jastrzebska, A. Kocot and L. Tajber, *J. Photochem. Photobiol. B: Biology*, **66**, 201–206 (2002).

51 P. Meredith, J. Riesz, C. Giacomantonio, S. Subianto, G. Will, A. Micolich and B. Powell, *International Congress on Synthetic Metals*, Wollongong (2004).

52 S. Subianto, G. Will and P. Meredith, *International Congress on Synthetic Metals*, Wollongong (2004).

# 4
# The Design of Natural Photosynthetic Antenna Systems

*Nancy E. Holt, Harsha M. Vaswani, and Graham R. Fleming*

## 4.1
## Introduction

The nanoscale dimensions of the photosynthetic complexes are of critical importance for their effective function. Ultrafast spectroscopic studies by ourselves and by other groups, advances in obtaining high-resolution structural models, and more realistic light-harvesting models have revealed that the electronic interactions between the light-absorbing components in photosynthetic antenna complexes and reaction centers (RCs) are both qualitatively and quantitatively different than in chemical model systems where chromophores are spaced by distances that are large compared with their size. As a result, there are only a few cases in which the energy transfer within natural photosynthetic antenna can be correctly characterized by conventional Förster theory.

In order to obtain a detailed understanding of the interactions between photosynthetic pigments embedded in protein complexes, two additional photosynthetic design features must be taken into account. One is that they utilize confined molecular geometries in which antenna molecules are spaced by distances that are small compared with their size. Such intermolecular distances enable transitions and energy levels that would be ineffective or even inoperative in systems with widely spaced chromophores to play crucial roles in light harvesting and energy transfer. The second is that they exploit energetic disorder in order to improve spectral coverage, reduce energy mismatches, and broaden excitonic manifolds to make the system exceedingly robust with respect to temperature variation and fluctuations in site energy and electronic coupling. Paradoxically, the same pigment interactions that can be used to optimize photosynthesis can also generate efficient photoprotection. Each of these points will be described and exemplified by theoretical and experimental work done in our laboratory on bacteria, cyanobacteria, and plants.

---

*Artificial Photosynthesis: From Basic Biology to Industrial Application*
Edited by Anthony F. Collings and Christa Critchley
Copyright © 2005 WILEY-VCH Verlag GmbH & Co. KGaA, Weinheim
ISBN: 3-527-31090-8

## 4.2
### Confined Geometries: From Weak to Strong Coupling and Everything in Between

Standard linear optical spectroscopy defines a length scale on which molecular transitions are characterized. Specifically, the photon characterizes the molecular transition density in the far field, thus averaging over the entire molecular dimension. In the confined geometries of photosynthetic light-harvesting complexes, neighboring molecules may sense each other's shape on a much finer scale. For example, consider the LH2 antenna complexes of purple bacteria (Fig. 4.1). LH2 has two rings of BChls that absorb at 800 nm (B800) and 850 nm (B850) [1, 2]. There is also a minor absorption by carotenoids (Cars) between 450 nm and 550 nm. The energy is first transferred within the light-harvesting complexes (LHC) and then, ultimately, to the reaction center (RC). The antennas thus enable the RCs to operate efficiently by collecting more light from a broader spectral range.

The structure of LH2 from *Rps. acidophila* was determined at a resolution of 2.5 Å [2]. The complex has nine BChls in the B800 ring, 18 BChls in the B850 ring, and nine Cars, which in this structure are rhodopsin glucoside. The energy transfer dynamics within LH2 can be separated into four processes, three of which occur between BChls. These processes are, in order of decreasing distance between the donor(s) and acceptor(s), B800 to B800, B800 to B850, and B850 to B850. The last process involves Car to BChl energy transfer. Using the different energy transfer pathways in LH2 as the basis for discussion, the successes and failures of conventional Förster theory (CFT) are outlined. To overcome the shortcomings of the theory, alternate methods, such as generalized Förster theory (GFT) and modified Redfield theory (MRT) are introduced.

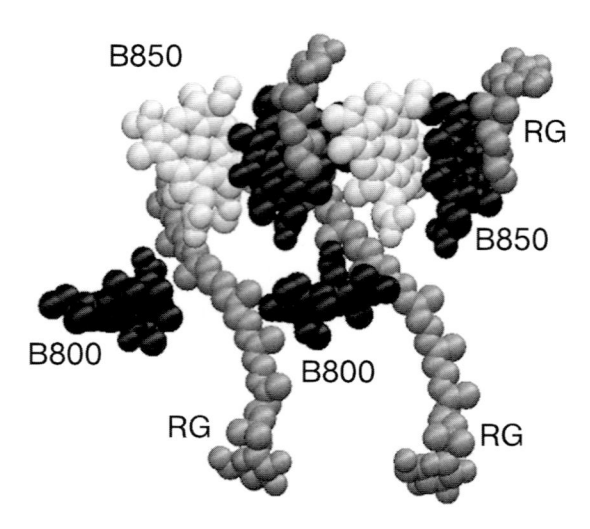

**Fig. 4.1** Van der Waals representation of the pigments from a portion of the LH2 ring. Rhodopsin glucoside (RG) is in gray and the lower BChls (black) are part of the B800 ring. The upper BChls (alternating black and light gray) belong to the B850 ring.

## 4.2.1
### Conventional Förster Theory: B800 to B800 Intra-band Energy Transfer

The fundamental equation for energy transfer between molecules is the Golden Rule. CFT uses this expression in the context of the dipole-dipole approximation under the assumption that the pigments are only weakly (incoherently) coupled to each other [3]. The rate of energy transfer in CFT is expressed using experimental parameters:

$$k = \frac{9000 \ln 10}{128 \pi^5 Nn^4} \cdot \frac{\kappa^2 \phi_D}{R^6 \tau_D} \int_0^\infty \frac{f_D(\nu)\varepsilon_A(\nu)}{\nu^4} d\nu \tag{1}$$

where $\kappa$ is the orientation dependence of the dipole-dipole coupling, $R$ is the distance between the donor and acceptor pigments, $\varphi_D$ is the donor fluorescence quantum yield, $\tau_D$ is the donor excited-state lifetime, $n$ is the index of refraction of the medium, and $N$ is Avogadro's number. The integral in the equation is the overlap of the donor emission and the acceptor absorption spectra.

In *Rps. acidophila*, the B800 pigments are separated by a distance of 21 Å [2]. At intermolecular separations this large with respect to the BChl size ($\sim 10 \times 10$ Å), CFT should be sufficient to explain the energy transfer between the B800 pigments. To test this hypothesis, we utilized an ultrafast, nonlinear spectroscopic technique, the three-pulse photon echo peak shift (3PEPS; for a more detailed explanation of this method, please see Ref. [4]), that can monitor energy transfer between pigments that cannot be spectroscopically distinguished from each other within the bandwidth of the laser pulses. We used the following Hamiltonian to simulate the photon echo data of the B800 band in *Rps. acidophila* [5, 6]:

$$H^0 = H^{el} + L + E^{ph} + H^{el\text{-}ph}; \quad H^1 = H^{Coul} \tag{2}$$

where $H^{el}$ is the electronic Hamiltonian, $L$ is the reorganization energy, $E^{ph}$ is the phonon energy, and $H^{el\text{-}ph}$ is the electron-phonon coupling. The Coulomb term (obtained from Eq. 1) is treated as a perturbation, a good approximation in the case of weak coupling. From these calculations, we concluded that B800 to B800 energy transfer occurs by incoherent hopping on a timescale of $\sim 500$ fs [5]. The excellent agreement between experiment and theory demonstrates that CFT provides an accurate description of energy migration within the weakly coupled, essentially monomeric B800 chromophores of LH2.

## 4.2.2
### Generalized Förster Theory: B800 to B850 Inter-band Energy Transfer

CFT often underestimates the rate of energy transfer. One specific example where conventional approaches fail completely is in the prediction of the energy transfer timescales from the B800 to B850 rings in LH2; experimentally it occurs in 800 fs [7] and 650 fs [4, 8, 9] for *Rps. acidophila* and *Rb. sphaeroides*, respectively, while CFT predicts $\sim 2$ ps [10, 11]. Analysis of the two rings provides some insight into the reasons for the disparities between the calculated and measured energy transfer timescales. The energy donor from the B800 ring is a single BChl pigment that absorbs at 800 nm and is located $\sim 18$ Å from the energy acceptor, the B850 ring [2]. However, the energy acceptor is not a single B850 BChl, but rather 18 acceptor states that are a product of strong coupling ($\sim 300$ cm$^{-1}$ [12]) among the 18 B850 BChls. Accordingly, a correct description of the energy transfer must include each of the individual acceptor states and a means to assess whether the distance between the donor and the acceptor is small or large with respect to the size of the acceptor.

GFT incorporates a number of features necessary to describe the energy transfer from B800 to B850 in LH2. Namely, it enables structural information to be retained in the model for energy transfer, it allows donor(s) and acceptor(s) to be properly specified in systems where the donors and acceptors are weakly coupled while the collection of donors and/or acceptors may not be, and it enables the energetic disorder universally present in photosynthetic systems to be properly included in the calculation of observables [13–16]. By using ab initio techniques to calculate the electronic coupling between B850 pigments along with GFT, we could quantitatively describe the energy transfer rate [14, 17]. Furthermore, the approach allowed us to explain two experimental observations that resisted elucidation by conventional methods: the remarkable insensitivity of the energy transfer rate to temperature (transfer timescales from both LH2 strains mentioned above increased to only 1.2 ps at 77 K [7, 18]) and its low-temperature, single-molecule spectra [14].

### 4.2.3
**Generalized Förster Theory with the Transition Density Cube Method: Car to Bchl Inter-pigment Energy Transfer**

In designing photosynthetic antenna, it seems counterintuitive to choose chromophores that have a low-lying excited state that completely lacks a transition dipole moment. However, nature has found an effective way to utilize such states for light harvesting. Cars, essentially substituted conjugated polyenes, have two well-characterized singlet excited states in the visible region of the spectrum. The higher-energy state ($S_2$) is one-photon allowed and absorbs in a region where BChls do not ($\sim 450$–550 nm). The other state ($S_1$) is one-photon forbidden by absorption and emission from the ground and excited states, respectively. Energy transfer from the Car $S_2$ to BChl in *Rps. acidophila* is extremely rapid (50–100 fs) [19, 20]; however, the overall efficiency of energy transfer from the Car to BChl is higher than can be explained by the Car $S_2$ transfer pathway alone. An alternate pathway via $S_1$ is involved (Fig. 4.2A) [21]. In order to unequivocally prove that there is $S_1$ to BChl energy transfer in photosynthetic antenna, we developed various forms of two-photon-excited, ultrafast spectroscopy that prepared the $S_1$ state directly from the ground state (i.e., not by $S_2$ internal conversion, as was done previously) [22–24]. We found the $S_1$ to BChl energy transfer timescale in bacterial LH2 from *Rb. sphaeroides* to be ~ 3 ps [23] and determined the spectrum and dynamics of Cars (Fig. 4.2 B).

The involvement of the $S_1$ state in light harvesting highlights a spectacular failure of the dipole model in characterizing photosynthetic pigment interactions: since $S_0$–$S_1$ has a very small net transition dipole moment because the photon averages over the entire molecule, the calculated rate would be zero. However, the Coulomb interaction between two regions of transition density, i and j, is given by $(q_i q_j/r_{ij})$ (Fig. 4.2 C). Therefore, the overall Coulombic coupling, which is the sum of the interactions between all such elements, depends on the full shape of the donor and acceptor transition density on a length scale much smaller than the molecular size. As a result, we developed a highly accurate

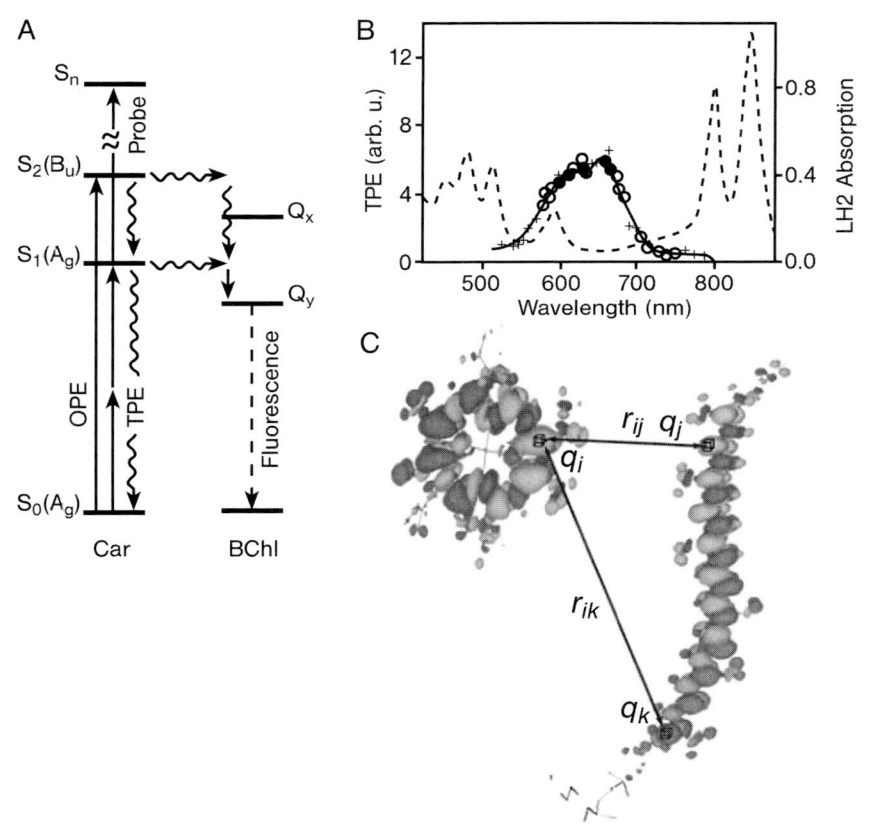

**Fig. 4.2** (A) The energy levels, excitation, and probing steps for Car-BChl interactions. (B) The two-photon excitation spectrum of spheroidene obtained by detecting fluorescence from BChl in LH2 of *Rb. sphaeroides*. (C) Transition densities calculated for a BChl molecule and a Car. Density elements, containing charges $q_i$ and $q_j$, are depicted together with their corresponding separation $r_{ij}$.

method, called the transition density cube (TDC) method, to calculate the full Coulombic coupling between two molecules [11]. In short, this is done by dividing the transition densities of the states involved into small cubes and computing the coupling by summing over all of the cube-cube interactions (Fig. 4.2 C). TDC combined with time-dependent density functional theory [25] gives values of the full Coulombic $S_1$-BChl coupling that are in excellent agreement with experimental estimates [26]. When incorporated into a Generalized Förster theory model, the calculated energy transfer rates both agree with experiment and quantitatively account for the large difference in $S_1$ to BChl transfer efficiency in the LH2 complexes of *Rps. acidophila*, *Rs. molischianum*, and *Rb. sphaeroides* [23, 26]. Furthermore, we showed that when the LH2 structure is modeled with a strictly symmetric polyene whose transition dipole is rigorously zero, sizeable Coulomb coupling to BChl can still occur simply from spatial proximity and the asymmetric arrangement of donors and acceptors [26].

As an aside, we note that singlet-singlet energy transfer involving transitions with little or no oscillator strength is often discussed in the context of electron exchange interactions between donor and acceptor – the Dexter mechanism [27]. While this form of coupling is undoubtedly crucial in triplet-triplet transfers, in all of the singlet-singlet transfer cases we have examined, the Coulombic coupling is much larger than the exchange coupling and the energy transfer is fully describable with Coulombic coupling alone.

### 4.2.4
### Modified Redfield Theory: Intra-band B850 Exciton Dynamics

In the case of the B850 molecules of LH2, the situation is more complex. The pigments are only 9 Å apart and the arrangement of the chromophores in a ring leads to strong interactions between them [2]. When the Coulombic coupling is large with respect to the pigment's reorganization energies and static variations in energy levels, Redfield theory is more appropriate than weak-coupling CFT approaches. Here, the Coulomb term is incorporated into the specification of the eigenstates, and the electron-phonon coupling is the perturbation that induces the energy transfer:

$$H^0 = H^{el} + L + E^{ph} + H^{Coul}\ ;\ \ H^1 = H^{el\text{-}ph} \tag{3}$$

Traditional Redfield theory does not interpolate correctly to Förster theory for weak coupling. Recently, Zhang et al. [28] introduced another version of the Redfield equation to describe population relaxation in disordered excitonic systems. Yang and Fleming [29] explored this modified form of Redfield theory for energy transfer and showed that it provides accurate results for both weakly and strongly coupled systems. Instead of taking the whole electron-phonon coupling as a perturbation, the molecular Hamiltonian is divided as follows:

$$H^0 = H^{el} + L + E^{ph} + H^{Coul} + \sum_k |k\rangle H^{el\text{-}ph}_{kk} \langle k| \; ; \quad H^1 = \sum_{\substack{k,k' \\ k \neq k'}} |k\rangle H^{el\text{-}ph}_{kk'} \langle k'| \qquad (4)$$

Consequently, even if the electron-phonon coupling strength is not particularly weak, a perturbative approach with respect to the off-diagonal Hamiltonian will be valid. The modified version of Redfield theory is thus able to describe energy transfer over a wide range of parameters.

Using calculated electronic couplings, measured disorder, and electron-phonon coupling strength as the input parameters for modified Redfield theory, we were able to describe the photon echo signal and absorption spectrum of B850 in the LH2 complex without adjustable parameters [6, 12]. The photon echo signal contains exciton relaxation timescales ranging from 100 fs to a few picoseconds. We showed that within 50 fs the exciton became delocalized between two to four molecules and that a mechanism in which excitation "hops" between adjacent groups of two to four molecules provides a realistic physical picture of the energy transfer within B850. The delocalization leads to the phenomenon of exchange narrowing [30], whereby the distribution of site energies is apparently narrowed by the averaging effect of the delocalized states. Exchange narrowing reduces the effective electron-phonon coupling and allows very rapid energy migration in a moderately disordered system [6].

## 4.3
### Energetic Disorder Within Light-harvesting Complexes

Photosynthetic complexes are, in general, both spatially and energetically disordered, and experimental techniques used to study them should ideally be able to separate these various types of disorder. We developed the three-pulse photon echo peak shift (3PEPS) method into an incisive tool for the study of energetically disordered systems, exploiting the disorder to determine energy transfer timescales, electron-photon coupling, and the degree to which the disorder is correlated within individual complexes. Master equation and response function calculations, employing either the Förster or the modified Redfield theory, were used to analyze 3PEPS data.

### 4.3.1
#### From Isolated Complexes to Membranes: Disorder in LH2

Using the coupled master equation–response function approach to analyze 3PEPS data, we determined the energy transfer timescale within B800 (500–600 fs), the coupling between B800 pigments (30 cm$^{-1}$), the energetic disorder within each complex (90 cm$^{-1}$), and the energetic disorder of the entire ensemble (103 cm$^{-1}$) [5, 6, 31]. Since the coupling between B800 chromophores is small compared to the disorder and thermal energy ($k_B T$), energy transfer within B800 occurs via incoherent hopping.

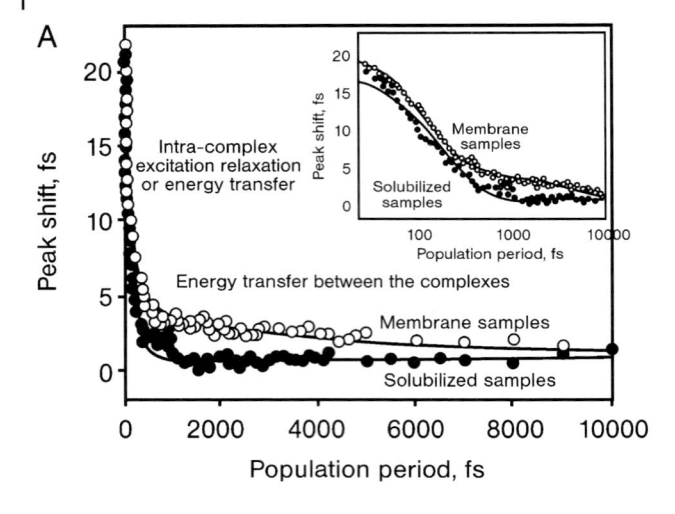

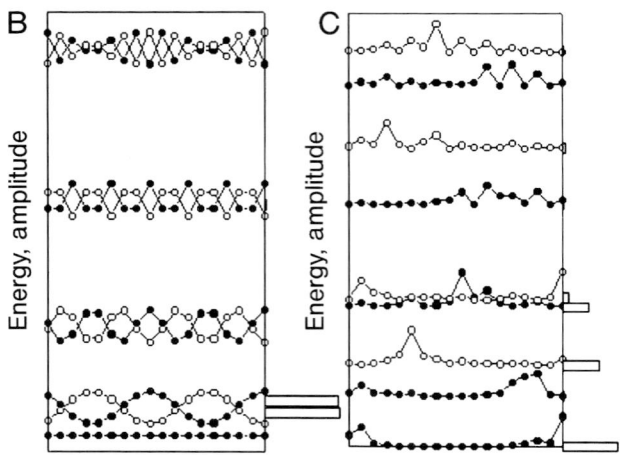

**Fig. 4.3** (A) 3PEPS experimental and simulated data for solubilized (circles) and membrane (squares) samples of the B850 band of LH2. The inset shows the same data on a logarithmic scale. Exciton wave functions calculated for the B850 band of LH2 for (B) a completely ordered model, with $J_{intra} = J_{inter} = 320$ cm$^{-1}$, and (C) for a statically disordered model, with $J_{intra} = 320$ cm$^{-1}$, $J_{inter} = 255$ cm$^{-1}$, $\sigma$ (disorder) $= 150$ cm$^{-1}$, and an energy offset between $\alpha$ and $\beta$ BChls of 530 cm$^{-1}$. The length of the horizontal gray bars to the right of each figure is proportional to the oscillator strength of the corresponding exciton states.

We turn our attention to the B850 ring in LH2 where the BChls are closer together. Now the coupling resembles the site energy disorder and $k_B T$, and thus the electronic states will be partially delocalized. Therefore, the intra-complex disorder associated with the delocalized states becomes dependent on the electronic coupling, whereas the inter-complex disorder does not. To study the intra-

and inter-complex energy transfer processes, we obtained 3PEPS data for the complex both in solution and in native membrane samples that contain LH2 as the sole BChl protein complex [12]. Both samples revealed an ultrafast decay on a 100-fs timescale that was attributed to exciton relaxation dynamics within individual aggregates. The samples in native membranes also contained a 5-ps decay component that was assigned to inter-complex energy transfer (Fig. 4.3 A). The observation of the 5-ps component in the membrane samples confirmed the presence of two levels of energetic disorder in the system: an intra-complex disorder of 150 cm$^{-1}$ and an inter-complex disorder of 65 cm$^{-1}$ [12]. It also confirmed the ability of 3PEPS to measure energy transfer rates between LHCs. The disorder causes redistribution of oscillator strength between optically allowed and forbidden transitions, thereby increasing spectral coverage (Fig. 4.3). In summary, our results on bacterial systems have helped us to better understand energy transfer and trapping in the entire bacterial photosynthetic unit and have encouraged us to study more-complex photosystems.

### 4.3.2
**Photosystem I**

PSI is even more spatially and energetically disordered than LH2, with 96 nonequivalent chlorophyll (Chl) molecules (Fig. 4.4 A) [32]. In our studies, we addressed the following problems: whether there were new design principles at work in PSI; how the system was optimized to focus and trap excitation at the primary electron donor (P700); identifying the rate-determining steps; and determining the viability of a first-principles calculation for quantitative predictions [33–36].

Calculations from first principles are challenging because of the presence of both moderately or strongly coupled pigments and pigments spaced at distances that make a weak coupling picture appropriate. For these types of calculations, modified Redfield theory was utilized. We first calculated the excitation energies of and couplings between each of the 96 Chls in the protein environment and obtained a good fit to the absorption spectrum [36]. Unlike bacterial light-harvesting antenna, PSI does not appear to have an energetic funnel, i.e., the outermost pigments do not, on average, have higher excitation energies than the pigments in the reaction center (Fig. 4.4 B).

We used our calculated excitation energies, modified Redfield theory, and a master equation calculation of the excitation transfer kinetics in PSI to calculate time-dependent fluorescence spectra and population kinetics. We found that the time constants are clustered into four groups: sub-100 fs, 200–300 fs, 2–3 ps, and 35–40 ps [35]. The global fits to experimental fluorescence data gave time constants of 360 fs, 3.6 ps, 9.6 ps, and 38 ps [37]. The experiments did not have sufficient time resolution to determine sub-100 fs time constants. Aside from this, our calculated time constants were in remarkable agreement with experimental values, with the exception of the 9.6-ps component that has been suggested to be associated with the trimer interface of PSI [38]. Our calculations were made for monomeric PSI and thus should not have this component.

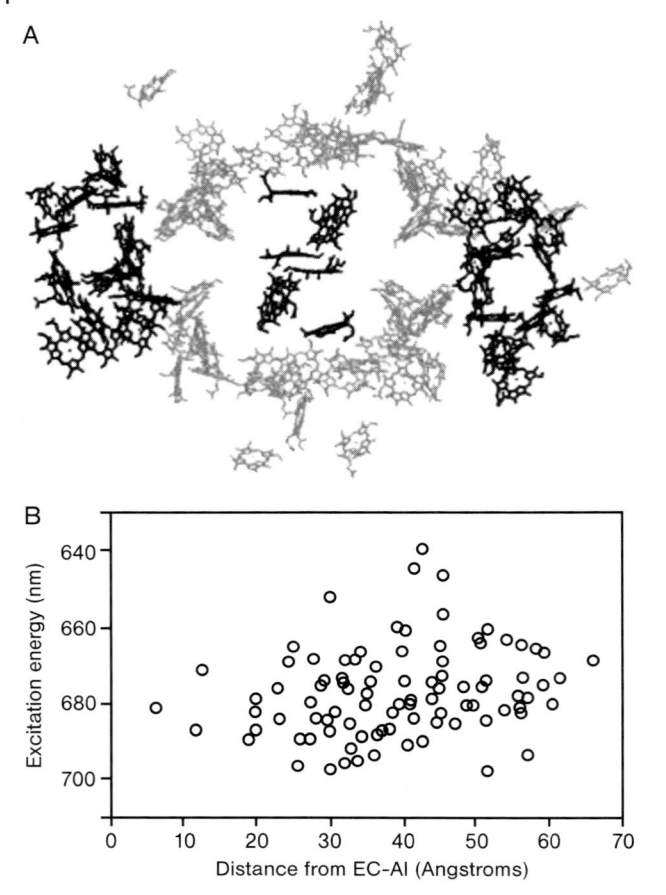

**Fig. 4.4** (A) The organization of pigments in PSI from *Synechococcus elongatus*. The RC is centrally located (P700 is seen from the side). The six RC Chls and the two linker Chls are shown in black. (B) Calculated $Q_y$ excitation energies of PSI Chls as a function of distance from the trap, delineated EC-A1.

Having developed a model of PSI that reproduced key experimental data, we explored the influence of the energy landscape [33, 35] on the trapping time and performed a detailed analysis of the rate-determining steps [34]. In a flat energy landscape, where all the Chls had identical lineshapes and transition energies and the spatial structure was held fixed, the calculated trapping time of PSI was actually shorter (25 ps vs. 38 ps). However, due to protein constraints and the proximity of the chromophores (which induces excitonic interactions), a flat energy landscape cannot be achieved in a realistic PSI complex. The high spatial connectivity of PSI makes it very robust to energy variation; however, the energies of the six RC Chls influence the overall trapping time significantly.

Along with the two linker Chls, the six RC Chls form a quasi-funnel structure. Fig. 4.5 A shows the distribution of trapping times for 1000 random permutations of the energies in PSI while keeping the energies fixed for only P700. In Fig. 4.5 B, we show the same calculations with the energies of the other four RC pigments and the two linker Chls also fixed. In the first case, the distribution of trapping time is broad; in the second case, the distribution is very narrow. The findings suggest that the energies of the RC and linker Chls are optimized for efficient transfer of excitation to the P700, although the precise energies of the other 88 antenna pigments are not crucial to the near-unit quantum efficiency of PSI.

Studies of the energy distribution in PSI after initial excitation of two different Chl molecules show that energy flow through the antenna contributes significantly to the overall trapping time [33]. Further analysis shows that there are three major contributions and that energy transfer around the reaction center is, in fact, dominated by entropic rather than enthalpic (i.e., energy funnel) considerations. Less evident is that back transfer, including transfer back from the RC Chls to the bulk antenna, is also a substantial component of the overall trapping time.

In summary, the transfer of excitation to P700 of PSI is an exceedingly efficient process. Overall, it differs markedly from the same process in the well-studied purple bacteria and in PSII, where atomic resolution structural data is not yet available. The two-dimensional projection of PSI seems to imply that the isolation of the antenna from the electron transfer components is the same as in purple bacteria (RC1 + LH1). Three important differences between the two systems illustrate that this is not true: (1) PSI is quasi-three-dimensional, while RC/LH1 is quasi-one-dimensional; (2) all six members of the RC Chls are energetically accessible from the antenna system in PSI, whereas only the pri-

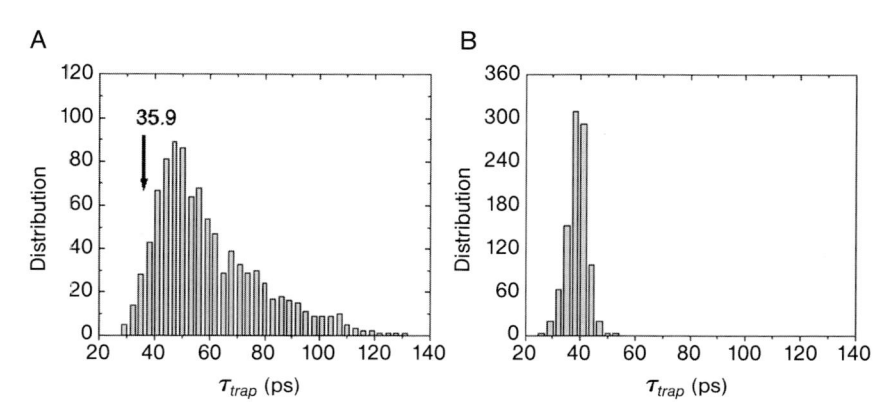

**Fig. 4.5** Distributions of trapping times (1000 samples) calculated by (A) a model that includes random shuffling of excitation energies in all the pigments except those of P700 and (B) random shuffling of excitation energies except those of the six RC Chls and the two linker Chls.

mary electron donor is accessible in purple bacteria; and (3) only PSI has linker Chls. These two systems efficiently solve the same problem in two different ways by exploiting the confined geometry and spatial and energetic disorder.

## 4.4
**Photochemistry and Photoprotection in the Bacterial Reaction Center**

Employing the information and methods discussed above, we can provide an answer to two well-known but mechanistically poorly understood problems concerning the reaction center of purple bacteria (Fig. 4.6). We focus on devising an accurate picture to describe known ultrafast energy transfer timescales from the accessory BChls (B) and the bacteriopheophytin (H) to the special pair (P). The difficulty arises because the lower excitation state of P ($P_-$) carries 88% of the dipole strength and is therefore strongly coupled to B according to the dipole approximation. However, $P_-$ has very poor spectral overlap with B. The upper exciton state ($P_+$) overlaps significantly with B emission but is predicted to be too weakly coupled to B to be an effective acceptor. Yet energy transfer from B to P proceeds on a 100-fs timescale [39–41]. Such a finding is completely paradoxical in the context of Förster theory within the reaction center and led to suggestions that a new energy transfer mechanism was at work in the RC. The answer to this conundrum lies in the idea that the absorption spectrum of the P acceptor does not contain the relevant information for predicting the electronic coupling between B and $P_+/P_-$. Using GFT [13, 15] and the TDC [11], we find that the magnitude of the coupling of B to $P_+$ and $P_-$ is almost identical – in stark contrast to the dipole-dipole result, which gives a 20-fold stronger coupling to $P_-$. Numerical calculations (with no adjustable parameters) described the absorption spectrum, the circular dichroism spectrum, and the energy transfer timescales well (Fig. 4.6) [13, 15, 16].

The second puzzling fact about the RC of purple bacteria is that when the special pair is oxidized (to $P^+$) the absorption band of P disappears, yet the decay of the excited B remains ultrafast ($\sim$150 fs) [42]. The consequences of this observation are physiologically significant because they provide purple bacteria, and most probably PSI that shows similar RC behavior, with an essential photoprotective mechanism that allows for excess energy dissipation in conditions where the amount of input light exceeds photosynthetic capacity. CFT generates a timescale for B to $P^+$ energy transfer that is five orders of magnitude too slow. In collaboration with J. Reimers, we calculated the electronic structure of the radical dimer cation and found that the energy transfer dynamics involve mixing of the strongly allowed transitions of $P^+$ with a manifold of exotic lower-energy transitions to enable energy transfer on a 150-fs timescale. Again, GFT and the TDC are used to identify the states of $P^+$ involved in the quenching of B in the oxidized RC, and we believe that a long-standing mystery has been solved. The dimer structure of the P allows for efficient trapping of excitation at the RC, ultrafast charge separation, and photoprotection against excess light in a single compact molecular framework.

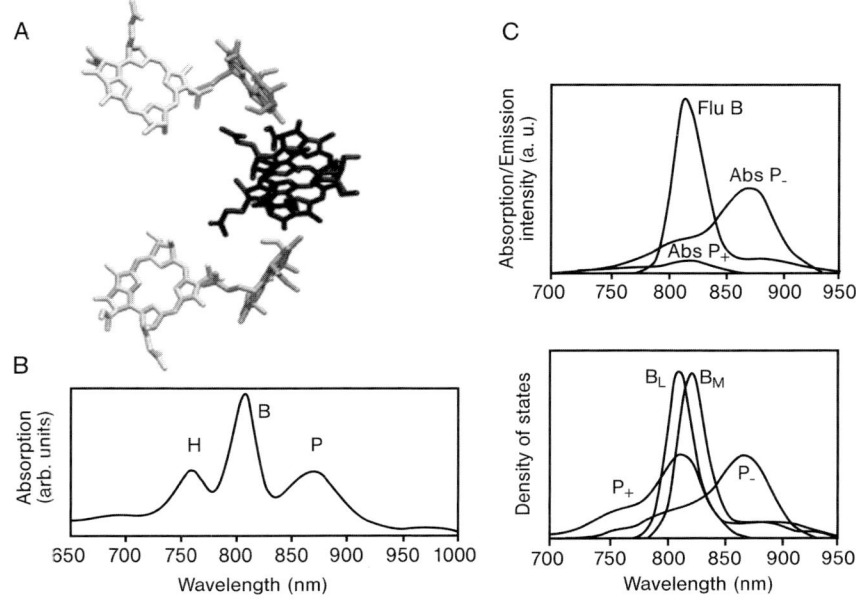

**Fig. 4.6** (A) Arrangement of the special pair (black), accessory BChls (gray), and bacteriopheophytins (light gray) in the reaction center of *Rb. sphaeroides*. (B) Absorption spectrum of the RC, with the features attributed to the pigments H, B, and P indicated. (C) Top panel: the absorption spectra of the special pair acceptor states $P_+$ and $P_-$ plotted over the fluorescence of the donor, B. Lower panel: the density of states calculated for the donors and acceptors in B to P energy transfer.

## 4.5
## The Regulation of Photosynthetic Light Harvesting

A green plant's ability to remove electrons from water requires that the PSII-RC have the highest oxidizing potential found in nature. The consequences of this process have generated unique sensitivities for green plants, with respect to their anoxygenic photosynthetic counterparts, under conditions in which photon fluxes exceed photosynthetic capacity. Specifically, the production of harmful photo-oxidative species – which cause destruction of vital proteins such as the PSII D1 reaction center protein, lipid bilayers, and pigments [43, 44] – is exacerbated during high light exposure. As a result, plants employ a regulatory mechanism that allows them to both utilize and dissipate solar energy with high efficiency in response to the incident light intensity.

The process that protects plants under most short-term light stress conditions occurs in PSII and is called feedback de-excitation quenching (qE) ([45, 46]; for a review of this process, see [47]). During qE, a nonradiative deactivation channel for singlet Chl molecules ($^1$Chl*) is formed that can harmlessly emit the excess absorbed energy as heat. The process significantly decreases the probability that the $^1$Chl* molecules will form triplet Chl molecules ($^3$Chl*), a species that reacts with ground-state oxygen ($^3$O$_2$) to form the strongly photo-oxidative singlet oxygen ($^1$O$_2$*) in addition to other highly reactive oxygen species [48]. In green plants and algae, qE can quench up to 80% of the $^1$Chl* [49–51]. Three elements are so far known to be necessary for qE: (1) the Car zeaxanthin (Zea), which is formed during qE by conversion from violaxanthin (Vio) via reversible reactions known as the xanthophyll (Xan) cycle [52]; (2) the 22-kDa subunit of PSII, PsbS [51]; and (3) a low-thylakoid lumen pH [45, 53].

The key unresolved issues concerning the mechanism of $^1$Chl* de-excitation during qE are the identity of the quenching species, in terms of both pigment composition and location, and the means by which the excess energy is dissipated. Based on the strong correlations between qE and Zea in plants, it was suggested that $^1$Chl* de-excitation might involve direct energy transfer from Chl to Zea [54–56]. As a result of this transfer, Zea would be excited to its first excited singlet state ($S_1$), which would decay rapidly ($\sim$ 9 ps) to the ground state by nonradiative dissipation. In order for this mechanism to be efficient, the energy of the $S_1$ state of Zea must lie below the state corresponding to $^1$Chl* ($Q_y$). Since the $S_1$ states of Cars are, for all intents and purposes, optically dark, obtaining an accurate value for the energy of this state has proven to be a formidable task. While there is no general consensus on the energy of the Zea $S_1$ state, a number of measurements suggest that its energy is comparable to the value of the Chl $Q_y$ band [56–58]. Alternatively, $^1$Chl* de-excitation may be due to an increased rate of internal conversion of Chl to the ground state by the formation of an exclusively Chl-based quenching species, a process that would be indirectly enhanced by the presence of Zea [59, 60].

In order to gain further insight into the quenching mechanism, we performed transient absorption (TA) measurements on intact thylakoid membranes with fully functional qE capability [61]. The measurements probed changes in the region of the Car $S_1 \rightarrow S_n$ transition after excitation of the Chl $Q_y$ band. Kinetic differences were evaluated when the system had no qE and when maximum, steady-state qE had been induced. The results for spinach are summarized in Fig. 4.7. They show an additional transient component with the lifetime and spectral characteristics of Zea $S_1 \rightarrow S_n$ absorption selectively under conditions of qE.

To test whether the changes were correlated to qE, we performed additional experiments on transgenic *A. thaliana* plants, two completely deficient and one enhanced in qE with respect to the wild type (WT), but all normal in their ability to carry out direct light-induced photochemistry (Fig. 4.8). The *WT + PsbS* mutant displays approximately 2.5 times more qE than the WT [46], and the amplitude of the new kinetic component is increased by roughly this amount compared to the WT. The lower plots in Fig. 4.8B show the results for two mutants, *npq4-1* and *npq4-E122QE226Q*, that are completely qE-deficient. The kinetics observed for both *npq4-* mutants were independent of high light illumination, showing that fully functional qE is required for the observation of the new transient. Additional experiments on *A. thaliana* mutants with distinct Car com-

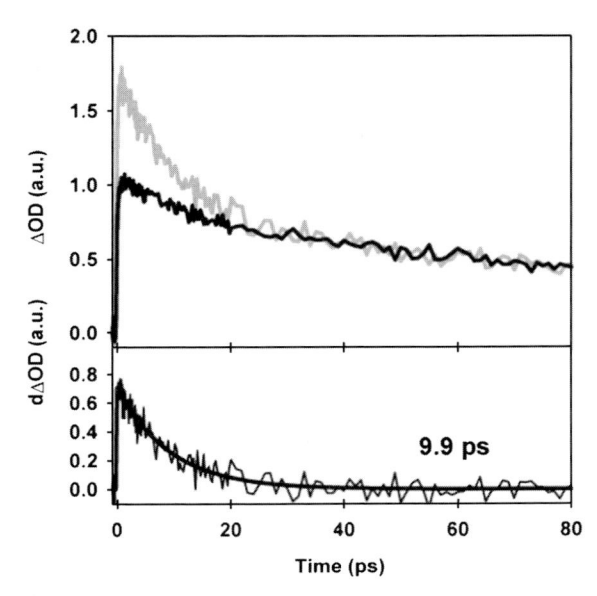

**Fig. 4.7** Top: The scaled TA kinetics probed at 540 nm for spinach thylakoids under conditions of no qE (black line) and maximum, steady-state qE (gray line) upon excitation at 664 nm. The profiles were normalized to 1.0 at the maximum amplitude of the kinetics without qE. Bottom: The difference between the no-qE and the maximum, steady-state qE curves and the corresponding monoexponential fit, which has a lifetime of 9.9 ps.

positions showed that Zea is also necessary to produce the ultrafast kinetic differences (data not shown).

By using the mutants and transgenic plants described above, we were able to identify a PsbS- and Zea-dependent kinetic difference in TA during qE [61]. The Zea excited state may be populated via energy transfer from the Chl $Q_y$ band or by the formation of a Zea-Chl heterodimer, which is more likely because of the characteristics of the kinetic difference. The heterodimer species may be the qE quencher. Furthermore, the formation of a heterodimer during qE may imply an electron transfer quenching mechanism. Recent calculations by Dreuw et al. [62], using a hybrid theoretical approach involving time-dependent density functional theory and configuration interaction singles that was able to capture the salient features of charge-transfer states, found that a heterodimer between Zea and a model Chl *a* chromophore, which contained the chlorin ring but lacked the phytyl chain, would result in Car cation and Chl anion radicals for Zea-Chl separations of less than 5.5 Å. While the mechanism of qE is still unclear, these TA measurements and theoretical calculations may provide another case in which nature has utilized optically dark states and confined geometries as a means to a very efficient end.

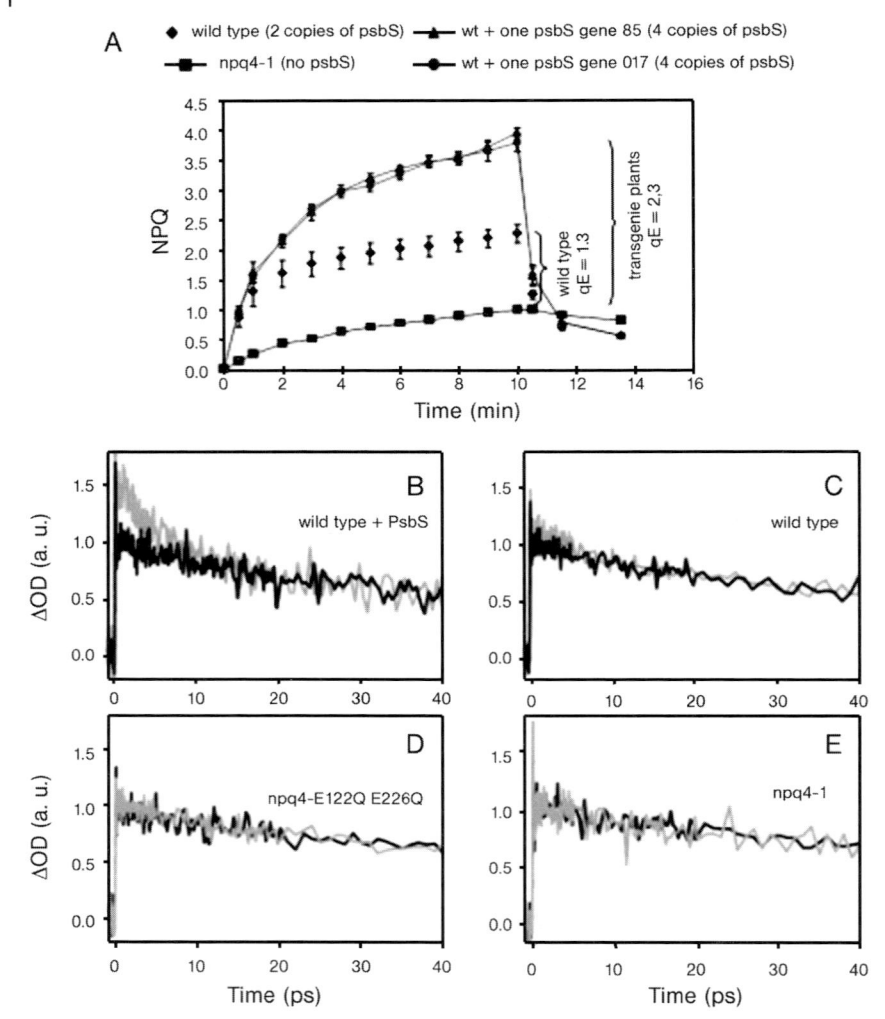

**Fig. 4.8** (A) NPQ induction curves in wild type, *npq4*, and two independent PsbS-overexpressing transgenic lines (No. 5 and 17). Plants were dark-adapted overnight prior to exposure to actinic light of 2000 μmol photons m$^{-2}$ s$^{-1}$. Data are shown as the mean ± standard error ($n=3$). The scaled TA kinetics measured at 540 nm upon excitation at 664 nm for *A. thaliana* plants: (B) *wild type + PsbS*, (C) *wild type*, (D) *npq4-E122QE226Q*, and (E) *npq4-1*. The gray and black lines indicate maximum steady-state and no-qE conditions, respectively, for the samples with qE capability, which corresponds to ON and OFF, respectively, for high light illumination in the qE-deficient mutants. The decay profiles were normalized to 1.0 at the maximum amplitude of the no-qE (light OFF) kinetics.

## 4.6
## Concluding Remarks

In many of the crucial energy transfer processes of the photosynthetic apparatus, nature has simply ignored the principle of maximum overlap of donor emission and acceptor absorption spectra. Instead, confined geometries in conjunction with energetic disorder are used to create ultrafast energy transfer pathways that would not be predicted from the linear spectra of the system. Application of these same design motifs in artificial light-harvesting systems may prove worthwhile.

## Acknowledgments

The work at Berkeley was supported in its entirety by the Director, Office of Science, Office of Basic Energy Sciences, Chemical Sciences Division, of the U.S. Department of Energy under Contract DE-AC03-76SF00098. We are grateful to Dr. Mary Gress for her support of our work. The ideas and results described here represent contributions from many colleagues and coworkers. We are indebted to the following people: Greg Scholes, Brent Krueger, Peter Walla, Chao-Ping (Cherri) Hsu, Jenny Yom, Xanthipe Jordanides, Ana Damjanovic, Ying-Zhong Ma, Xiao-Ping Li, Patricia Linden, John Kennis, Ritesh Agarwal, Bradley Prall, Abbas Rizvi, Mino Yang, Martin Head-Gordon, Petra Fromme, Krishna Niyogi, and Jeffrey Reimers.

## References

1 Koepke, J., Hu, X., Muenke, C., Schulten, K., and Michel, H. (1996) *Structure* **4**, 581.
2 McDermott, G., Prince, S. M., Freer, A. A., Hawthornthwaite-Lawless, A. M., Papiz, M. Z., and Cogdell, R. J. (1995) *Nature* **374**, 517.
3 Förster, T. (1948) *Ann. Phys.* **6**, 55.
4 Joo, T., Jia, Y., Yu, J.-Y., Lang, M. J., and Fleming, G. R. (1996) *J. Chem. Phys.* **104**, 6089.
5 Agarwal, R., Yang, M., Xu, Q.-H., and Fleming, G. R. (2001) *J. Phys. Chem. B* **105**, 1887.
6 Yang, M., Agarwal, R., and Fleming, G. R. (2001) *Photochem. and Photobiol. Part A* **142**, 107.
7 Ma, Y.-Z., Cogdell, R. J., and Gillbro, T. (1997) *J. Phys. Chem. B* **101**, 1087.
8 Pullerits, T., Hess, S., Herek, J. L., and Sundstrom, V. (1997) *J. Phys. Chem. B* **101**, 10560.
9 Jimenez, R., Dikshit, S. N., Bradforth, S. E., and Fleming, G. R. (1996) **100**, 6825.
10 Fowler, G. J. S., Visschers, R. W., Grief, G. G., Grondelle, R. v., and Hunter, C. N. (1992) *Nature* **355**, 848.
11 Krueger, B. P., Scholes, G. D., and Fleming, G. R. (1998) *J. Phys. Chem. B* **102**, 5378.
12 Agarwal, R., Rizvi, A. H., Prall, B. S., Olsen, J. D., Hunter, C. N., and Fleming, G. R. (2002) *J. Phys. Chem. A* **106**, 7573.
13 Scholes, G. D., Jordanides, X. J., and Fleming, G. R. (2001) *J. Phys. Chem. B* **105**, 1640.

14 Scholes, G. D., and Fleming, G. R. (2000) *J. Phys. Chem. B 104*, 1854.
15 Jordanides, X. J., Scholes, G. D., and Fleming, G. R. (2001) *J. Phys. Chem. B 105*, 1652.
16 Jordanides, X., Scholes, G. D., Shapley, W. A., Remers, J. R., and Fleming, G. R. (2004) *J. Phys. Chem. B 108*, 1753.
17 Krueger, B. P., Scholes, G. D., Gould, I. R., and Fleming, G. R. (1999) *Phys. Chem. Comm.*, 8.
18 Shreve, A. P., Trautman, J. K., Frank, H. A., Owens, T. G., and Albrecht, A. C. (1991) *Biochemica et Biophysica Acta 1058*, 280.
19 Macpherson, A. N., Arellano, J. B., Fraser, N. J., Cogdell, R. J., and Gillbro, T. (2001) *Biophys. J. 80*, 923.
20 Krueger, B. P., Scholes, G. D., Jimenez, R., and Fleming, G. R. (1998) *J. Phys. Chem. B 102*, 2284.
21 Zhang, J.-P., Fujii, R., Qian, P., Inaba, T., Mizoguchi, T., Koyama, Y., Onaka, K., Watanabe, T., and Nagae, H. (2000) *J. Phys. Chem. B 104*, 3683.
22 Krueger, B. P., Yom, J., Walla, P. J., and Fleming, G. R. (1999) *Chem. Phys. Lett. 310*, 57.
23 Walla, P. J., Linden, P. A., Hsu, C.-P., Scholes, G. D., and Fleming, G. R. (2000) *Proc. Natl. Acad. Sci. USA 97*, 10808.
24 Walla, P. J., Yom, J., Krueger, B., and Fleming, G. R. (2000) *J. Phys. Chem. B 104*, 4799.
25 Petersilka, M., Gossmann, U. J., and Gross, E. K. U. (1996) *Phys. Rev. Lett. 76*, 1212.
26 Hsu, C.-P., Walla, P. J., Head-Martin, M., and Fleming, G. R. (2001) *J. Phys. Chem. B 105*, 11016.
27 Dexter, D. L. (1953) *J. Chem. Phys. 21*, 836.
28 Zhang, W. M., Meier, T., Chernyak, V., and Mukamel, S. (1998) *J. Chem. Phys. 108*, 7763.
29 Yang, M., and Fleming, G. R. (2002) *Chem. Phys. 282*, 161.
30 Knapp, E. W. (1984) *Chem. Phys. 85*, 73.
31 Agarwal, R., Yang, M., and Fleming, G. R. (2001) in *Ultrafast Phenomena XII*, pp. 653, Springer.
32 Jordan, P., Fromme, P., Witt, H. T., Klukas, O., Saenger, W., and Krauss, N. (2001) *Nature 411*, 909.
33 Vaswani, H. M., Yang, M., Damjanovic, A., and Fleming, G. R. (2004) in: *Femtochemistry and Femtobiology: Ultrafast Events in Molecular Science* (Martin, M. M., and Hynes, J. T., Eds.), pp 401, Elsevier.
34 Yang, M., and Fleming, G. R. (2003) *J. Chem. Phys. 119*, 5614.
35 Yang, M., Damjanovic, A., Vaswani, H., and Fleming, G. R. (2003) *Biophys. J. 85*, 140.
36 Damjanovic, A., Vaswani, H. M., Fromme, P., and Fleming, G. R. (2002) *J. Phys. Chem. B 106*, 10251.
37 Kennis, J. T. M., Gobets, B., Stokkum, I. H. M. v., Dekker, J. P., Grondelle, R. v., and Fleming, G. R. (2001) *J. Phys. Chem. 105*, 4485.
38 Holzwarth, A. R., Schatz, G., Brock, H., and Bittersmann, E. (1993) *Biophys. J. 64*, 1813.
39 Jia, Y. W., Jonas, D. M., Joo, T. H., Nagasawa, Y., Lang, M. J., and Fleming, G. R. (1995) *J. Phys. Chem. 99*, 6263.
40 Jonas, D. M., Lang, M. J., Nagasawa, Y., Joo, T., and Fleming, G. R. (1996) *J. Phys. Chem. 100*, 12660.
41 Haran, G., Wynne, K., Moser, C. C., Dutton, P. L., and Hochstrasser, R. M. (1996) *J. Phys. Chem. 100*, 5562.
42 Jonas, D. M., Lang, M. J., Nagasawa, Y., Bradforth, S. E., Dikshit, S. N., Jiminez, R., Joo, T., and Fleming, G. R. (1995) in: *Proceedings of the Feldafing III Workshop, Munich*.
43 Barber, J., and Andersson, B. (1992) *Trends Biochem. Sci. 17*, 61.
44 Niyogi, K. K. (1999) *Annu. Rev. Plant Physiol. Plant Mol. Biol. 50*, 333.
45 Müller, P., Li, X.-P., and Niyogi, K. K. (2001) *Plant Physiol. 125*, 1558.
46 Li, X.-P., Müller-Moulé, P., Gilmore, A. M., and Niyogi, K. K. (2002) *Proc. Natl. Acad. Sci. USA 99*, 15222.
47 Holt, N. E., Fleming, G. R., and Niyogi, K. K. (2004) *Biochemistry 43*, 8281.
48 Havaux, M., and Niyogi, K. K. (1999) *Proc. Natl. Acad. Sci. USA 96*, 8762.

49 Demmig-Adams, B., III, W. W. A., Barker, D. H., Logan, B. A., Bowling, D. R., and Verhoeven, A. S. (1996) *Physiol. Plant. 98*, 253.
50 Bassi, R., and Caffarri, S. (2000) *Photosynth. Res. 64*, 243.
51 Li, X.-P., Björkman, O., Shih, C., Grossman, A. R., Rosenquist, M., Jansson, S., and Niyogi, K. K. (2000) *Nature 403*, 391.
52 Demmig-Adams, B., and III, W. W. A. (1996) *Trends Plant Sci. 1*, 21.
53 Horton, P., Ruban, A. V., and Walters, R. G. (1996) *Annu. Rev. Plant Phys. 47*, 655.
54 Demmig-Adams, B. (1990) *Biochim. Biophys. Acta 1020*, 1.
55 Owens, T. G., Shreve, A. P., and Albrecht, A. C. (1992) Dynamics and mechanism of singlet energy transfer between carotenoids and chlorophylls: light harvesting and nonphotochemical fluorescence quenching, Vol. 4, Kluwer Academic, Dordrecht, The Netherlands.
56 Frank, H. A., Cua, A., Chynwat, V., Young, A., Gosztola, D., and Wasielewski, M. R. (1994) *Photosynth. Res. 41*, 389.
57 Frank, H. A., Bautista, J. A., Josue, J. S., and Young, A. J. (2000) *Biochemistry 39*, 2831.
58 Josue, J. S., and Frank, H. A. (2002) *J. Phys. Chem. A 106*, 4815.
59 Horton, P., Ruban, A. V., and Walters, R. G. (1996) *Annu. Rev. Plant Phys. 47*, 655.
60 Horton, P., Ruban, A. V., and Wentworth, M. (2000) *Phil. Trans. R. Soc. Lond. B 355*, 1361.
61 Ma, Y.-Z., Holt, N. E., Li, X.-P., Niyogi, K. K., and Fleming, G. R. (2003) *Proc. Natl. Acad. Sci. USA 100*, 4377.
62 Dreuw, A., Fleming, G. R., and Head-Gordon, M. (2003) *J. Phys. Chem. B 107*, 6500.

# 5
# Identifying Redox-active Chromophores in Photosystem II by Low-temperature Optical Spectroscopies

*Elmars Krausz and Sindra Peterson Årsköld*

## 5.1
## Introduction

Photosystem II (PSII) is a complex pigment-protein assembly, partially embedded in the thylakoid membrane of plants, algae, and cyanobacteria. PSII provides the primary bioenergetic driving force for the electron transfer chain of the photosynthetic light-driven reactions. Light absorption by PSII leads to the most energetic process in biology: the extraction of electrons from water on the lumenal side of the thylakoid membrane along with the reduction of quinones on the stromal side. This electron transfer results in the creation of a proton gradient across the membrane, as well as the toxic byproduct of molecular oxygen. The proton gradient provides chemical energy to power the subsequent dark reactions of photosynthesis in which carbon dioxide is fixed [1, 2].

At the heart of PSII is the reaction center assembly (Fig. 5.1). Seminal to the reaction center are the D1 and D2 protein subunits, each being comprised of five membrane-spanning helices. D1 and D2 each contain three chlorophyll (Chl) $a$'s, one pheophytin (Pheo) $a$, and one plastoquinone. A non-heme Fe ion is present between the plastoquinones $Q_A$ and $Q_B$, at the D1-D2 interface. The tyrosine protein residues D1-161 ($Y_Z$) and D2-160 ($Y_D$) are redox-active. There are two $\beta$-carotenes (car) in the reaction center. Crystallographic evidence points to both carotenes being within D2 [3]. The Mn cluster, site of catalytic water oxidation, lies at the lumenal edge of the D1 protein. This crucial complex is stabilized by a number of extrinsic proteins [3–7].

The D1 and D2 subunits are themselves invariably associated with the smaller cytochrome (cyt) $b_{559}$ protein subunit, which forms an integral part of the reaction center. Cyt $b_{559}$ has two membrane-spanning helices. The redox-active heme unit of cyt $b_{559}$ is located between these helices and coordinated by histidine residues on each. Even the most strongly solubilized PSII preparations (D1/D2/$b_{559}$ particles) retain the cyt $b_{559}$ subunit [8], but its role is not fully understood.

*Artificial Photosynthesis: From Basic Biology to Industrial Application*
Edited by Anthony F. Collings and Christa Critchley
Copyright © 2005 WILEY-VCH Verlag GmbH & Co. KGaA, Weinheim
ISBN: 3-527-31090-8

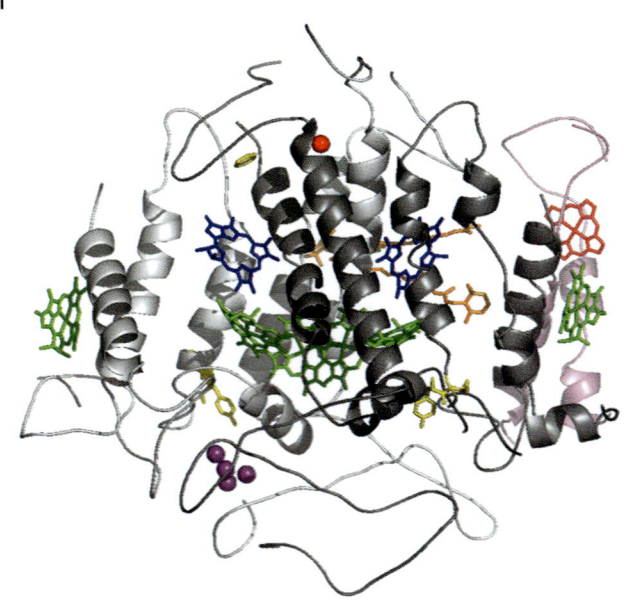

**Fig. 5.1** The X-ray crystal structure, at 3.7-Å resolution, of the photosystem II reaction center region, derived from *Thermosynechococcus vulcanus* cores. The cofactors, as fitted to the electron densities, are color coded as follows: green chlorophyll chlorin rings, blue pheophytin chlorin rings, orange carotenes, red non-heme iron, pink cyt $b_{559}$ heme, purple Mn cluster, yellow quinone and redox-active tyrosines (D1-161 and D2-160). The protein subunit D1 is light gray, D2 dark gray, and cyt $b_{559}$ subunits pink. Structural data from Kamiya and Shen [3], PDB ID code 1IZL.

Intimately associated with the reaction center of PSII are the inner antenna proteins, CP43 and CP47. A functionally active PSII core preparation retains these proteins and other less strongly bound manganese-stabilizing proteins. The core is the minimal PSII assembly able to efficiently oxidize water. In the thylakoid membrane, a PSII core is supplemented by large light-harvesting assemblies, which are not (nominally) involved in redox chemistry. These outer light-harvesting assemblies differ significantly between plants and cyanobacteria [1]. By contrast, the fundamental core PSII machinery functions with a remarkable similarity across a wide variety of organisms. In this work we show that low-temperature spectroscopy is able to identify variations between PSII cores sourced from plants and those from cyanobacteria. We are also able to compare spectra of active cores with those of the pigment-containing protein subunits from which they are constructed: D1/D2/$b_{559}$ particles, isolated CP43, and isolated CP47. The whole is thus compared to the sum of the parts, indicating changes wrought by progressive detergent solubilization of a PSII core.

Fundamental questions remain regarding critical aspects of PSII. One question is the nature of P680, the light-activated electron donor in PSII. Charge separation in PSII occurs by spontaneous ionization of the excited state of P680, often labeled

P680*. There is ongoing discussion as to the extent of involvement of the different reaction center pigments in P680* and about the rate of primary charge separation [9]. Many critical experiments pertinent to these deliberations have been performed on D1/D2/$b_{559}$ preparations. We further put forward the case in this chapter that the reaction center proteins become disrupted and inhomogeneous upon solubilization and that spectra from D1/D2/$b_{559}$ particles do not accurately reflect the properties of native P680 present in active PSII.

As with the analogous L and M reaction center proteins in the bacterial photosystem, there is crude mirror symmetry between the D1 and D2 proteins. There are other parallels with the bacterial reaction center. Photoactivated electron flow occurs through just one branch, mainly located on the D1 protein. Charge separation of P680 to form P680$^+$ is accompanied by reduction of the Pheo $a$ on D1 (Pheo$_{D1}$). This anion intermediate then goes on to reduce the plastoquinone $Q_A$, bound near Pheo$_{D1}$ by the D2 protein. However, the donor side of PSII is uniquely different from the bacterial system. P680$^+$ rapidly oxidizes $Y_Z$, which in turn oxidizes the manganese cluster. This cluster cycles through four sequential one-electron oxidative steps, described by the Kok cycle [10, 11], each step driven by an oxidation of P680. PSII is advanced from the lowest state, $S_0$, progressively to the $S_1$, $S_2$, and $S_3$ and $S_4$ states, and is then spontaneously re-reduced to the $S_0$ state as oxygen is evolved. In the dark, the cluster rests in the $S_1$ state.

A PSII preparation can be poised in a particular metastable S-state ($S_0$–$S_3$) by appropriate illumination protocols and then be trapped by freezing [12–14]. Detailed spectra of these species can then be obtained, particularly via EPR spectroscopy at low temperatures. If illumination of a PSII sample is performed below 130 K, manganese oxidation is inhibited and alternative electron donors are sought by the system. These pathways can be quantitatively monitored using precision optical spectroscopies [15, 16]. The disappearance of chromophores is paralleled by the appearance of oxidized species, which most often absorb in a different part of the spectrum. The investigation of electrochromic shifts associated with $Q_A^-$ radical anion formation is very conveniently performed by very low-temperature (1.7 K) illumination of a sample poised in the $S_1$ state. The $S_1(Q_A^-)$ state is formed with a quantum efficiency approaching 1 and is stable over a period of hours at helium temperatures. Electrochromic shifts smaller than 1 cm$^{-1}$ can be routinely determined in this way [17].

## 5.2
## Experimental Methods

### 5.2.1
### Sample Preparation

PSII membrane fragments (BBYs) were prepared from spinach as previously described [18]. Fully active, detergent-solubilized PSII core complexes (cores) were prepared from BBYs using the recent improvement [19] on the original method from van Leeuwen et al. [20]. Solubilized protein preparations of CP43, CP47, and D1/D2/$b_{559}$ preparations containing five and six Chl $a$'s were also prepared by a previously described procedure [8, 21–23]. *Synechocystis* (*Syn*.) 6803 core complexes were prepared from a glucose-tolerant, HT-3A histidine-tagged strain of *Syn*. 6803 [24].

A prerequisite for our low-temperature spectroscopies is the availability of large-aperture, uniform, highly transparent, and strain-free optical samples. This requires the use of an effective glassing medium (cryoprotectant), which was either 45% 1:1 glycerol:ethylene glycol or 60% glycerol (final concentrations). All sample handling was performed in subdued, indirect green light and performed as quickly as practicable. Active PSII samples were dark-adapted for 5 min at room temperature before glassing into He($l$).

## 5.2.2
### Illumination

Sample illuminations were performed with a stabilized 150-W quartz halogen lamp. The lamp was imaged $\sim$1:1 onto the sample through a 10-cm water filter to remove unwanted infrared light. Light was further passed through a bank of glass filters that transmitted in the 500–600 nm region. This is a region of uniform and relatively low absorbance of PSII samples, allowing a uniform illumination of the sample through its thickness. The intensity of light at the sample was 1–2 mW cm$^{-2}$. This corresponds to $\sim 3$–$6\times10^{15}$ photons per second. We estimate our typical core sample to contain $\sim 5\times10^{-10}$ moles PSII, i.e., $\sim 30\times10^{13}$ PSII centers.

BBYs and PSII cores were efficiently converted to the $S_2Q_A^-$ state by illumination as described above, at 260 K for 10–30 s. The samples were then quenched to 4 K over a period of 30–60 s. In cores, $Q_B$ is absent, limiting the photochemistry to this single-electron movement. In BBY samples, DCMU was added to block electron transfer from $Q_A^-$ to $Q_B$, leading to an analogous $S_2Q_A^-$ formation as in cores. Other illuminations were performed with the sample at 1.7 K. These create non-physiological ($S_1Q_A^-$) states in which electron donation from manganese is inhibited.

## 5.2.3
### Spectra

The sample cells used are specially constructed, strain-free, cylindrical fused quartz-windowed assemblies. Their diameter is 12 mm and the path length ranges from 0.1 mm to 1 mm. The cell was mounted onto a low-thermal mass sample rod with a facility for rapid cell interchange. The sample rod was introduced into an Oxford Instruments SM4 superconducting cryostat through a helium lock system, the latter fitted with sample observation/illumination windows. Absorption and CD/MCD spectra were collected simultaneously on a spectrometer designed and constructed in our laboratory [25]. The entire system was built on an optical table and constructed for good mechanical stability.

The system is built around a 0.75-m Spex Czerny-Turner single monochromator using a 1200-L mm$^{-1}$ grating blazed at 500 nm and a SM4 cryostat modified for use with metalloenzymes. Critical aspects of the performance of this system with respect to PSII samples were that the wavelength reproducibility between scans was better than 0.002 nm and that the signal-to-noise in absorption, CD, and MCD spectra were often close to the theoretical maximum for a given light level ("shot noise" limited). The small spectral shifts sometimes observed were not caused by wavelength jitter or drift in the monochromator, nor were they masked by excess noise.

Electrochromism spectra presented were recorded with 50-μm slit widths (0.05 nm resolution) in order to minimize light-induced photochemistry associated with measurement (actinic) light from the monochromator. Adequate sensitivity in CD and MCD spectra could be obtained only with significantly wider slits (500 μm) and, consequently, 100 times higher light levels. For CD/MCD measurements on active PSII preparations, samples were pre-illuminated with green light at 1.7 K (locking the sample in the $S_1Q_A^-$ state as described above) so that CD/MCD spectral measurements did not lead to accumulating changes in absorption spectra. Dichroic data was acquired with a magnetic field of +5T or −5T placed over the sample. MCD data were extracted by subtracting the obtained traces, CD data by adding them.

## 5.3
## Results and Discussion

### 5.3.1
### Absorption and CD Signatures: Plant PSII Cores and BBYs

Table 5.1 provides some critical properties of our PSII BBYs and ensuing core preparations. The catalytic activity of the cores, scaled per PSII center, matches that of BBYs. We found 32 Chl $a$'s per PSII in our plant cores. This Chl $a$ content is near the number inferred from crystallographic studies on cyanobacterial core complexes (CP47: 16 chls, CP43: 14 chls, D1/D2: 6 chls; total 36 Chl $a$'s) [7]. By low-temperature MCD, we found one cyt $b_{559}$ per 32 Chl $a$'s in the core material, again consistent with crystallographic and other studies [26]. The MCD of PSII membrane fragments shows one cyt $b_{559}$ per two pheophytins, thus establishing that there is a single cyt $b_{559}$ per PSII in this material. Cyt $b_{559}$ is >95% oxidized in the core samples used for optical study, but its redox state in BBYs can vary, depending on the history of the sample and the particular cryoprotectant used.

Fig. 5.2 provides a comparative overview of the low-temperature absorption spectra of BBYs and cores in the near-UV/visible region. The dominant absorption feature (435 nm) is due to the B-bands of Chl $a$ and Chl $b$. Being a compo-

**Table 5.1** PSII-enriched membrane fragments and core complexes from spinach: oxygen-evolving activities and cofactor component stoichiometries [19].

|  | BBY[a] | PSII Cores |
|---|---|---|
| *Activity* | | |
| μmol $O_2$/mg chl/h [b] | 600–800 | 3700–4300 |
| kmol $O_2$/mol PSII/h | 120–160 | 120–140 |
| *Components* [c] | | |
| Pheophytin *a* | 2.00 | 2.00 |
| Chlorophyll *a* | 134 ± 2 | 32 ± 1 |
| Chlorophyll *b* | 62 ± 1 | 0.8 ± 0.2 |
| β-carotene | 9 ± 1 | 7 ± 1 |
| Mn [d] | ~ 5 | 4.5 ± 0.5 |
| cyt $b_{559}$ [e] | 1.0 ± 0.2 | 1.0 ± 0.05 |

a) PSII-enriched membrane fragments.
b) Oxygen-evolving activities of cores and BBYs were converted to a per PSII basis by using the chl *a* and *b* contents established below.
c) All component stoichiometries are given per 2.00 pheophytin *a*, assumed to represent one PSII center.
d) Includes bound Mn only.
e) Determined by analysis of MCD spectra and gel band intensities.

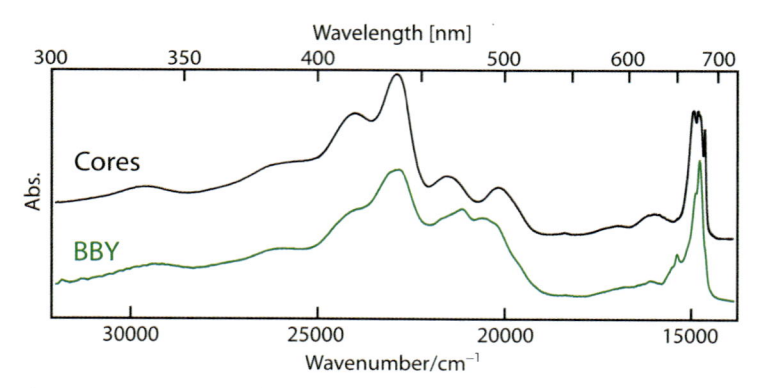

**Fig. 5.2** Low-temperature absorption spectra of PSII cores (upper trace, 1.7 K) and membrane fragments (lower trace, 10 K) derived from spinach. The spectra have been scaled to similar peak absorption intensities and offset for clarity.

nent of the light-harvesting antenna, Chl *b* is absent in cores but shows a distinct $Q_y$ feature at 650 nm in the BBY spectrum. Prominent β-carotene absorption peaks are seen in the 450–520 nm region, and the $Q_y$ bands of Chl *a* and Chl *b* dominate the red region (650–700 nm). Weaker bands near 540–560 nm, visible upon magnification, are identifiable as the distinctive $Q_x$ bands of Pheo *a*.

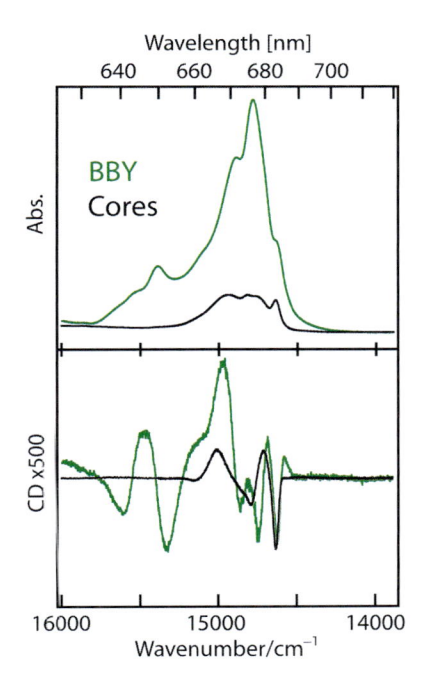

**Fig. 5.3** Absorption (top panel) and CD (bottom panel), recorded simultaneously at 1.7 K, of BBY (green) and PSII cores (black). The data is scaled per PSII: the BBY absorption has an area corresponding to 172 Chl $a$ equivalents, and that of the cores corresponds to 33 Chl $a$ equivalents. The CD traces have been scaled by the same factors and multiplied by 500.

Also present in this region of the BBY spectrum is the characteristic Q-band of the reduced form of cyt $b_{559}$.

The Chl $Q_y$ band shows quite well-developed structure in both preparations. Notable is a sharp feature near 683.5 nm (see Fig. 5.3) with a full width at half maximum of 50 cm$^{-1}$, and an intensity corresponding to 2.2 Chl $a$'s in cores, that we have associated with P680 [19, 27]. A number of vibrational sidebands associated with this peak appear weakly in the 16,000 cm$^{-1}$ region. The unstructured bands in the 17,000 cm$^{-1}$ region can be attributed to $Q_x$ transitions of Chl $a$. This assignment is strongly supported by their characteristic negative MCD signature (not shown). In Fig. 5.3, the absorption and CD of BBYs and cores in the $Q_y$ region are displayed on a per-PSII scale. The 683.5-nm band is present in both BBY and core spectra with virtually the same position, width, and intensity, in both absorption and CD. We take the qualitative and quantitative retention of this feature as direct evidence that our isolated, fully active PSII cores have the same spectral characteristics as the CP43/CP47/D1/D2/$b_{559}$ pigment assembly within BBYs. Illumination-induced changes, presented in Fig. 5.8, also point to the quantitative retention of PSII core spectral features between BBYs and our core preparations. Further on (Fig. 5.5), we show that this spectral homology of reaction center pigments is disrupted when further solubilizing cores to D1/D2/$b_{559}$ particles.

## 5.3.2
**Absorption and CD Signatures: Plant and Cyanobacterial PSII Cores**

As mentioned in the Introduction, PSII sourced from cyanobacteria has many properties in common with that obtained from plants. This is despite the fact that these organisms experience entirely different environmental conditions and have very different light-harvesting systems. Fig. 5.4 compares absorption and CD spectra of active cores prepared from spinach and Syn. 6803 [28]. Both preparations have a number of spectral features with line widths in the range of 50–80 cm$^{-1}$. However, there is no pronounced feature in Syn. 6803 corresponding to the 683.5-nm absorption feature in spinach. The CD spectra of both systems display some typical characteristics of coupled pigment systems, having a magnitude far larger than that of isolated monomeric Chl $a$ and displaying a dispersive and conservative lineshape: the CD shows positive and negative features, and the area of negatively and positively signed CD features balance. The single strong negative feature at 683.5 nm in spinach appears as a doublet in Syn. 6803, with features at 681 nm and 684 nm. This split is also clearly visible in the absorption spectra. The band corresponding to the positive CD feature at 665 nm in spinach also appears to be split in the cyanobacterial system.

We have suggested that the well-developed structure in the red region of the $Q_y$ band in spinach PSII, 675 nm and above, is associated with reaction center pigments [17, 19]. There may also be some structural contribution from pigments in the inner antenna, particularly CP43, in this region. The significant differences in spectral structure seen in Fig. 5.4 immediately point to the possibility that the reaction center is somewhat differently organized in the two organisms.

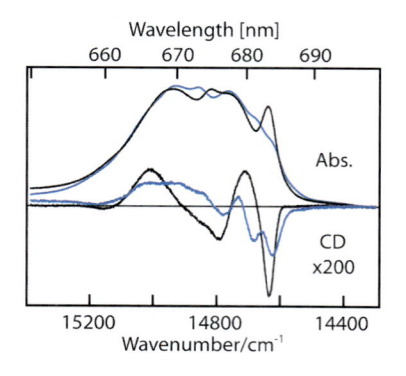

**Fig. 5.4** Simultaneously acquired absorption and CD spectra at 1.7 K, from spinach cores (black) and Synechocystis 6803 (blue). The spectra have been normalized to the same absorption peak area, and the CD has been multiplied by 200.

## 5.3.3
**Absorption Signatures: The Native and Solubilized Reaction Center**

The solubilized D1/D2/$b_{559}$ complex has been extensively studied since the original description of its preparation by Namba and Satoh [8]. This preparative breakthrough allowed the PSII reaction center pigments and the dynamical processes within PSII to be studied without spectral masking and interference from light-harvesting pigments. These preparations have proven to be somewhat inhomogeneous and variable [21], but preparations can be made that contain either five or six Chl $a$'s. D1/D2/$b_{559}$ particles also contain two Pheo $a$'s and $\beta$-carotenes. The $\beta$-carotene content is lower in the 5-Chl preparation [29].

All the pigments in PSII cores have their $Q_y$ absorption contained within the 650–700 nm spectral region. If we attribute equal oscillator strength to each Chl $a$ or Pheo $a$, regardless of its exact spectral position or degree of electronic coupling with other pigments, and use the known ratio of oscillator strengths of Pheo $a$ to Chl $a$ in the $Q_y$ region (0.7), it is straightforward to use the scaled spectra from isolated subunits to "synthesize" a PSII core spectrum, as is done in Fig. 5.5. This figure also presents appropriately scaled spectra of each of the isolated subunits, 6-Chl D1/D2/$b_{559}$, CP43, and CP47 (also prepared from spinach), from which this synthesis was constructed, as well as the spectrum of active PSII cores. All spectra were taken at the same temperature and in the same cryoprotectant medium and were recorded on the same instrument. The number of Chl $a$ equivalents used to scale each spectrum is given in Fig. 5.5.

The synthesized spectrum (gray) has an overall shape similar to that of the actual spectrum (black) of PSII cores, but it lacks much of the structure, particularly the prominent 683.5-nm feature. In a previous analysis of this type, data from different workers recorded at different temperatures on different spectrometers were compared and combined [27]. A similar discrepancy was noted and attributed primarily to changes in the spectrum of D1/D2/$b_{559}$ reaction center pigments upon rupture of the core complex. The improved data in Fig. 5.5 fully support this interpretation. Although there is sharp structure in CP43 near 683 nm, its magnitude does not account for the prominent 683.5-nm feature in the PSII cores. If we assume that differences between the synthesized

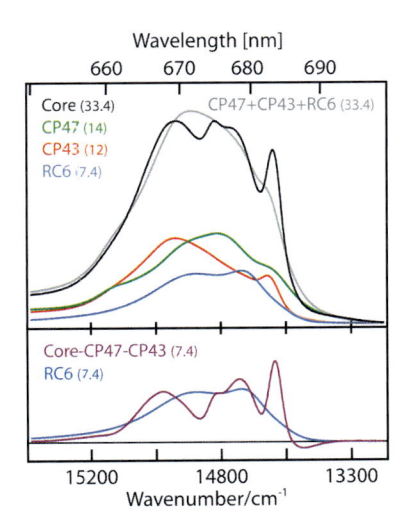

**Fig. 5.5** Absorption spectra at 1.7 K of spinach PSII cores (black), isolated CP47 (green), isolated CP43 (red), and solubilized D1/D2/$b_{559}$ (blue). Each spectrum has been scaled so that the area matches the known pigment content of that fragment, with each Pheo counting as 0.7 of a Chl. The number of Chl $a$ equivalents used for the scaling is given. The sum of the fragments is displayed in gray. The intact core absorption minus that of isolated CP43 and CP47, displayed in purple, represents the intact reaction center. The vertical scaling is the same in both panels.

subunit spectrum and the assembled core spectrum are entirely due to variations of pigments within the D1/D2/$b_{559}$ proteins, then we can obtain the "spectrum" of the native reaction center by subtraction of the CP43 and CP47 spectra from that of the active cores. This is also shown in Fig. 5.5 (purple). Comparing this to the solubilized D1/D2/$b_{559}$ spectrum, the latter shows a severe collapse of structure and a blue shift of 2–3 nm of the region of dominant absorption.

### 5.3.4
### MCD Signatures: P680 and Chl$_Z$

The room-temperature MCD of a D1/D2/$b_{559}$ preparation was reported some time ago as indicating that the red-most pigments associated with P680 had a significantly reduced MCD amplitude compared to normal Chl $a$. There have been other reports of reduced MCD of reaction centers and aggregated pigment systems [30–32], and we have reported that the 683.5-nm feature in PSII cores in particular shows a 50% MCD reduction [19, 27].

Fig. 5.6 shows the absorption (blue) and MCD spectra (green) in the $Q_y$ region of D1/D2/$b_{559}$ preparations containing six or five chls, measured at 1.7 K. The absorption spectra have been scaled per PSII. The MCD spectrum of a Chl $a$ in the $Q_y$ region is dominated by a Faraday $B$ term [33]. $B$ terms have the same lineshape as absorption spectra, are temperature independent, and can have positive or negative amplitude. The $B$ term of $Q_y$ of Chl $a$ is positive [32].

The MCD spectra in Fig. 5.6, obtained as the difference between +5 T and −5 T data, are presented per T and PSII and then multiplied by a scaling factor. The scaling factor was chosen so that, for an isolated Chl $a$, the scaled MCD $B$ term and the absorption spectra would overlap. Such an overlap is indeed seen for the single Chl $a$ bound to the Cyt $b_6f$ complex (unpublished results).

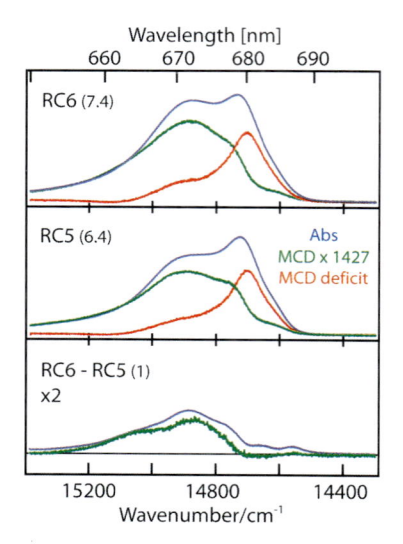

**Fig. 5.6** Absorption and MCD data at 1.7 K from D1/D2/$b_{559}$ fragments containing 6 Chl $a$'s (top panel) and 5 Chl $a$'s (middle panel). The absorption traces (blue) are scaled per PSII, with the number of Chl $a$ equivalents as specified. The MCD (green) is given per T and PSII and is multiplied by 1427. This scaling factor was chosen to produce overlapping absorption and MCD spectra for isolated Chl $a$. The MCD deficit (red) is the difference between the absorption and the scaled MCD, blue minus green. The bottom panel shows the difference in absorption and MCD of the 6-Chl and 5-Chl D1/D2/$b_{559}$ materials, doubled in amplitude for clarity. The difference in MCD deficit between the two samples is negligible. The vertical scaling is the same in all panels.

For both 5- and 6-Chl preparations, the overlap of absorption and scaled MCD is reasonable at the high-energy side of the band, but there is a dramatic reduction of the MCD relative to the absorption in the red region. This discrepancy peaks around 680 nm, where the MCD B term reduction observed is $\sim 80\%$. Fig. 5.6 plots the difference between the absorption and the scaled MCD, thus displaying the reduction, or "MCD deficit," throughout the $Q_y$ band (red). This deficit spectrum is almost identical in the two preparations. Interestingly, it bears a strong resemblance to reports of transient changes in the absorption spectra upon intense short-pulse laser excitation [34], which bleaches P680. This correspondence points to the MCD deficit being a remarkable signature for P680.

The bottom panel of Fig. 5.6 shows the difference between the appropriately scaled (per PSII) spectra of the 6- and 5-Chl preparations, magnified twice for clarity. Upon removal of the sixth Chl, absorption intensity is lost on the high-energy side of the absorption, with the loss peaking at 672 nm. The Chl lost is thought to be the peripheral Chl in the D1 protein ($Chl_Z$). We note that the difference in scaled MCD is almost identical to the difference in absorption. Thus, $Chl_Z$ behaves as a "normal" Chl $a$ and shows no MCD deficit.

We have previously reported a $\sim 50\%$ MCD deficit in the 683.5-nm feature of active PSII cores [19, 27]. This characteristic was used to help establish this band as being associated with P680. In pioneering work [35], the assignment of a somewhat weaker absorption feature near 683 nm in the spectrum of a thermophilic cyanobacterial core preparation (*Synechococcus elongatus*) to P680 was discounted on the basis that there was a similar feature in the spectra of solubilized CP43 derived from this core preparation. At this time, the pigment contents of the different protein fragments were not well established. We have recorded the simultaneous absorption and MCD spectra of PSII cores, CP43, CP47, and D1/D2/$b_{559}$, all prepared from spinach and measured under identical conditions (absorption traces shown in Fig. 5.5). Scaling the MCD spectra using the same procedure as used in Figs. 5.5 and 5.6, we can subtract the MCD (scaled per PSII) of isolated CP43 and CP47 from that of the active cores to obtain the MCD associated with the native reaction center. Correspondingly, we can generate the MCD deficit spectrum of the native reaction center; this is presented, along with corresponding spectra of solubilized D1/D2/$b_{559}$ material, in Fig. 5.7.

A comparison between the native reaction center absorption and MCD deficit is both interesting and remarkable. The absorption peak at 667 nm, having an area corresponding to approximately two Chl $a$'s, has no MCD deficit. In this sense, it has the same property as the band associated with the Chl $a$ that was removed in passing from the 6-Chl to the 5-Chl D1/D2/$b_{559}$ preparation (Fig. 5.6). It is also in a similar spectral range and is twice as intense (per PSII). On this basis we assign this peak to the two peripheral chls in the native system. As noted previously [19], and our improved data confirm this, the region 686–700 nm also shows no MCD deficit in the cores (not shown). This region also has no absorption in the native reaction center and can be attributed to a "trap" pigment in CP47 in the PSII cores. Thus, these three weakly interacting, non-photoactive Chl $a$ have MCD B term magnitudes similar to that of isolated Chl $a$.

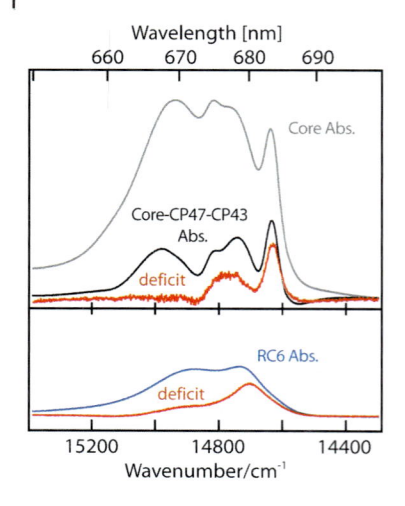

**Fig. 5.7** Top panel: absorption (black) and MCD deficit (red) of the intact PSII reaction center, deduced by subtracting the absorption and MCD deficit spectra of isolated CP43 and CP47 from those of intact PSII cores. The intact core absorption (gray) is shown for reference. Bottom panel: absorption (blue) and MCD deficit (red) of solubilized D1/D2/$b_{559}$ material. The vertical scaling is the same in both panels.

The MCD deficit of the 683.5-nm absorption peak in the native reaction center, as seen in Fig. 5.7, is greater than 80%. This is evident from the fact that the absorption and MCD deficit spectra have comparable amplitude. This corresponds to a greater reduction in the MCD for this peak than the 50% seen directly in the PSII core spectra [19, 27], as the latter contains a CP43 contribution in this region, which displays no MCD reduction (not shown). An extensive report of the CD and MCD of the isolated PSII subunits is underway [36].

The notion of a quasi-degeneracy of a weaker CP43 feature (with no MCD deficit) with a stronger P680 feature (with a strong deficit) at 683.5 nm in spinach cores is supported by the observation of two bands in the absorption and CD spectra of *Syn.* 6803 cores in the same region (Fig. 5.4). In the cyanobacterial system, the accidental degeneracy appears to be lifted. There are two distinct negative peaks in the CD, with corresponding absorption peaks. In the CD spectrum of purified spinach CP43, there is a strong negative CD feature near 683 nm [36, 37]. The combined amplitude of the two negative CD peaks in *Syn.* 6803 has a comparable strength to the single, intense peak in spinach. This additivity points to the CD not being associated with coupling between the native reaction center and pigments in CP43.

The area of the absorption of the solubilized D1/D2/$b_{559}$ and the deduced native reaction center are identical per definition, corresponding to 7.4 Chl *a* equivalents (Fig. 5.7). Interestingly, the corresponding MCD deficits are also of nearly identical magnitude. The area corresponds to a total of $\sim 3$ Chl *a*'s, equivalent to three chls having zero MCD. The collapse of the absorption spectral structure upon solubilization, with the dominant peak shifting to higher energy by 3–4 nm (80–100 cm$^{-1}$) and the Chl$_Z$ absorption to lower energy by a comparable amount, is mirrored by the MCD deficit spectra. This implies that the MCD deficit is an inherent property of the P680 pigments, maintaining its intensity although the couplings within the reaction center are changing.

### 5.3.5
### Electrochromic Signature: Pheo$_{D1}$ in Active PSII

Flash illumination of PSII preparations leads to a range of transient spectral changes. These occur over a wide spectral range and on timescales from picoseconds to seconds [1]. When experiments are performed under physiological conditions, light absorption leads to an electron being extracted from the manganese cluster on the lumenal side of the membrane, advancing the S-state of the Mn cluster, along with an electron being accepted by a quinone on the stromal side of the membrane. Experiments performed with modified samples or at low temperature lead to the system seeking additional or alternate electron transfer pathways.

Many spectral changes can be attributed to the creation of metastable species, such as P680$^+$, along with the parallel bleaching of the parent species, P680 in this case. As well as transient gain/loss of redox-active elements in the multi-step charge-separation process in PSII, pigments close to oxidized or reduced species can experience significant spectral shifts. Electrochromic (Stark) shifts are induced when the creation of a charge center close to a pigment shifts the excitation energy of the pigment via the interaction of the electric field of the charge with the change in dipole moment induced by electronic excitation of the pigment.

The "C550" signal, a distinctive transient optical feature of ca. millisecond duration seen near 550 nm in PSII at room temperature, is an example of an electrochromic shift. This signature was discovered early in the study of PSII preparations, but its origin was not understood. It has subsequently been shown to arise from a Stark shift to higher energy of the Pheo$_{D1}$ $Q_x$ transition, induced by the formation of $Q_A^-$. We have estimated the magnitude of the $Q_x$ shift from low-temperature spectroscopy to $\geq 80$ cm$^{-1}$ [17] (see below). The effect of $Q_A$ reduction on the $Q_x$ transition can be calculated by taking the electric field change experienced by Pheo$_{D1}$, now known to be 13 Å from $Q_A$, and by assuming a dielectric constant of $\varepsilon = 1$–2 (appropriate for the interior of a protein at cryogenic temperatures [38]). A shift of the observed magnitude can be accounted for if the change in dipole moment of the $Q_x$ excitation, $\Delta\mu_x$, is $\geq 0.5$ D [17].

In a relatively recent room-temperature study [39], transient absorption differences of pea PSII cores in the $Q_y$ spectral region were reported and modeled. The analysis concluded that there was a small electrochromic shift (relative to that of the $Q_x$ transition) of $\sim 6$ cm$^{-1}$ to high energy on the $Q_y$ transition on Pheo$_{D1}$ subsequent to $Q_A^-$ formation. In a related study [40] of Mn-depleted *Syn.* 6803 PSII preparations, a set of chemical reduction and illumination protocols were designed to create samples in which $Q_A$ was first neutral and then anionic, with other redox elements remaining unchanged. Low-temperature (77 K) absorption spectra of these samples were recorded. Upon subtraction of scaled spectra of the different samples, a "C550" signature analogous to that obtained in transient spectra was evident. Larger spectral changes in the $Q_y$ region were also present. These were interpreted in terms of $\sim 80$-cm$^{-1}$ $Q_y$ shifts of Pheo$_{D1}$

and comparable shifts in a number of other pigments. This interpretation is in marked contrast to the pea PSII core analysis mentioned above, and to our analysis below [17]. There have been a number of other low-temperature studies of illumination-induced changes in chemically modified PSII complexes at low temperatures [15, 41–43].

We take advantage of the improved resolution in active PSII core spectra recorded at 1.7 K. This enhances our ability to assign pigment band positions and to quantify electrochromic shifts. By using active PSII samples and measuring spectra before and after the efficient generation of $Q_A^-$ in a single sample by weak green-light illumination, reproducible spectral changes can be obtained. Illuminations performed at 260 K on PSII samples poised in the $S_1Q_A$ state move an electron from the Mn to the quinone, generating the physiological $S_2Q_A^-$ state. The efficiency of the conversion was confirmed by observation of the characteristic multiline pattern in EPR [19]. Low-temperature (1.7 K) illuminations generate a metastable $S_1Q_A^-$ state. We have seen that this state does not persist in cores above 100 K, returning to the $S_1Q_A$ state, but at $\leq 10$ K it is stable. The very low light levels used in the spectrometer eliminate measurement-induced changes in spectra.

In Fig. 5.8 we present absorption spectra (black) obtained from plant core and BBY preparations in both the $Q_x$ and $Q_y$ spectral regions, along with the differences in absorption induced by 260-K (red) and 1.7-K (blue) illuminations. For both materials, the changes induced are dominated by the electrochromic shifts of the $Pheo_{D1}$ bands at 545 nm and 685 nm, induced by $Q_A$ reduction.

Small changes in the overall $Q_y$ absorption profile occur for the BBY samples upon freeze/thaw cycles, and are evident in Fig. 5.8 (dotted red trace). These are not dependent on illumination and never show sharp features such as the illumination-induced Stark shifts. Apart from the changes associated with freeze/thaw cycles mentioned above, $Q_x$ and $Q_y$ electrochromic shifts upon 260-K and 1.7-K illumination are very similar to the changes seen in PSII cores. This further supports our contention that spectral features of the PSII cores are fully maintained between the solubilized and membrane-bound materials.

Interestingly, both illumination temperatures give rise to essentially identical electrochromic shifts, although the electron donors differ. In the case of illumination at 260 K, the electron donor for both materials is the Mn cluster, moving from $S_1$ to $S_2$ in the Kok cycle. For the 1.7-K illumination experiment, the absence of any change in the integrated Chl $a$ $Q_y$ absorption establishes that the donor is not Chl, to a precision of <0.1 Chl per PSII. This is confirmed by NIR spectra (not shown), where the Chl radical absorbs. Spectral measurements near 450 nm, where $\beta$-carotene absorbs, and in the 960-nm region (not shown), where the $\beta$-carotene radical cation absorbs, establish that in PSII cores, $\beta$-carotene is a dominant electron donor upon 1.7-K illumination.

PSII in BBYs and cores differ in a respect that is apparent in the $Q_x$ spectra of Fig. 5.8: cyt $b_{559}$ rests in the reduced form in BBYs but is fully oxidized in cores. The reduced form has distinct absorption in the 555–560 nm region [26] and its absorption band is evident in the spectrum of BBYs. This assignment is

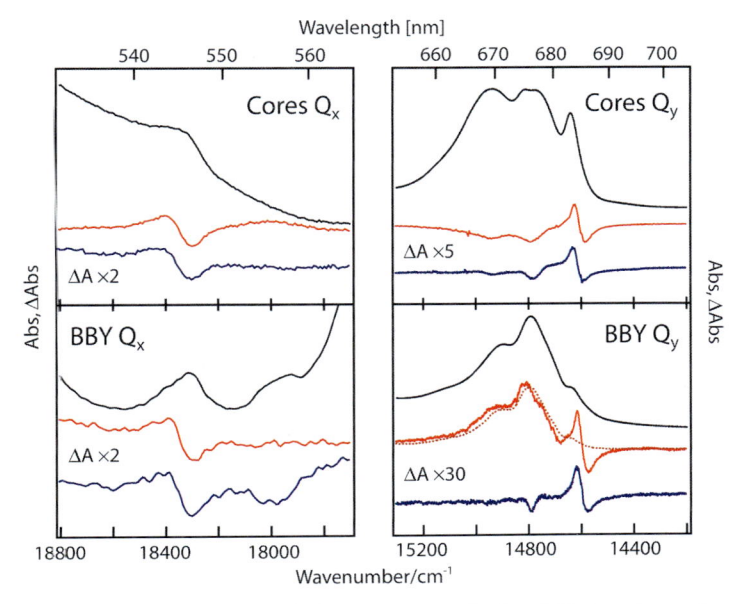

**Fig. 5.8** Absorption spectra (black traces) and differences in absorption induced by illuminating spinach PSII cores and BBY at 1.7 K (blue traces) and 260 K (red traces). For the $Q_x$ region, the differences are doubled in size compared to the absorption spectra; for the $Q_y$ region, the BBY differences are multiplied by 30 and the core differences by 5. The core $Q_x$ data are displayed 20 times larger than the core $Q_y$ data, and the BBY $Q_x$ data are displayed 100 times larger the BBY $Q_y$ data. The difference obtained in the $Q_y$ region from bringing the BBY sample to 260 K and refreezing it without illumination is displayed (dotted red trace). The BBY $Q_x$ data have been adjusted linearly and smoothed for clarity. The data were accumulated at 1.7 K.

verified by a characteristic MCD A term signature (not shown). The band is bleached by the low-temperature illumination that induces electrochromic shifts at 1.7 K. This shows that in BBYs, cyt $b_{559}$ is the preferred electron donor at 1.7 K, although the relevant electron transfer distance from P680 to cyt $b_{559}$ is ~ 30 Å. In a BBY sample in which cyt $b_{559}$ is pre-oxidized, $Q_A^-$ formation remains efficient, with the electron donor becoming $\beta$-carotene as in cores.

Despite these different electron sources, the Stark patterns induced in the different materials and at different illumination temperatures are very similar in both spectral regions. This points to the shift patterns being dominated by the formation of $Q_A^-$ and not being significantly influenced by potential charges associated with the different electron donors.

Quantification of the Pheo $Q_x$ shifts is difficult because of the composite and asymmetric nature of the bands in this region. $Pheo_{D1}$ and $Pheo_{D2}$ have overlapping absorptions in this region, together with both reduced and oxidized forms of cyt $b_{559}$ [26]. A lower limit of the Stark shift, based directly on the change of the $Q_x$ peak maximum in BBYs, the system in which spectral features are best resolved, is 80 cm$^{-1}$ [17].

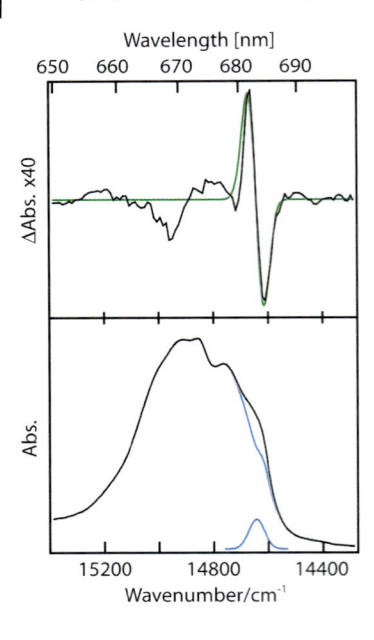

Fig. 5.9 Bottom panel, black trace: *Syn.* 6803 absorption prior to illumination. Blue traces: a Gaussian centered at 683.0 nm, with the magnitude of one Pheo *a* (0.7 Chl *a* equivalents) and the *Syn.* 6803 absorption minus this Gaussian. Top panel, black trace: The difference induced in the Syn. 6803 absorption by illumination at 260 K (black trace). The green trace was generated by shifting the Gaussian (bottom panel, blue trace) 0.2 nm to the blue. The top-panel traces are multiplied by 40.

Analogous electrochromism data were obtained from *Syn.* 6803 PSII cores, utilizing the same 260-K illumination protocol used for plant PSII samples. Fig. 5.9 shows data in the $Q_y$ region: the dark spectrum (black, bottom panel) and the changes induced by 260-K illumination (black, top panel). A strong derivative-shaped feature is seen at 683 nm in the difference spectrum, of similar magnitude to that seen in plants at 685 nm (Fig. 5.8). This shift is well described (green trace) by introducing a Gaussian with an area corresponding to one Pheo *a*, centered at 683 nm (blue trace), and translating it 0.2 nm to higher energy. This corresponds to a $\sim 5$ cm$^{-1}$ blue shift [17].

These data show a Pheo$_{D1}$ Stark shift in $Q_x$ that is 10 times larger than that in $Q_y$ ($\geq 5$ cm$^{-1}$ and $\geq 80$ cm$^{-1}$, respectively). If the $\Delta\mu$'s of the two states are nearly perpendicular, and of comparable magnitude, this points to an unexpected orientation of the transition dipoles of Pheo$_{D1}$, with $\Delta\mu_x$ nearly parallel to the $Q_A$-Pheo$_{D1}$ inter-distance vector and with $\Delta\mu$ magnitudes near 0.5 D [17]. Alternatively, the $\Delta\mu$ of the $Q_x$ state must be unexpectedly large. A large value (5 D) has indeed been reported by recent external-field electrochromism experiments on D1/D2/$b_{559}$ material [44].

In cyanobacteria, the (inferred) spectral position of Pheo$_{D1}$ (683 nm) is markedly different from the absorptions at 681 nm and 684 nm that give rise to strong CD signals. We have associated these with P680 and CP43. Very recent hole-burning experiments [45–47] point to the 684-nm feature being associated with CP43. These experiments also point to the possibility of a weak, lower-energy absorption of P680 near 690 nm, not previously detected as it underlies the CP47 low-energy absorption.

## 5.4
## Conclusions

### 5.4.1
### Low-temperature Precision Polarization Spectroscopies

In this chapter, we have outlined an approach that seeks to establish clear spectral signatures of critical components of the complex machinery of fully active PSII via the application of polarized steady-state optical spectroscopies. Spectra are recorded on a single spectrometer and over a wide spectral range (300–1100 nm). Measurements are made with high reproducibility, sensitivity, and precision, allowing remarkably small spectral shifts and other changes to be directly assessed from differences in static spectra rather than by transient techniques. The stability and sensitivity of the spectrometer are such that changes smaller than $10^{-4}$ $\Delta A$ can be addressed, but with both "before" and "after" spectra available at high precision, more information is obtained. Furthermore, light fluences and powers at the sample are kept below levels that significantly influence the sample and are far lower than those used in transient spectroscopies.

Spectra are taken at very low temperatures, which significantly enhances the resolution. This, along with the ability to see both weak and strong spectral features with useful signal-to-noise ratios in the same sample, helps unravel the spectral congestion in active PSII. This congestion arises inevitably from the large number of redox-inactive chromophores masking the already formidable number of redox-active elements in PSII. Consequently, subtle but fundamental differences between PSII preparations prepared differently or sourced from different organisms become immediately apparent.

Absorption, CD, and MCD all have an inherent advantage in that through Beer's law they allow quantitative estimations of concentrations and changes. Both paramagnetic and diamagnetic species appear in spectra. Species such as chlorophyll and $\beta$-carotene radicals, which are difficult to distinguish through EPR, absorb in different spectral ranges and are entirely distinct. CD is known to be very sensitive to inter-pigment coupling. MCD arises fundamentally from the Zeeman effect and has an entirely different electronic basis than CD. Thus it is an invaluable addition to absorption and CD, particularly when measured simultaneously. MCD can immediately distinguish between diamagnetic and paramagnetic chromophores [33], which is a significant advantage in PSII, which encapsulates a range of active redox centers.

### 5.4.2
### Signatures of P680 and Chl$_Z$

Quantitative comparisons of spectra taken of active PSII cores with those of the solubilized protein subassemblies D1/D2/$b_{559}$, CP43, and CP47 indicate a significant change in P680 upon disassembly. By contrast, absorption, CD, and MCD spectral signatures of plant BBYs are quantitatively retained in core preparations. Together with maintained enzymatic activity, this establishes cores as the minimal assembly containing "native" PSII reaction centers. We extract spectral information of the native reaction center by subtracting the spectra of isolated CP43 and CP47 from that of active PSII cores. In this way, we hope to pinpoint the characteristics of P680 as it occurs in nature.

D1/D2/$b_{559}$ preparations containing five and six chls show a strong MCD "deficit" region in the Chl $Q_y$ band, peaking at 680 nm. The profile of this deficit is very similar to transient bleaches attributed to P680 in these materials, indicating that the MCD deficit is associated with the P680 chls. By contrast, the Chl missing in the 5-Chl preparation shows no deficit. This pigment has the MCD of a normal Chl *a*. The MCD of the native reaction center shows a pattern similar to that seen in D1/D2/$b_{559}$, although shifted, and with several sharp features. There is dramatic reduction of the MCD of the 683.5-nm feature, attributed (in part) to P680, but no reduction for the high-energy pigments.

From the MCD deficit patterns in native reaction centers and solubilized D1/D2/$b_{559}$, we infer that in plant cores, P680 has a resonance at 683.5 nm, which overlaps with a weaker CP43 excitation in spinach. From CD spectra, we conclude that this accidental degeneracy is lifted in *Syn.* 6803, where two weaker components are discerned at 681 nm and 684 nm. In solubilized D1/D2/$b_{559}$, P680 appears to peak at 680 nm, as judged by the MCD deficit spectrum. By comparing 5- and 6-Chl preparations, the peripheral $Chl_Z$ is identified at 672 nm in this material. This shows full MCD intensity, as does a resonance in the intact reaction center spectrum at 667 nm, which we assign to the two $Chl_Z$'s in the intact system.

A strong MCD deficit appears to be a signature of P680. New theoretical developments are looking into the origin of this effect [48].

### 5.4.3
**Electrochromism Signature of Pheo$_{D1}$**

Illumination-induced formation of $Q_A^-$ leads to spectral changes that pinpoint the Pheo$_{D1}$ excitations and their Stark shifts. The shift on $Q_x$ is 10-fold greater than that on $Q_y$, indicating unexpected characteristics of the Pheo$_{D1}$ $\Delta\mu$'s. Cyanobacterial and plant PSII differ significantly in the energies of the electronic excitation of Pheo$_{D1}$, with the shift centered at 685 nm in spinach material and at 683 nm in *Syn.* 6803.

### 5.4.4
**Coupling and Robustness in P680 and Biomimetic Systems**

As mentioned in the introduction, there is ongoing debate with respect to the coupling between the reaction center pigments in PSII and, consequently, the electronic makeup of P680. The pair-wise center-to-center inter-pigment distances for the four Chl *a*'s and two Pheo *a*'s are comparable, with the central Chl's ($P_{D1}$ and $P_{D2}$) closer to one another than the rest. The distances allow a maximal interaction of $\sim 50$–$100$ cm$^{-1}$, stronger between the central Chl's. We have seen very well-structured features in plant PSII cores with integrated intensity both greater than and far less than a single Chl $a$. We take this as prima facie evidence that inter-pigment coupling is larger than spectral linewidths. The structure seen in spinach is not reproduced in *Syn.* 6803, in which a number of equally narrow features are present, but none that has the prominence of the 683.5-nm feature in spinach. The signatures of illumination-induced electrochromic shifts, CD, and MCD each present a different projection of the reaction center pigments and their interactions.

From this multidimensional set of spectroscopic data, corresponding excitations can be tracked from one organism/preparation to another and a broader understanding of P680 can evolve. A comparison between spinach and *Syn.* 6803 indicates that the electronic coupling within P680 changes significantly. As both species are successful, the function of P680 is clearly not inhibited. Our studies on other organisms confirm a strong variability in P680 [49]. The observation of a narrow and single dominant electrochromic signature in Q$_y$ for both plants and cyanobacteria suggests that the excitation on Pheo$_{D1}$ is *not* strongly delocalized in active PSII. An analysis of weaker electrochromic structure, along with MCD and CD, may help quantify inter-pigment interactions in various organisms further. In isolated D1/D2/$b_{559}$ preparations, in which the stabilizing influence of CP43 and CP47 proteins is absent, no well-resolved structure is seen, although the MCD deficit signature of P680 is maintained.

Our approach is helping to establish a more detailed picture of the electronic structure of active PSII along with its fundamental charge-separation and electron transfer mechanisms. The development of clear spectral signatures for components may also assist in the design of genetically modified PSII and in the study and design of biomimetic systems.

## Acknowledgments

We thank Barry J. Prince, Joseph L. Hughes, Paul J. Smith, and Ron J. Pace for fruitful discussions. S.P.Å. gratefully acknowledges support from the Wenner-Gren Foundations.

## References

1 Blankenship, R.E. (2002) *Molecular Mechanisms of Photosynthesis*, 1 ed., Blackwell Science Ltd.
2 Rutherford, A.W. (1989) *Trends in Biological Sciences* 14, 227–232.
3 Kamiya, N., and Shen, J.-R. (2003) *Proceedings of the National Academy of Science USA* 100, 98–103.
4 Kamiya, N., and Shen, J.-R. (2001) in *PS2001: 12th International Congress of Photosynthesis*, pp S5-002, Brisbane, Australia.
5 Zouni, A., Witt, H.T., Kern, J., Fromme, P., Krauss, N., Saenger, W., and Orth, P. (2001) *Nature* 409, 739–743.

6 Svensson, B., Etchebest, C., Tuffery, P., van Kan, P., Smith, J., and Styring, S. (1996) *Biochemistry 35*, 14486–14502.

7 Ferreira, K. N., Iverson, T. M., Maghlaoui, K., Barber, J., and Iwata, S. (2004) *Science 303*, 1831–1838.

8 Namba, O., and Satoh, K. (1987) *Proceedings of the National Academy of Science USA 84*, 109–112.

9 Dekker, J. P., and van Grondelle, R. (2000) *Photosynthesis Research 63*, 195–208.

10 Kok, B., Forbush, B., and McGloin, M. (1970) *Photochemistry and Photobiology 11*, 457–475.

11 Forbush, B., Kok, B., and McGloin, M. (1971) *Photochemistry and Photobiology 14*, 307–321.

12 Siderer, Y., and Dismukes, G. C. (1981) *Proceedings of the National Academy of Science USA 78*, 274–278.

13 Åhrling, K. A., Peterson, S., and Styring, S. (1997) *Biochemistry 36*, 13148–13152.

14 Styring, S., and Rutherford, A. W. (1988) *Biochemistry 27*, 4915–4923.

15 Hanley, J., Deligiannakin, Y., Pascal, A., Faller, P., and Rutherford, A. W. (1999) *Biochemistry 38*, 8189–8195.

16 Tracewell, C. A., Cua, A., Stewart, D. H., Bocian, D. F., and Brudvig, G. W. (2001) *Biochemistry 40*, 193–203.

17 Peterson Årsköld, S., Masters, V. M., Prince, B. J., Smith, P. J., Pace, R. J., and Krausz, E. (2003) *Journal of the American Chemical Society 125*, 13063–13074.

18 Smith, P. J., Åhrling, K. A., and Pace, R. J. (1993) *Journal of the Chemical Society, Faraday Transactions 89*, 2863–2868.

19 Smith, P. J., Peterson, S., Masters, V. M., Wydrzynski, T., Styring, S., Krausz, E., and Pace, R. J. (2002) *Biochemistry 41*, 1981–1989.

20 van Leeuwen, P. J., Nieveen, M. C., van de Meent, E. J., Dekker, J. P., and van Gorkom, H. J. (1991) *Photosynthesis Research 28*, 149–153.

21 Seibert, M. (1993) in *The Photosynthetic Reaction Center* pp 319–356, Academic Press, Inc.

22 Alfonso, M., Montoya, G., Cases, R., Rodriguez, R., and Picorel, R. (1994) *Biochemistry 33*, 10494–10500.

23 den Hartog, F. T. H., Vacha, F., Lock, A. J., Barber, J., Dekker, J. P., and Völker, S. (1998) *J. Phys. Chem. B 102*, 9174–9180.

24 Hillier, W., Hendry, G., Burnap, R. L., and Wydrzynski, T. (2001) *Journal of Biological Chemistry 276*, 46917–46924.

25 Stranger, R., Dubicki, L., and Krausz, E. (1996) *Inorganic Chemistry 35*, 4218–4226.

26 Stewart, D. H., and Brudvig, G. W. (1998) *Biochimica et Biophysica Acta 1367*, 63–87.

27 Masters, V., Smith, P., Krausz, E., and Pace, R. (2001) *Journal of Luminescence 94/95*, 267–270.

28 Peterson Årsköld, S., Krausz, E., Smith, P. J., Pace, R. J., Picorel, R., and Seibert, M. (2004) *Journal of Luminescence 108*, 97–100.

29 Newell, W. R., van Amerongen, H., Barber, J., and van Grondelle, R. (1991) *Biochimica et Biophysica Acta 1057*, 232–238.

30 Nozawa, T., Kobayashi, M., Wang, Z.-Y., Itoh, S., Iwaki, M., Mimuro, M., and Satoh, K. (1995) *Spectrochimica Acta 51A*, 125–134.

31 Kobayashi, M., Wang, Z.-Y., Yoza, K., Umetsu, M., Konami, H., Mimuro, M., and Nozawa, T. (1996) *Spectrochimica Acta 51A*, 585–598.

32 Zevenhuijzen, D., and Zandstra, P. J. (1984) *Biophysical Chemistry 19*, 121–129.

33 Piepho, S. B., and Schatz, P. N. (1983) *Group Theory in Spectroscopy with Applications to Magnetic Circular Dichroism*, John Wiley & Sons.

34 van Kan, P. J. M., Otte, S. C. M., Kleinherenbrink, F. A. M., Nieveen, M. C., Aartsma, T. J., and van Gorkom, H. J. (1990) *Biochimica et Biophysica Acta 1020*, 146–152.

35 Breton, J., and Katoh, S. (1987) *Biochimica et Biophysica Acta 892*, 99–107.

36 Peterson Årsköld, S., Picorel, R., Seibert, M., Smith, P. J., Pace, R. J., and Krausz, E. (2005) *in preparation*.

37 Groot, M.-L., Frese, R. N., de Weerd, F. L., Bromek, K., Pettersson, A., Peterman, E. J. G., van Stokkum, I. H. M., van Grondelle, R., and Dekker, J. P. (1999) *Biophysical Journal 77*, 3328–3340.

38 Steffen, M. A., Lao, K., and Boxer, S. G. (1994) *Science 264*, 810–816.
39 Mulkidjanian, A. Y., Cherepanov, D. A., Haumann, M., and Junge, W. (1996) *Biochemistry 35*, 3093–3107.
40 Stewart, D. H., Nixon, P. J., Diner, B. A., and Brudvig, G. W. (2000) *Biochemistry 39*, 14583–14594.
41 van Mieghem, F., Brettel, K., Hillmann, B., Kamlowski, A., Rutherford, A. W., and Schlodder, E. (1995) *Biochemistry 34*, 4798–4813.
42 Hillmann, B., Brettel, K., van Mieghem, F., Kamlowski, A., Rutherford, A. W., and Schlodder, E. (1995) *Biochemistry 34*, 4814–4827.
43 Hillmann, B., and Schlodder, E. (1995) *Biochimica et Biophysica Acta 1231*, 76–88.
44 Frese, R. N., Germano, M., de Weerd, F. L., van Stokkum, I. H. M., Shkuropatov, A. Y., Shuvalov, V. A., van Gorkom, H. J., van Grondelle, R., and Dekker, J. P. (2003) *Biochemistry 42*, 9205–9213.
45 Hughes, J. L., Prince, B. J., Krausz, E., Smith, P. J., Pace, R. J., and Riesen, H. (2004) *Journal of Physical Chemistry B 108*, 10428–10439.
46 Hughes, J. L., Prince, B. J., Peterson Årsköld, S., Krausz, E., Pace, R. J., Picorel, R., and Seibert, M. (2004) *Journal of Luminescence 108*, 131–136.
47 Prince, B. J., Krausz, E., Peterson Årsköld, S., Smith, P. J., and Pace, R. J. (2004) *Journal of Luminescence 108*, 101–105.
48 Hughes, J. L., Pace, R. J., and Krausz, E. (2004) *Chemical Physics Letters 385*, 116–121.
49 Peterson Årsköld, S., Smith, P. J., Shen, J.-R., Pace, R. J., and Krausz, E. (2005) *Photosynthesis Research, in press*.

# 6
# The Nature of the Special-pair Radical Cation Produced by Primary Charge Separation During Photosynthesis

*Jeffrey R. Reimers and Noel S. Hush*

## 6.1
## Introduction

Today's photovoltaic devices are designed around the use of semiconductor materials made either from inorganic semiconductors such as silicon or from organic polymers. These devices harvest light, initiate primary charge separation, and provide carrier diffusion to external electrodes. Photosynthesis in plants and bacteria involves in a key step an analogous photovoltaic mechanism that oversees the harvesting of optical energy and its conversion into electrical energy, en route to long-term storage as chemical energy. This naturally occurring process operates using single-molecule, nominally insulating devices rather than the 100-nm or larger semiconducting objects found in applications photovoltaics. A major goal of the current research of the Molecular Electronics Group at the University of Sydney is to understand the processes of the natural light-harvesting systems. Special emphasis is placed on the mechanisms of primary and secondary charge separation, with the long-term goal being the design of new biomimetic photovoltaic technologies that directly emulate elements of natural photosynthetic systems. To this end, artificial antenna systems and reaction centers have been designed by Crossley [1–3], while details of the mechanism of primary charge separation have been elucidated by Reimers and Hush [4–17]. Herein, this later work is reviewed, building a thorough picture of the central role of the primary electron donor.

## 6.2
## The Special Pair

During natural photosynthesis, light energy is harvested in manifold types of antenna systems and funneled to the *special pair* dimer P of chlorophyll or bacteriochlorophyll molecules that initiate primary charge separation, ejecting an electron to become a dimer radical cation. Significant differences in this process

*Artificial Photosynthesis: From Basic Biology to Industrial Application*
Edited by Anthony F. Collings and Christa Critchley
Copyright © 2005 WILEY-VCH Verlag GmbH & Co. KGaA, Weinheim
ISBN: 3-527-31090-8

occur between the major classes of photosynthetic proteins – those found in purple bacterial systems as well as in the PSI and PSII systems of cyanobacteria and green plants. To date, very much more spectroscopic and structural information is available concerning the reaction centers of purple bacteria. While our studies have concentrated mainly on the purple bacteria *Rhodobacter (Rb.) sphaeroides*, we believe that the results obtained can readily be generalized to account for the differences observed in PSI and PSII. Much more research is required in order to establish this, however, and the story told here is largely that of understanding the nature of the special pair in bacterial photosynthesis.

The most interesting feature of the special pair is actually that it is a dimer. The two halves of the dimer, known as $P_L$ and $P_M$, interact with each other through an electronic coupling $J$, and this coupling is very important in determining the chemical and spectroscopic properties of the system and hence is central to the charge-separation process. It takes on slightly different values before and after charge separation, but this variation is small and not of concern here. Hence, by examining the properties of the final charge-separated state $P^+$, much is learned about the processes that control the charge separation. $P^+$ is a dimer radical cation, an unusual chemical species whose properties are unlike those of typical organic or biological molecules, being in fact more similar to those of inorganic charge-transfer complexes such as the famous Creutz-Taube ion [18].

A characteristic and very striking feature of $P^+$ is the appearance [19–21] of dominant transitions in the vibrational infrared spectrum that do not correspond to transitions observed in either the infrared spectrum of the neutral or radical-anion state of the bacteriochlorophyll monomers or the spectrum of the neutral special pair. Shown in Fig. 6.1 is the observed [21] infrared difference spectrum of the cation radical of *Rb. sphaeroides* less the spectrum of the neutral species. Only vibrational modes whose spectrum differs upon oxidation appear in this difference spectrum. Some modes simply change in frequency, giving rise to sharp positive peaks (at the vibration frequency of the cation radical) with close-lying neighboring sharp negative peaks (at the vibration frequency of the neutral species). However, most of the signal observed in the difference spectrum below 1800 $cm^{-1}$ is associated with new transitions that occur only in the cation-radical species.

Vibrational infrared absorptions are associated with molecular motions that rearrange the internal charge distribution, hence producing a fluctuating dipole moment that interacts with electromagnetic radiation. The new transitions in $P^+$ arise from vibrations that cause the +ve charge (the *electron hole*) to be moved between the two monomers, producing large changes in dipole moment and hence very intense infrared absorptions. They are known in the inorganic electron transfer literature as *phase phonon* lines [22–27]. Our calculations [5] have shown that the most intense of these lines are associated with antisymmetric combinations of local motions on each bacteriochlorophyll, like those shown in Fig. 6.2, that involve expansion and contraction of the macrocyclic rings. These motions are activated when a monomeric bacteriochlorophyll is oxidized. Bacteriochlorophylls are related to porphyrin, a molecule that has a

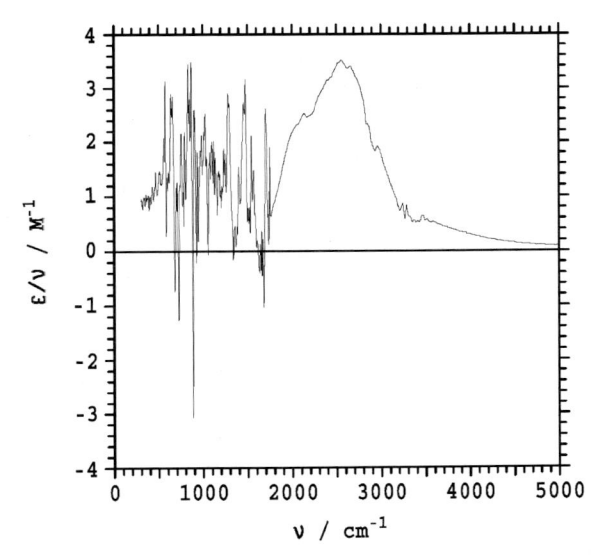

**Fig. 6.1** Observed [21] FTIR difference spectrum ($\varepsilon$ is the molar extinction coefficient, $\nu$ is frequency) of P$^+$ minus P for *Rb. sphaeroides* showing a broad electronic absorption among the sharp vibrational transitions.

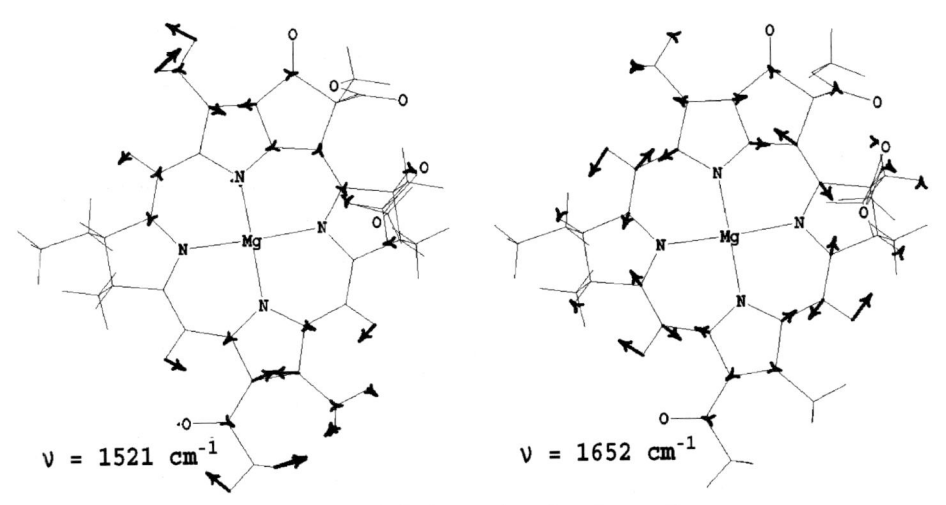

**Fig. 6.2** Two monomeric bacteriochlorophyll vibrational modes whose anti-symmetric combinations are strongly coupled to the hole-transfer process in the special-pair radical cation, as obtained from density-functional calculations [5].

center of symmetry allowing its vibrations to be classified as of type g (symmetric) or type u (antisymmetric). It is well known that only u modes can give rise to infrared absorptions, and hence the absence of these types of modes in the spectrum of monomers and neutral special pair is readily understood. The most active mode involves the local motion at 1652 cm$^{-1}$, shown in Fig. 6.2, and is of g type; that such a mode should dominate the spectrum of the radical cation is quite remarkable.

Another striking feature of the spectrum of P$^+$ shown in Fig. 6.1 is the appearance of the broad electronic transition centered at 2700 cm$^{-1}$ among the vibrational infrared spectrum. The lowest-energy electronic transition in most organic molecules, including chlorophylls and bacteriochlorophylls, occurs either in the visible region of the spectrum or beyond into the ultraviolet, and it is typical to scan for such spectra in the 800–200 nm region. For P$^+$, however, the lowest-energy electronic transitions occur in the 5000–2000 nm region, more usually depicted as the infrared wavenumber range of 2000–5000 cm$^{-1}$. Further intense transitions are also found in the 1250 nm range near 8000 cm$^{-1}$. All of

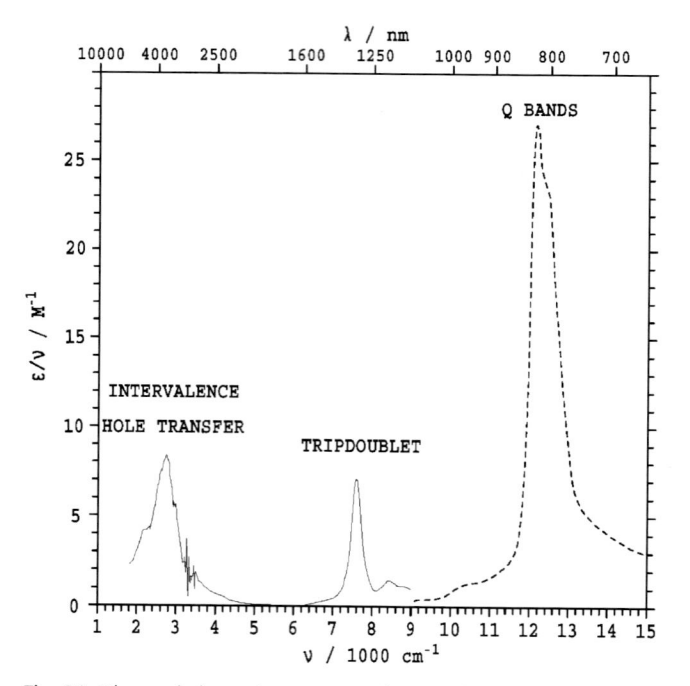

**Fig. 6.3** Observed absorption spectrum from the IR region to the visible region for *Bl. viridis* ($\varepsilon$ is the molar extinction coefficient, $v$ is frequency, $\lambda$ is wavelength). The Q bands are distributed over various chromophores, while the hole-transfer and trip-doublet bands are localized on the special pair (solid line) from FTIR difference spectroscopy [19] (with normalization taken from [28]), with the noise around 3400 cm$^{-1}$ attributable to modulation of water OH-stretch vibrations (dashed line) (from [28]).

these transitions are unprecedented in the spectroscopy of monomeric or neutral dimeric species. Some characteristic spectra are shown in Fig. 6.3. In this figure, the spectrum is not shown in the typical fashion as a plot of the extinction $\varepsilon$ as a function of wavelength $\lambda$ but rather as a plot of $\varepsilon/\nu$ as a function of frequency $\nu$. In this fashion, all influences of the properties of light on the measured absorption are removed, revealing the inherent molecular properties. The areas and bandwidths apparent in this representation provide true indications of the relative intensities and the vibrational relaxation processes that dominate the spectrum and reveal the photophysical and chemical properties of $P^+$.

The absorption at 8000 cm$^{-1}$ was first observed [29] in 1969 but was never assigned. We have shown that this absorption is due to a *trip-doublet* transition [14]. This is in itself a rather unique transition that corresponds not to the excitation of a single electron, as is the norm in molecular spectroscopy, but rather to the synchronous excitation of two electrons. In effect, the transition can be described as a singlet-to-triplet transition on one of the halves of the dimer cation radical assumed to be in its neutral state, intensified by the presence of the nearby other half, assumed to be in its cation-radical state. This process is analogous to the well-known effect caused by the presence of molecular oxygen in facilitating singlet-to-triplet absorptions in organic molecules. While two-electron excitations are formally forbidden in single-photon spectroscopy, they become allowed due to mixing with singly excited states facilitated by the intermolecular coupling. As the two halves of the special pair are close together and strongly coupled, the intensity of the trip-doublet bands in bacterial photosynthetic systems becomes substantial, making prominent these rather exotic transitions.

## 6.3
## The Hole-transfer Band

The broad absorption in the 2000–4000 cm$^{-1}$ region derives its intensity from the *intervalence hole transfer* band of the dimer cation radical. In one limit known as the *localized diabatic* description [4, 5, 30], this transition can be thought of as taking the electron hole from one of the monomers in the dimer and transferring it to the other. This band is characteristic of mixed-valence inorganic complexes and provides critical information concerning the nature of the system, in this case the nature of the interactions within the special pair as well as key properties associated with the asymmetry of the dimer.

Three different ways in which the hole-transfer absorption may be perceived are shown in Fig. 6.4, a figure that depicts molecular potential energy surfaces for the two lowest-energy electronic states as a function of a generalized nuclear coordinate. The relative intensities of the phase-phonon lines discussed previously are indicative of the projections of this generalized coordinate onto the normal modes of vibration of the radical cation. One way of perceiving the problem is via the localized diabatic picture referred to previously. This involves

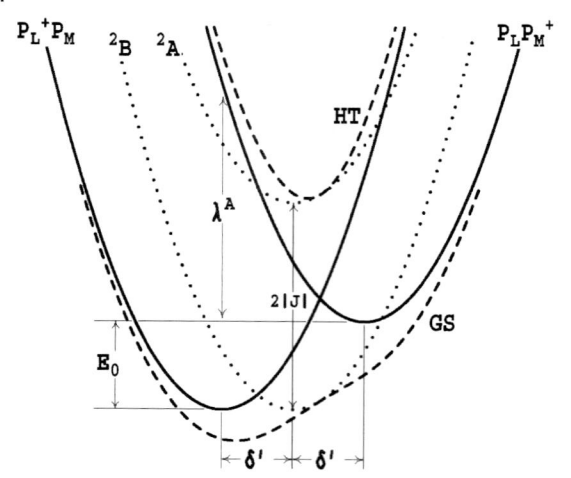

**Fig. 6.4** Three different representations of the ground-state and first-excited-state electronic potential-energy surfaces for the special-pair radical cation as a function of a generalized antisymmetric nuclear coordinate. Solid lines: localized diabatic surfaces; dots: delocalized diabatic surfaces depicting $C_2$ symmetry; dashed lines: Born-Oppenheimer adiabatic surfaces (named GS and HT) obtained by parametric diagonalization of either set of diabatic states. $\lambda$ is the reorganization energy, $\delta$ is related to the geometrical displacements between monomeric neutral and cation-radical bacteriochlorophylls, and $J$ is the electronic coupling between the two chromophores.

states described as $P_L^+P_M$ and $P_LP_M^+$, representing assemblies of neutral and cationic individual dimer halves. It is a simplistic picture as it ignores the effects of the electronic coupling $J$ that acts between the dimer halves. The effect of this coupling is to attempt to enforce symmetry on the electronic wavefunctions, in much the same way that bonding and antibonding molecular orbitals arise from atomic orbitals for, e.g., the hydrogen molecule. For bacterial photosynthesis, calculations indicate that the bonding orbital has B symmetry in the pseudo $C_2$ point group of the dimer, while the antibonding orbital has A symmetry. Opposing the electronic coupling, the reorganization energy $\lambda$ shown in Fig. 6.4 is associated with the desire of the neutral and cationic monomers to adopt different molecular geometries as a function of the generalized nuclear coordinate. Hence the states $P_L^+P_M$ and $P_LP_M^+$ are modeled as (hyper)parabolic surfaces with different equilibrium geometries. Ignoring completely the reorganization energy allows surfaces for the A and B states to be constructed: this is known as the *delocalized diabatic* picture of the interaction, and these are also shown in Fig. 6.4. In this picture, the charge is equally shared (or delocalized) over both halves of the dimer rather than localized on one half only.

The final way of perceiving the coupling is through the introduction of the Born-Oppenheimer adiabatic approximation [31], as is nearly universally applied in chemistry, to eliminate all direct electronic interactions by, in this case, diagonalizing the Hamiltonian obtained using either pair of diabatic states parametrically as a function of the nuclear coordinates. This produces the adiabatic surfaces shown in Fig. 6.4. These are highly anharmonic and it is thus quite difficult to determine the vibrational frequencies and intensities required in order to predict absorption spectra. Further, the key element of the Born-Oppenheimer approximation is the separation of nuclear and electronic motions. Basically, this approach is sensible as the light electrons travel much faster than the heavy nuclei; thus it is possible to determine the nuclear motion in an averaged field of the electrons. However, from Fig. 6.3 it is clear that the energies involved in vibrational excitation are equivalent to those involved in hole-transfer electronic motion, and hence the nuclear and electronic motions are tightly coupled. As a result, the Born-Oppenheimer approximation is a poor starting point for understanding intervalence transfer problems, and Born-Oppenheimer breakdown effects must explicitly be included in any spectral simulation. Born-Oppenheimer-based methodologies are thus poorly suited, with the electronic-nuclear (vibronic coupling) interactions being more readily treated using either diabatic approach.

In the diabatic approaches, one starts by ignoring either the electronic coupling $J$ or the reorganization energy $\lambda$ in order to generate a basis set of mixed vibrational and electronic states that are used to depict the full Hamiltonian of the dimer cation radical. This Hamiltonian must then be constructed and its spectrum determined and compared to the observed absorption spectrum. Naively, the most efficient computational approach is the one whose initial assumptions most closely resemble the system being studied – i.e., if the charge in reality is delocalized, then the delocalized diabatic basis should be used to expand the full Hamiltonian operator, etc.

A third key molecular property is the asymmetry parameter $E_0$. Physically, asymmetry arises as the geometries of the two halves of the dimer are not identical owing to asymmetric interactions with the surrounding protein, while asymmetric electric fields also create a preference for the positive charge to be localized on one side; in the case of bacterial photosynthesis, this side is $P_L$. In our treatment, these manifold effects are incorporated into one parameter that simply indicates the difference in redox potential for oxidation of charge-localized dimer halves. By considering electrochemical measurements for heterodimer mutants in which one bacteriochlorophyll in the special pair is replaced with a bacteriopheophytin, we have recently argued [4] that the experimental value for this quantity is $E_0 = 0.058 \pm 0.020$ eV, though a very wide range of prior estimates have been made. This redox asymmetry aids the reorganization energy in favoring charge localization rather than delocalization.

## 6.4
## Initial Investigations of the Hole-transfer Band

The first efforts [13, 19] made at interpreting the observed intervalence hole-transfer spectrum in terms of molecular properties led to the conclusion that the charge is 95–97% localized on $P_L$. These analyses were based on analytical formulae rather than full spectral simulation. However, spin-density measurements made at around the same time depicted the hole in *Rb. sphaeroides* as being actually localized 68% on $P_L$, in the region midway between full localization (100%) and full delocalization (50%). This provided a major disparity that required explanation as the two scenarios depicted really quite different chemical species. We argued [30] that the analytical expressions used in interpreting the spectra were inadequate in this application as they involved the use of one-electron theory to depict the observed transition moment. According to simple one-electron theory, trip-doublet bands should be forbidden as they correspond to two-electron excitations, and so the very observation of intense trip-doublet absorptions demonstrates the inadequacy of this approximation. Using the localized diabatic approach, we performed [30] numerical spectral simulations to map out likely solution spaces for $J$, $\lambda$, and $E_0$, concluding that, provided intensity information was ignored, interpretations with as little as 68% of the charge localized on $P_L$ were feasible. The challenge then became to perform quantitative simulations from which accurate values of $J$, $\lambda$, and $E_0$ could be extracted.

The next approach was to make a priori predictions of the spectrum, performing full spectral simulations. This required independent determination of the values of $J$, $\lambda$, and $E_0$, as well as the vibrational frequencies and the projections of the generalized nuclear coordinate used in the sketch of Fig. 6.4 onto all of these modes. Alternatively [4, 30], these projections may be represented as vibronic coupling constants, and this is the approach that was finally adopted. All of the vibrational parameters were evaluated [5] using density-functional calculations either for the bacteriochlorophyll monomers, in their neutral and cation-radical states, or for the dimer cation radical, in its ground and first-excited states. The relative values of the vibronic coupling constants were obtained by analysis of the calculated infrared transition moments, which were expressed as an expected term, taken from the calculated spectrum of the neutral dimer, plus a vibronic contribution originating from the coupling between the ground state and hole-transfer excited state of the dimer cation radical. Finally, the absolute values of the vibronic coupling constants were obtained by constraining the total reorganization energy to the previously determined value of $\lambda$.

Various attempts were made at the calculation of the main controlling parameters $J$, $\lambda$, and $E_0$, with varying degrees of success. The reorganization energy $\lambda$ relates to geometry changes on oxidation and is the easiest quantity to determine reliably. However, quantitative error estimates [4, 5] indicated that the likely accuracy in these calculations is on the order of 0.02 eV or about 10%. The electronic coupling $J$ is difficult to calculate because it is very sensitive to the spacing between the bacteriochlorophylls in the special pair, and the coupling contains a significant dispersive component that, at the present time, is not computationally feasible to evaluate accurately [10]. Early computations appeared to underestimate $J$ by a factor of two, although in light of our more authoritative estimates of the experimental quantity, this factor now appears to be only on the order of 30%. Finally, the redox asymmetry $E_0$ is extremely difficult to calculate, as it is sensitive to long-range, variably screened electrostatic potential influences that require detailed knowledge of the protein structure well beyond that provided by X-ray analysis tools. While local modulation of $E_0$ due to site-directed mutagenesis can be modeled [8], a full, accurate, a priori calculation of the intrinsic quantity is not feasible.

We hence sought to determine $J$, $\lambda$, and $E_0$ from analysis of experimental data. A method for extracting $J$ and $E_0$ from the combination of observed electron-spin distributions and redox potentials for a series of mutants of *Rb. sphaeroides* had been generated by Allen et al. [32–34], but the results obtained were inconsistent with the observed hole-transfer band maximum. We improved [9] the theory used to interpret this data and added to it allowance for the reorganization energy $\lambda$. Unfortunately, the available experimental data was insufficient to allow for the determination of $\lambda$, but we determined the somewhat realistic values of $J=0.18\pm0.03$ eV and $E_0=0.136\pm0.030$ eV.

It was clear that only couplings at the lower limit of the deduced range could possibly be consistent with the observed FTIR spectra, and hence we performed

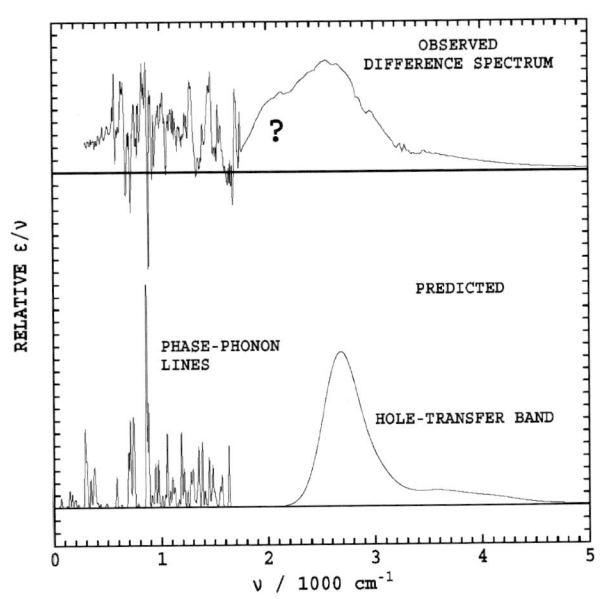

**Fig. 6.5** Observed [21] difference spectrum and the predicted [7] hole-transfer electronic band contour and phase-phonon line distribution. The predictions cannot account for the feature observed at 2200 cm$^{-1}$.

preliminary spectral simulations [7] using $J=0.15$ eV, $E_0=0.14$ eV, and the B3LYP-calculated value [5] of $\lambda=0.20$ eV. It is necessary to restrict the number of vibrational modes used in the simulations, and we initially chose the 91 most active modes of the bacteriochlorophyll dimer. It was clear [7] that the calculations had converged as a function of the number of modes included and the degree of excitation allowed in each mode. The calculated spectrum (normalized to match the observed intensity) is shown in Fig. 6.5, where it is compared to the observed FTIR difference spectrum. Major success was achieved in that the location of the band maximum near 2700 cm$^{-1}$, the width of the band, and the details of the high-frequency tail were well reproduced by these a priori calculations. Indeed, all of these properties are very sensitive to $J$, $\lambda$, and $E_0$ as well as the nature of the partitioning of the coupling among the vibrational modes, and it was clear that, for the first time, an authoritative interpretation of the spectrum could be advanced. However, this preliminary simulation provided no indication of the origin of the shoulder observed near 2200 cm$^{-1}$ to the red of the main hole-transfer absorption band. Proper quantitative analysis of the spectrum demanded a proper explanation of this feature.

## 6.5
## Identification of the SHOMO to HOMO Band

Previously [12], we had argued that the 2200 cm$^{-1}$ shoulder was not part of the same spectral system as the main hole-transfer band. This conclusion was based on the observation that as the chromophore is varied and as mutations are introduced in the surrounding protein, the spacing between the shoulder and main band, as well as the relative intensity ratio, changes erratically. Initial suggestions [12] as to the origin of this band proved [6] incorrect. Recently [6], we have assigned this shoulder to a second-highest occupied molecular orbital (SHOMO) to highest-occupied molecular orbital (HOMO) transition. This assignment is based on ab initio and density-functional calculations of the energy of this transition as a function of chromophore leading to qualitative interpretations of the shape of the intervalence hole-transfer bands for a variety of bacterial, PSI, and PSII photosystems [6], in conjunction with quantitative calculations of the band intensity [4]. The possibility of such a transition contributing to the spectrum had long been considered, but the possibility had been eliminated, as this transition would also occur at about the same energy in monomeric cation radicals, yet absorption is registered only in dimeric cation radicals. Our interpretation of this is that the SHOMO to HOMO transition is intrinsically very weak and hence is not observed in monomeric cation radicals, but in the special pair it is vibronically coupled to the hole-transfer excited state (with which it is nearly degenerate) and its entire energy originates from this coupling. Subsequently [5], we evaluated the vibronic coupling constants between these states and showed [4] that the predicted values were consistent with the observed stolen intensity.

## 6.6
## Full Spectral Simulations Involving all Bands

A further complication demanded by this picture is that there must exist a fourth low-lying electronic state obtained by combining the SHOMO to HOMO excitation with the hole-transfer excitation, and this state must be strongly coupled to the other states. Hence, quantitative analysis of the FTIR difference spectrum of $P^+$ requires at least the inclusion of these four electronic states. Also, our preliminary calculations [7] included explicitly only antisymmetric modes, the modes that are actively coupled to the electron transfer, but the shape of the spectrum is affected also by symmetric modes as these broaden the spectrum due to the Franck-Condon principle. We thus generated a revised computational procedure [4] that included, at a mutually consistent level, four electronic states, 50 antisymmetric modes, and 20 symmetric modes. A new representation of the vibronic Hamiltonian for electron transfer problems was then introduced [4] using the delocalized diabatic representation, and a feasible computational scheme based on time-dependent quantum mechanics was developed to evaluate the absorption spectrum. In all, up to $4 \times 10^9$ individual vibrational levels were included in these spectral simulations. The calculated spectrum was expressed as a function of seven adjustable parameters, and the observed spectrum was fitted using this model.

Only one region of the parameter space gave acceptable solutions, and the final calculated spectrum is compared to the observed one in Fig. 6.6. Note that in this figure both spectra have been broadened to a resolution of 300 cm$^{-1}$ in order to eliminate from the observed FTIR difference spectrum the contributions from vibrational modes whose frequency simply changes upon oxidation of the special pair. In this way, all of the intensity observed at frequencies below 1800 cm$^{-1}$

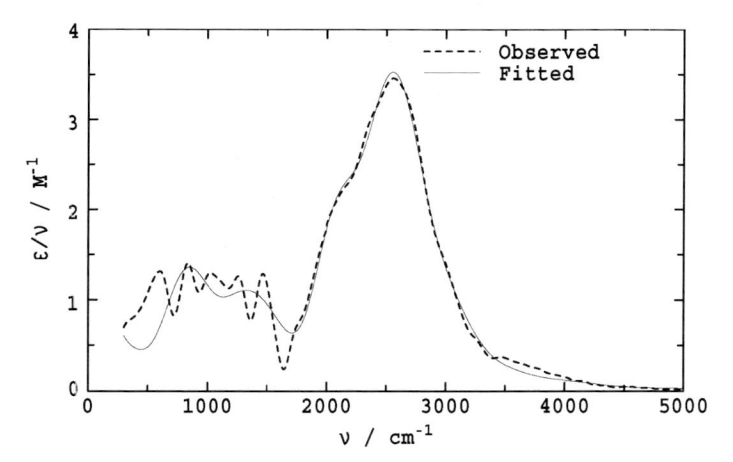

**Fig. 6.6** Observed [21] FTIR difference spectrum $P^+$ minus P for *Rb. sphaeroides* compared to the fitted spectrum [4], all broadened to 300 cm$^{-1}$ resolution.

can be attributed to phase-phonon activity, allowing for direct comparison with the calculated enhancement of the vibrational spectrum upon oxidation. Very good agreement is found between the calculated and observed spectra.

The fitted parameters are given in Table 6.1, where they are compared to alternate estimates. In all cases, the fitted parameters appear to be highly realistic estimates. This indicates that the solution obtained does not arise from some overdetermined fit but actually does depict the physical reality.

For the most important parameters $J$, $\lambda$, and $E_0$, the best current alternate estimates come from the recent analysis of extensive ENDOR and redox data of Müh et al. [35]. This work takes our equations [9] for the analysis of data of this type and applies them to a vastly enhanced experimental dataset. Unfortunately, unique values for $J$, $\lambda$, and $E_0$ still cannot be obtained, and in Table 6.1 two indicative solutions are provided, labeled No. 1 and No. 2. Our fitted values for $J$ and $\lambda$ are within the range suggested by the ENDOR/redox analysis, while our value of $0.069 \pm 0.002$ eV is outside the expected range of 0.09–0.13 eV. However, by analyzing [4] redox data [36] for some heterodimer mutants, we have obtained the much more reliable alternative estimate of $E_0 = 0.058 \pm 0.020$ eV, in good agreement with the spectroscopically fitted value.

**Table 6.1** Properties for *Rhodobacter sphaeroides* obtained by fitting [4] the observed spectra [21] of the special-pair radical cation and by other means

| Property [a] | Fitted | ENDOR/redox [b] | | Calculated [c] | | | Other |
|---|---|---|---|---|---|---|---|
| | | No. 1 | No. 2 | B3LYP | RB3LYP | PW91 | |
| $J$/eV | $0.126 \pm 0.002$ | 0.12 | 0.16 | 0.098 | 0.077 | 0.078 | 0.088–0.105 [d] |
| $\lambda$/eV | $0.139 \pm 0.003$ | 0.1 | 0.2 | 0.17 | 0.168 | 0.14 | |
| $E_0$/eV | $0.069 \pm 0.002$ | 0.09 | 0.13 | | | | $0.058 \pm 0.020$ [e] |
| $\lambda^S$/eV | $0.0027 \pm 0.0011$ | | | 0.004 | 0.014 | 0.023 | |
| $\lambda^{HT\text{-}SH}$/eV | $0.071 \pm 0.004$ | | | | 0.054 | 0.04 | |
| $E_{SH}$/eV | $0.180 \pm 0.001$ | | | $0.40 \pm 0.25$ | $0.37 \pm 0.25$ | | |
| $M$/eÅ | 1.53 | | | | | | 1.3 [f] |
| $\rho_L$/e | $0.674 \pm 0.007$ | 0.678 | 0.678 | | | | |

a) In addition to the primary parameters $J$, $E_0$, and $\lambda$, the remaining adjustable parameters are: $\lambda^S$: the total reorganization energy associated with symmetric vibrations of wavenumber $>200$ cm$^{-1}$; $\lambda^{HT\text{-}SH}$: the total antisymmetric-mode reorganization energy associated with the vibronic coupling between the hole-transfer and SHOMO to HOMO electronic states; $E_{SH}$: the energy of the SHOMO to HOMO state relative to the ground state; and $M$: the allowed electronic transition moment connecting the ground state to hole-transfer state. $\rho_L$ is the charge density localized on $P_L$.
b) From ENDOR and redox midpoint potential data [35] supporting a range of solutions varying from No. 1 to No. 2.
c) From [5] and [6].
d) DFT calculations from [5] and [16].
e) From [4] HL(M202)+LH(L131) mutant [36].
f) From INDO/S calculations [14].

The other parameters that are adjusted to fit the spectroscopic data are all fitted to values that are close to those obtained from independent estimates. These parameters are the symmetric-mode reorganization energy $\lambda^S$, the reorganization energy that reflects the total magnitude of the vibronic coupling between the SHOMO to HOMO and hole-transfer electronic states $\lambda^{HT-SH}$, the energy offset of the SHOMO to HOMO state $E_{SH}$, and the hole-transfer transition moment $M$.

As we have previously reasoned [30], the value of the transition moment is much smaller than one would expect for a transition involving significant delocalization (full delocalization is naively expected to produce $M \sim 8$ eÅ) due to configuration interaction between the hole-transfer and trip-doublet states. Hence, simple theories [14, 19], when applied to interpret the observed spectrum in terms of the degree of charge delocalization, have produced qualitatively incorrect results. The fraction of the charge localized on $P_L$ as deduced from the full spectral simulation is $0.674 \pm 0.007$ e, a value that is in excellent agreement with the experimental spin-density localization deduced from ENDOR data [32] of 0.678 e. Our spectral simulations have thus resulted in the development of a unified picture that describes a wide range of the observed physical, chemical, and spectroscopic properties of the photosynthetic reaction centers of purple bacteria.

## 6.7
### Predicting Chemical Properties Based on the Spectral Analysis

The parameters extracted from the spectral analysis are all possibly subject to modification through mutation of the reaction centre, and provided a means is available for monitoring these changes, the changes in chemical properties of the mutants can be predicted from the model. Shown in Fig. 6.7 is the observed redox midpoint potential for the special pair as a function of the observed ENDOR fractional charge on $P_L$ obtained for four related series of mutant reaction centers of *Rh. sphaeroides* [35]. As the location of the mutation site are known, our four-state model can be used to predict these correlations, and the results are also shown in the figure. Indeed, a different correlation is expected for each mutant series owing to the differing relative location of the perturbation to the two halves $P_L$ and $P_M$. In this approach [9, 33, 34], the sole effect of mutation is assumed to be the independent, electrostatically induced variation of the midpoint potentials of the two halves of the dimer, with the difference in this variation thus acting to modulate $E_0$ and hence the charge distribution and intervalence hole-transfer spectrum. What is thus of most importance is the ratio $\eta$ of the modulation of the midpoint potential of the distant bacteriochlorophyll to that of the nearer one [9, 35]. As all of the mutations depicted in Fig. 6.7 arise from variations at either the symmetrically related M160 or L131 sites, we assume that the ratio is the same for all of the mutant series. Hence $\eta$, along with the redox midpoint potential of the wild type, is used as an adjustable parame-

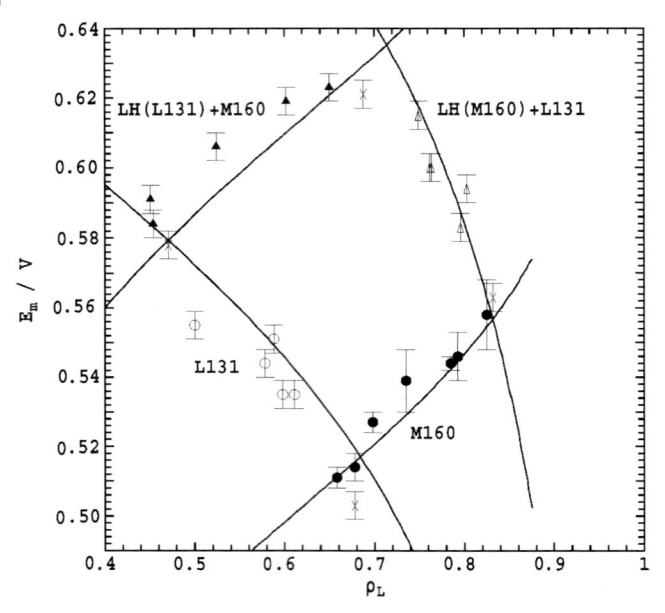

**Fig. 6.7** The observed midpoint potential $E_m$ as a function of the observed charge density on $P_L$ for the M160, L131, LH(M160)+L131, and LH(L131)+M160 series of mutants of Rb. sphaeroides [35] is compared to the expected correlation curves for each series determined [17] from our four-state interaction model. Experimental error bars are shown, while species marked × belong to each of two mutant series.

ter to fit the predicted correlations to the observed ones in Fig. 6.7. The optimized value of $\eta$ for these mutant series is 0.18, in excellent agreement with prior independent estimates [35] of $\sim 0.2$. Also shown in Fig. 6.7 are the experimental error bars in the observed midpoint potentials [35]; the root mean square error between the observed and calculated data points of 0.008 V is of similar magnitude to these. Also, from the figure it is clear that some of the experimental data points do not smoothly correlate with the others. These deviations are most likely associated with effects of mutation outside of the simple electrostatic perturbations assumed on the interpretive model. Because simulations of the structures of some of these and other reaction centers indicate that typically the effects of site-directed mutagenesis are quite complex [8], the agreement between the predicted and observed data in Fig. 6.7 is as good as can be expected. We have also investigated other mutant series [37], obtaining similar results [17].

From the previous analysis of the ENDOR/redox data, values are obtained for the redox asymmetry $E_0$ of each mutant reaction center. Our four-state model for the properties of the special-pair radical cation, again assuming that the only effect of mutation is the modulation of electrostatic interactions, can be used to predict the vertical transition energy of the hole-transfer band [37]. This vertical transition energy is formally defined as the average excitation energy of the band, a quantity that, for high-energy, symmetric-shaped bands, is readily approximated by the energy of maximum absorption [38]. For low-energy, strikingly asymmetric bands such as those observed for the hole-transfer band of photosynthetic special-pair radical cations, this approximation is invalid, and hence we explicitly evaluate [38] the actual average transition energy from the observed hole-transfer band contours for the wild type and a variety of mutant reaction centers [19, 37, 39]. The observed change in this vertical transition energy [17] is plotted against the observed [32–34, 37] charge distribution on $P_L$ in Fig. 6.8. Also shown in the figure is the correlation expected based on our four-state model of the special-pair radical cation. The observed and calculated data both show a minimum in the vertical excitation energy associated with mutants for which the charge is completely delocalized over both halves of the dimer. For such mutants, $E_0 \sim 0$ so that symmetry is maximized; symmetry reduction always acts to increase the transition energy. Quantitatively, the observed correlation falls somewhat outside the predicted one; this is most likely due to breakdown of the assumption used in this analysis that the sole effect of mutation is variation of $E_0$. One point that is clearly established, however, is that the mini-

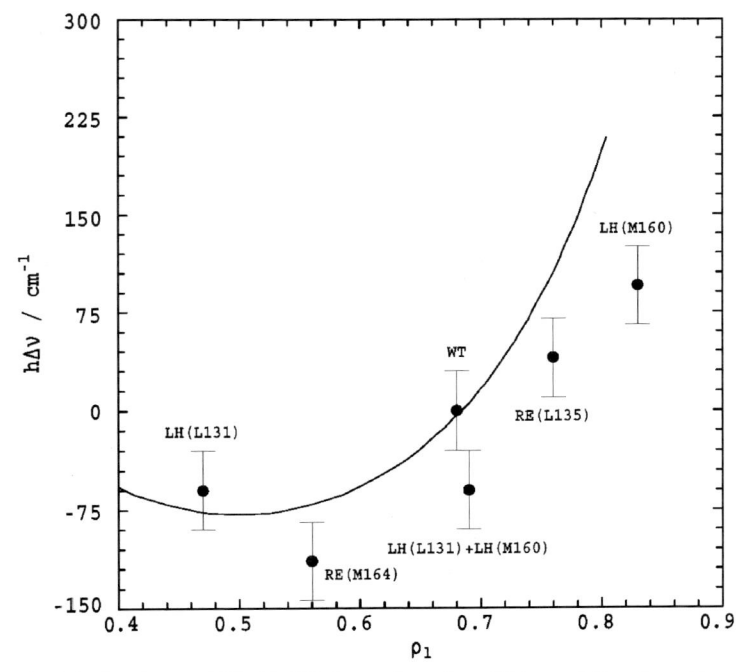

**Fig. 6.8** The observed [19, 37, 39] change from the wild type of *Rb. sphaeroides* in vertical excitation energy of the hole-transfer band of the special-pair radical cation for various mutants is correlated with the observed [32–34, 37] spin density on $P_L$. The solid line indicates the correlation expected from our four-state analysis [4] of the hole-transfer absorption band of the wild type.

mum hole-transfer band energy lowering that can be achieved by electric field-induced Stark modulation is on the order of 100 cm$^{-1}$. Hence, this effect cannot account for the observation of the 2200 cm$^{-1}$ shoulder shown in Figs. 6.1 and 6.6, as the required lowering would be on the order of 500 cm$^{-1}$.

The response of the hole-transfer absorption band to small changes in the electric field across the special pair has been determined directly by Treynor and Boxer [40] and is shown in Fig. 6.9 as the fractional change in absorption observed for an applied external electric field of magnitude 0.6 MV cm$^{-1}$. For the Stark modulation of spectra not dominated by vibronic coupling, the theory of Liptay [41, 42] is usually used to interpret this. Applied to intervalence spectroscopy [43, 44], qualitatively this theory stipulates that if the charge is localized on one half of the special pair, then the response will be strong and contain two nodes, while if the charge is delocalized, then the response will be weak and contain only one node. As the spectrum of the special-pair radical cation is dominated by vibronic coupling effects, simple Liptay theory is inapplicable and full spectral simulation is required [45]. The observed spectrum is shown as recorded and after broadening to 100 cm$^{-1}$ resolution along with that predicted directly based on our four-state model analysis [4] of the absorption spectrum [17].

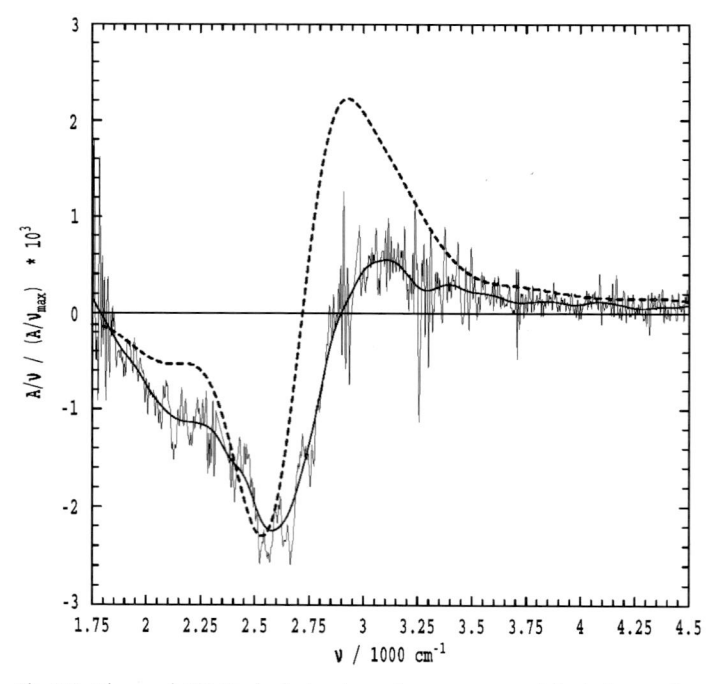

**Fig. 6.9** Observed [40] Stark electroabsorption spectrum of the hole-transfer band of the special-pair radical cation of Rh. sphaeroides, additionally broadened to 100 cm$^{-1}$ resolution, and that as calculated [17] from our four-state model [4].

The observed and calculated spectra are strikingly similar, with both being of the same strength, both containing just one node, and both displaying long, high-frequency tails and a shoulder at 2200 cm$^{-1}$. Quantitatively, the locations of the node differ somewhat, while the observed band is significantly more asymmetric about the node than is the calculated one. Such differences are common even for small molecules [44], for which very accurate calculations are possible.

## 6.8
## Conclusions

A comprehensive assignment of the spectrum of the special-pair radical cation of *Rb. sphaeroides* in the 2000–5000 cm$^{-1}$ region has been obtained using a four-state model [4]. The parameters in this model have been fitted to the observed infrared difference spectrum and shown to be adequate for the description of the complex redox, charge-distribution, and Stark-effect properties of the system [17]. This verifies the correctness of the basic model and the appropriateness of the quantitative solution obtained. The influence of the special pair on all aspects of the overall photosynthetic charge-separation process can be understood from the results of our analysis, with one of the most influential properties being the degree of charge localization within the dimer radical cation. In addition, the analysis obtained here for bacterial photosynthesis will prove to be central to the development of a proper understanding of the role of the special pair in algal and plant photosynthetic systems and to the design of artificial photovoltaic devices that mimic natural photosynthesis.

## Acknowledgments

We thank the Australian Research Council for providing 10 years of support for this project through Research Fellowships for J.R.R. and a variety of Large Research Grants; we also thank the Australian Partnership for Advance Computing for provision of computational resources.

## References

1 M.J. Crossley, P. Thordarson, *Angew. Chem. Int. Ed.* **2002**, *41*, 1709.
2 E.K.L. Yeow, P.J. Sintic, N. Cabral, J.N.H. Reek, M.J. Crossley, K.P. Ghiggino, *Phys. Chem. Chem. Phys.* **2000**, *2*, 4281.
3 E.K.L. Yeow, K.P. Ghiggino, J.N.H. Reek, M.J. Crossley, A.W. Bosman, A.P.K.J. Schenning, E.W. Meijer, *J. Phys. Chem. B* **2000**, *104*, 2596.
4 J.R. Reimers, N.S. Hush, *J. Chem. Phys.* **2003**, *119*, 3262.
5 J.R. Reimers, W.A. Shapley, A.P. Rendell, N.S. Hush, *J. Chem. Phys.* **2003**, *119*, 3249.
6 J.R. Reimers, W.A. Shapley, N.S. Hush, *J. Chem. Phys.* **2003**, *119*, 3240.
7 J.R. Reimers, A.P. Rendell, N.S. Hush, in: *12th Int Conf on Photosynthesis*, CSIRO, **2001**.

8  J. M. Hughes, M. C. Hutter, J. R. Reimers, N. S. Hush, *J. Am. Chem. Soc.* **2001**, *123*, 8550.
9  J. R. Reimers, J. M. Hughes, N. S. Hush, *Biochemistry* **2000**, 16185.
10 J. R. Reimers, M. C. Hutter, J. M. Hughes, N. S. Hush, *Int. J. Quant. Chem.* **2000**, *80*, 1224.
11 M. C. Hutter, J. M. Hughes, J. R. Reimers, N. S. Hush, *J. Phys. Chem. B* **1999**, *103*, 4906.
12 J. R. Reimers, M. C. Hutter, N. S. Hush, *Photosynth. Res.* **1998**, *55*, 163.
13 J. R. Reimers, N. S. Hush, *Chem. Phys.* **1995**, *197*, 323.
14 J. R. Reimers, N. S. Hush, *J. Am. Chem. Soc.* **1995**, *117*, 1302.
15 J. R. Reimers, N. S. Hush, *Inorg. Chim. Acta* **1994**, *226*, 33.
16 W. A. Shapley, J. R. Reimers, N. S. Hush, *Int. J. Quant. Chem.* **2002**, *90*, 424.
17 J. R. Reimers, N. S. Hush, *J. Am. Chem. Soc.* **2004**, *126*, 4132.
18 C. Creutz, H. Taube, *J. Am. Chem. Soc.* **1969**, *91*, 3988.
19 J. Breton, E. Nabedryk, W. W. Parson, *Biochem.* **1992**, *31*, 7503.
20 E. Nabedryk, J. P. Allen, A. K. W. Taguchi, J. C. Williams, N. W. Woodbury, J. Breton, *Biochem.* **1993**, *32*, 13879.
21 J. Breton, E. Nabedryk, A. Clérici, *Vibrational Spectrosc.* **1999**, *19*, 71.
22 M. J. Rice, V. M. Yartsev, C. S. Jacobsen, *Phys. Rev. B* **1980**, *21*, 3437.
23 M. J. Rice, *Phys. Rev. Lett.* **1976**, *37*, 36.
24 M. J. Rice, N. O. Lipari, S. Strässler, *Phys. Rev. Lett.* **1977**, *39*, 1359.
25 M. J. Rice, L. Pietronero, P. Brüesch, *Solid State Comm.* **1977**, *21*, 757.
26 M. J. Rice, *Solid. State Commun.* **1979**, *31*, 93.
27 S. B. Piepho, E. R. Krausz, P. N. Schatz, *J. Am. Chem. Soc.* **1978**, *100*, 2996.
28 R. K. Clayton, B. J. Clayton, *Biochim. Biophys. Acta* **1978**, *501*, 478.
29 D. W. Reed, *J. Biol. Chem.* **1969**, *244*, 4936.
30 J. R. Reimers, N. S. Hush, *Chem. Phys.* **1996**, *208*, 177.
31 M. Born, R. Oppenheimer, *Ann. Phys.* **1927**, *84*, 457.
32 X. Lin, H. A. Murchisson, V. Nagarajan, W. W. Parson, J. P. Allen, J. C. Williams, *Proc. Natl. Acad. Sci.* **1994**, *91*, 10265.
33 K. Artz, J. C. Williams, J. P. Allen, F. Lendzian, J. Rautter, W. Lubitz, *Proc. Natl. Acad. Sci. USA* **1997**, *94*, 13582.
34 A. Ivancich, K. Artz, J. C. Williams, J. P. Allen, T. A. Mattioli, *Biochem.* **1998**, *37*, 11812.
35 F. Müh, F. Lendzian, M. Roy, J. C. Williams, J. P. Allen, W. Lubitz, *J. Phys. Chem. B* **2002**, *106*, 3226.
36 J. P. Allen, K. Artz, X. Lin, J. C. Williams, A. Ivancich, D. Albouy, T. A. Mattioli, A. Fetsch, M. Kuhn, W. Lubitz, *Biochem.* **1996**, *35*, 6612.
37 E. T. Johnson, F. Müh, E. Nabedryk, J. C. Williams, J. P. Allen, W. Lubitz, J. Breton, W. W. Parson, *J. Phys. Chem. B* **2002**, *106*, 11859.
38 Z.-L. Cai, J. R. Reimers, *J. Phys. Chem. A* **2000**, *104*, 8389.
39 E. Nabedryk, J. Breton, J. Wachtveitl, K. A. Gray, D. Oesterhelt, *NATO ASI Ser., Ser. A* **1992**, *237*, 147.
40 T. P. Treynor, S. S. Andrews, S. G. Boxer, *J. Phys. Chem. B* **2003**, *107*, 11230.
41 W. Liptay, in: *Modern quantum chemistry*, Vol III (Ed. O. Sinanoglu), Academic Press, New York, **1965**, pp. 45.
42 W. Liptay, *Angew. Chem. Internat. Edit.* **1969**, *8*, 177.
43 J. R. Reimers, N. S. Hush, in: *Mixed Valence Systems: Applications in Chemistry, Physics, and Biology* (Ed. K. Prassides), Kluwer Acad. Publishers, Dordrecht, **1991**, p. 29.
44 J. R. Reimers, N. S. Hush, *J. Phys. Chem.* **1991**, *95*, 9773.
45 J. R. Reimers, N. S. Hush, in: *A.C.S. Advances in Chemistry Series 226: Electron Transfer in Biology and the Solid State* (Eds. M. K. Johnson, R. B. King, D. M. K. Jr, C. Kutal, M. L. Norton, R. A. Scott), A.C.S., Washington DC, **1990**, p. 339.

# 7
# Protein-based Artificial Photosynthetic Reaction Centers [1]

*Reza Razeghifard and Thomas J. Wydrzynski*

## 7.1
## Introduction

The biological conversion of solar energy into chemical energy is carried out by membrane-bound photosynthetic protein complexes. These large pigment-protein complexes have evolved into multicomponent systems, which efficiently capture light energy and promote the directional transfer of electrons. The light energy captured by (bacterio)chlorophylls (BChls) in the light-harvesting systems is channeled to special photo-oxidizable BChls in the reaction center (RC) where the photochemistry occurs. These special BChls are called primary electron donors (P) and are the key determinant for the unique function of each photosystem. They are often designated by their particular maximum absorption wavelength. Two photosystems (PSI and PSII) work in tandem in plants and cyanobacteria, while only one photosystem (the bacterial reaction center [BRC]) is found in anoxygenic photosynthetic bacteria.

## 7.2
## Natural Reaction Centers

### 7.2.1
### Structure and Function

The reaction center of *Rhodopseudomonas viridis* was the first membrane protein to be structurally characterized by high-resolution X-ray crystallography, winning the 1988 Nobel Prize in Chemistry for Deisenhofer, Michel, and Huber [1]. In more recent breakthroughs in this field, medium-resolution X-ray crystallographic information has now become available for the structures of dimeric PSII [2–4] and PSI

---

1) Abbreviations: Chl, chlorophyll; ET, electron transfer; Fe-S, iron-sulfur cluster; RC, reaction center; P, primary electron donor; Pheo, pheophytin; PSI, photosystem I; PSII, photosystem II.

*Artificial Photosynthesis: From Basic Biology to Industrial Application*
Edited by Anthony F. Collings and Christa Critchley
Copyright © 2005 WILEY-VCH Verlag GmbH & Co. KGaA, Weinheim
ISBN: 3-527-31090-8

[5, 6], revealing the spatial arrangement of the polypeptides and redox cofactors. The overall architecture of all known RCs is composed of two homologous, integral membrane proteins that form a heterodimer with a C2 symmetry axis. The polypeptides of the corresponding heterodimers are designated as L/M, D1/D2, and PsaA/PsaB for bacterial, PSII, and PSI RCs, respectively. The majority of the redox-active cofactors are organized in the dimeric cores into two possible symmetrically arranged electron transfer (ET) branches (Fig. 7.1). Even though the fundamental principles of photochemical reaction are conserved, each RC is unique according to the chemical nature of its electron acceptors and donors.

RCs can be divided into two categories based on the type of their electron acceptors. PSI and certain green sulfur bacteria use iron-sulfur clusters ($Fe_4S_4$) as electron acceptors and thus belong to the FeS-type RC, while the purple bacterial and PSII RCs use pheophytin/iron-quinone complexes as acceptors and are known as the FeQ-type RC. Each RC is equipped with a P that has its redox potential optimized to serve the particular photochemical function. In the purple BRC, $P_{870}$, is a special pair of BChl $a$ molecules with a redox midpoint potential ($E_m$) of +0.45 for the $P_{870}^+/P_{870}$ couple. The BChl molecules in this red-shifted dimer are separated by ~ 3 Å in the pyrrole ring I and by ~ 7 Å from the ring centers. The chlorin heads are inclined to each other by ~ 15° and are believed to be excitonically coupled. Upon light excitation, $P_{870}^*$ donates an electron to a ubiquinone (UQ) via a bacteriopheophytin $a$ (BPheo $a$) along the so-called active ET branch. The oxidized $P_{870}^+$ then receives an electron from the heme iron of cytochrome $c_2$. The heme, which is covalently attached to cysteine residues via thioether linkages, functions as the electron donor because it can alternate between the $Fe^{2+}$ and $Fe^{3+}$ oxidation states.

In PSII, the Chl $a$ molecules in the $P_{680}$ dimer are positioned about 10 Å apart from each other and can generate an extremely strong oxidant ($E_m$ ~ +1.2 V) that is needed to overcome the energy barrier required to oxidize water to molecular oxygen (+870 mV at pH 6) as the ultimate source of electrons in oxygenic photosynthesis. Water oxidation takes place at a manganese-containing catalytic site (Mn cluster) where the four required oxidizing equivalents provided by sequential photoactivations of $P_{680}$ are progressively accumulated by increasing the redox state of the Mn cluster. The photoactivated $P_{680}^*$ transfers an electron to the D2-bound plastoquinone ($Q_A$) via the D1-bound pheophytin $a$ (Pheo $a$). The electron hole on $P_{680}^+$ is then filled by extracting an electron from the Mn cluster via a redox-active tyrosine residue ($Y_Z$) acting as an intermediate electron carrier. In both BRCs and PSII RCs, the electron is passed on to a secondary quinone molecule ($Q_B$) on the acceptor side that becomes doubly reduced after two successive charge-separation events. The doubly reduced $Q_B^=$ is then released into the lipid phase of the membrane after taking up two protons and forming the quinol $PQH_2$.

In Chl $a$-type PSI RCs, P is a heterodimer of Chl $a$/Chl $a'$ with a $E_m$ ~ +0.5 V for $P_{700}^+/P_{700}$. The excited $P_{700}$ is much stronger reductant than $P_{680}^*$. The primary electron acceptor ($A_0$) is a monomeric Chl molecule, which transfers the electron from $P_{700}$ to a phylloquinone ($A_1$). After sequential reduction of several

Fe$_4$S$_4$ clusters (F$_X$, F$_A$ and F$_B$) by the reduced A$_1$, the electron is finally transferred to the ferredoxin protein (in some cases flavodoxin), which is used to reduce NADP$^+$ to NADPH$_2$, the reducing compound needed for CO$_2$ fixation. The Cu ion of plastocyanin (in some cases heme in cytochrome $c_6$) acts as the electron donor to P$_{700}^+$.

### 7.2.2
### Creation of a Charge-separated State

As a general principle, all photosynthetic RCs, regardless of their source, use light energy to create a charge-separated state by moving electrons along a chain of carriers away from the oxidized P$^+$, which is ultimately reduced by a neighboring electron donor rather than by a back reaction. The use of multiple cofactors separated by < 14–15 Å ensures that electron tunneling is a much more efficient mechanism of transfer than is catalytic turnover [7]. The protein matrix arranges the redox cofactors along the desired transfer pathway, with specific positions, orientations, and tuned redox potentials in order to induce directional electron transfer and avoid charge recombination. The result is a charge-separated state in which the oxidized and reduced cofactors are located on opposite sides of the protein (Fig. 7.1). Initial electron transfer is extremely fast (some tens of picoseconds), but it gets gradually slower as the chance of charge recombination becomes smaller. While the ET from Y$_Z$ to P$_{680}^+$ is completed in some nanoseconds in PSII, the ET from the Mn cluster to Y$_Z^\bullet$ depends on the redox state of the Mn cluster and can be three orders of magnitude slower [8]. It is a very special property of the natural system that the ET can vary from picoseconds to milliseconds. In PSII RC, the result of a successful photochemical event is the reduction of Q$_A$ on the acceptor side with a $E_m \sim -80$ mV, while on the donor side the oxidized species is the tyrosine radical Y$_Z^\bullet$ with a $E_m \sim +1.1$ V, which is capable of generating the oxidizing equivalents used by the water-oxidizing complex. In PSI, the electron is donated by plastocyanin with a $E_m \sim +350$ mV, while on the acceptor side a reduced Fe$_4$S$_4$ is generated with a reduction potential (–550 mV) high enough to reduce the Fe-S cluster in ferredoxin. The redox potential of the cofactors, and therefore the free energy gap required for the forward electron transfer, is entirely controlled by the protein environment.

### 7.2.3
### Mutational Studies

The difference in the functionality of RCs is the result of incorporating chemically different redox cofactors and the tuning of the redox activity by the protein. Therefore, how the protein environment controls the redox properties of the cofactors is fundamental to the understanding of ET. Several factors can be considered for controlling the redox activity. One important factor is metal coordination. By preparing a series of site-directed mutants at a histidine axial ligation site for P$_{700}$ in PSI, it was shown that the $E_m$ of P$_{700}^{\bullet+}$/P$_{700}$ can be increased, in some cases by 140 mV [9]. It is believed that the axial ligation of the $A_0$ Chl to the sulfur atoms of methionine rather than to the nitrogen atoms of histidine tunes its redox activity [6]. It also appears that histidines are not involved in the ligation of the accessory Chls in PSII and corresponding Chls in PSI [4, 6]. It is worth noting that the recruitment of Chl or Pheo into the reaction sequence is determined by the presence or the absence of a ligating residue. For the BRC, introduction of M-His202Leu and L-His173Leu mutations causes the replacement of BChls with two BPheo molecules, while introduction of the M-Leu214-His mutation favors the incorporation of a BChl into the BPheo binding site [10]. As the axial ligation with histidine is a common theme for binding of the primary electron donors, other factors must also be involved in tuning the redox activity, such as differences in the local dielectric or the H-bond network. For example, it was shown that the formation of up to four hydrogen bonds with the chlorin ring of the BChl special pair in a genetically modified *Rhodobacter sphaeroides* could additively increase the oxidation potential by a remarkable 355 mV [11, 12], enabling the mutated BRC to oxidize a nearby engineered tyrosine residue [13] and Mn ions [14]. Interestingly, the increase in the midpoint potential was found to be dependent on the chemical nature of the H-bond forming residue, with histidine being the most effective. Likewise, the $E_m$ could be changed, but to a lesser extent, by introducing ionizable residues around the BChl special pair [15, 16].

The organization of P as enforced by the protein backbone is another factor controlling redox activity. Thus, $P_{680}$ is more monomeric in nature than are $P_{700}$ and $P_{870}$, which could also help account for the very high $E_m$ of $P_{680}$. Likewise, it was found that when phylloquinone is replaced by plastoquinone in mutants of *Synechocystis* sp. PCC 6803, the forward ET still occurs with high efficiency [17]. In other cases, however, the redox potential of a cofactor depends on the chemical identity of the cofactor. The low redox potential of $P^+/P$ in Chl *d*-containing PSI in *Acaryochloris marina* (+350 mV) compared with that of $P_{700}$ (+500 mV) appears to be due to small structural differences between Chl *d* and Chl *a* molecules [18].

## 7.3
**Synthetic and Semi-synthetic Reaction Centers**

Regardless of the specific chemical reactions that are coupled to the ET, all RCs are efficient solar energy converters. This is because the primary radical pair is stabilized by a series of fast and efficient short-range electron transfer steps (picoseconds to nanoseconds) that lead to a stable charge separation across a distance of $\sim 30$ Å (see Fig. 7.1). This stable charge separation is a consequence of secondary electron transfer events that occur on the order of tens of microseconds and are unique to the protein environment [19]. The high quantum efficiency for charge separation has been an inspiration for creating artificial systems.

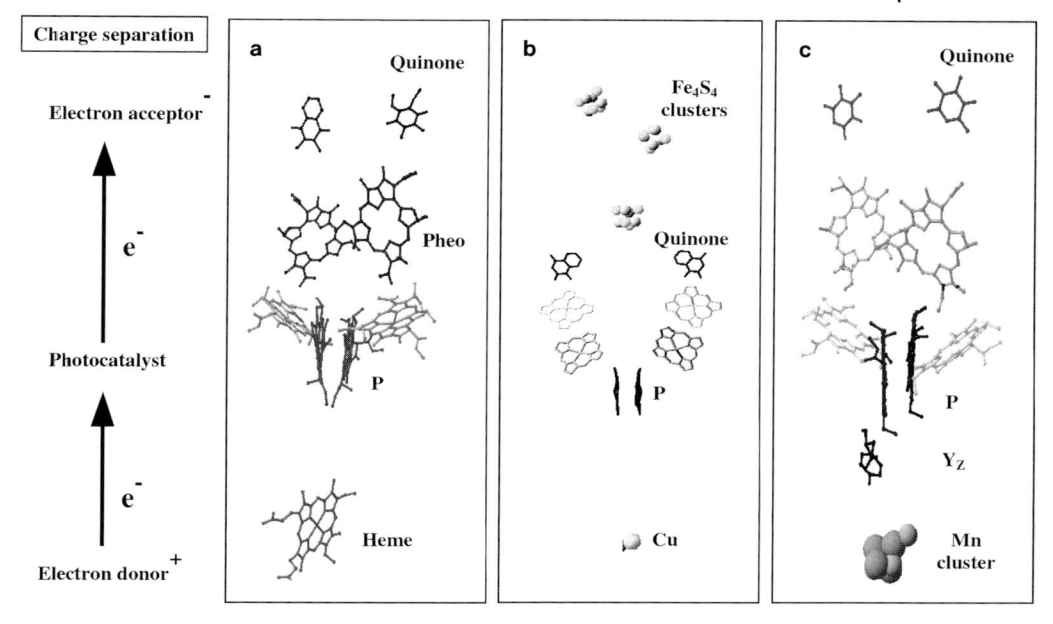

**Fig. 7.1** Protein bound cofactors of (a) purple bacterial, (b) PSI, and (c) PSII reaction centers. Phytyl tails of chlorophyll and quinones are removed for clarity. Figures were generated by Swiss-PdbViewer [79] from Protein Data Bank codes 1PRC, 1JB0, and 1S5L.

## 7.3.1
### Multi-layered Films

In the natural system, the ability of the photosystems to preferentially orient themselves in the membrane is a crucial factor for the high efficiency of the photosynthetic process. Various approaches to mimic this have been employed by depositing purified RCs into synthetic membranes in a desired geometric order and measuring net charge movement. Langmuir-Blodgett films of BRCs incorporated into planar phospholipid membranes or monolayers on solid supports were the first system to show that the natural photochemical reactions of the BRCs could be maintained in an artificial system [20–22]. Other techniques involving biochemical modification of the BRCs have since been applied to better control the directional alignment of proteins in order to improve photocurrent efficiency. For example, when biotinylated BRCs are immobilized onto $SnO_2$-coated glass plates by avidin-biotin interactions, a two to four times larger photocurrent is detected compared with BRCs adsorbed into membranes [23]. Alternatively, by preparing 24-layer BRC films, the overall response was increased by a factor of eight [24]. Even though these studies show major progress in creating semi-artificial photoactive membranes, the approach requires further improvement. In particular, preparing active biological samples with long-term stability is problematic. For example, the response of the BRC embedded membranes drops to one-fourth in a one-month period [23]. The gradual loss of activity could be due to conformational changes associated with such large biological complexes and/or to pigment degradation. In order to make a relatively more stable system, an abstracted version of the RC with photocatalysts more stable than chlorophylls can be used. In one approach, an elegant design has been described by Gust and Moore (see Chapter 10) to create a non-peptide artificial RC using triads of pure redox cofactors [25, 26]. In this design, a primary chromophore, an electron donor, and an electron acceptor are covalently linked. To increase the lifetime of the charge-separated state while maintaining fast forward ET, the conventional acceptors in the triads were replaced with fullerenes [27, 28] or were incorporated in part into a protein matrix from the donor end while the rest of the triad dangled in solution. In the latter system, it was shown that the lifetime of the charge-separated state could be increased for the donor (Zn-protoporphyrin)–sensitizer ($Ru^{+2}(bpy)_3$)–acceptor (cyclic viologen) triad from 300 ns to 600–900 ns (pH dependent) or 1.1–18 μs when the acceptor part was incorporated into cytochrome $b_{562}$ or myoglobin, respectively [29]. A long-lived charge-separated state (>2 ms) was also generated in a heme triad-myoglobin system [30].

### 7.3.2
**Synthetic Reaction Centers**

Recent advances in protein design now offer a new approach to constructing protein-based artificial reaction centers. Starting from scratch, a simple design would be a polypeptide scaffold with engineered binding sites for cofactors. Much is known about how to create such a simple design from the structural analysis of natural systems. The first step in constructing an artificial RC is to incorporate cofactors into a helical structure in order to modulate their redox properties and control their geometry. Directional ET can then be created by adjusting the location of the cofactors on the protein scaffold. Exceptionally simple de novo designs have been successfully used to create synthetic proteins with predicted four-α-helical structures [31, 32]. In these instances, α-helices are obtained by engineering hydrophobic residues to align onto one side of a helical-forming segment. This allows the hydrophobic residues to sequester themselves away from the aqueous solvent and to assemble a hydrophobic interior pocket suitable for cofactor binding. The exposed negatively and positively charged residues in the helical-forming segment provide for stability and water solubility. Iterative redesign of the first synthetic four-α-helical bundle proteins has successfully created native-like structures [33, 34]. By introducing favorable hydrophobic interactions, the synthetic peptides acquire a well-packed hydrophobic core that is needed to form native states from molten globules. For natural proteins, folding occurs in a cellular environment in the presence of cofactors, and their interactions appear to be critical for stabilizing the native state. This was also found to be the case for a de novo-designed heme peptide (Fig. 7.2). The

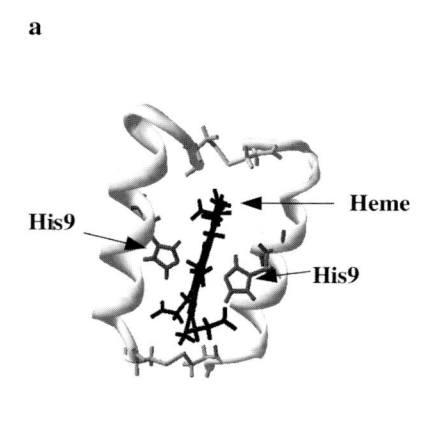

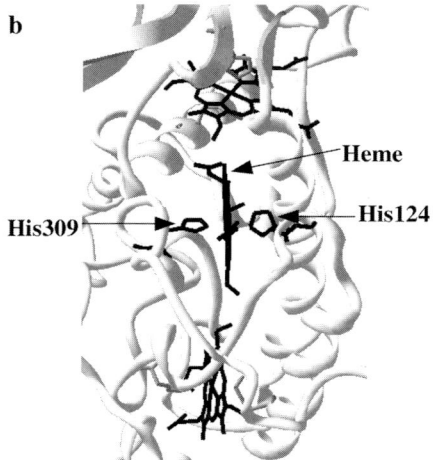

**Fig. 7.2** (a) *Bis*-ligated heme in a de novo-designed peptide and (b) the BRC from *Rhodopseudomonas viridis* [80]. The heme peptide structure represents a water-soluble [Co$^{III}$(coproporphyrinate)]$^{3-}$ bound to two monomeric 15-mer peptides linked at the termini by disulfide bridges. Figures were generated by Swiss-PdbViewer from Protein Data Bank codes 1PBZ and 1PRC.

synthetic peptide folded in the presence of the metalloporphyrin and gave rise to a single conformation in solution [35].

The advantage of the de novo helical bundle proteins is that they can be readily modified without loss of their structural integrity, as has been shown by NMR spectroscopy [36]. The robustness of the design thus allows incorporation of organic cofactors or transition metals into structurally related variants to create artificial redox-active proteins. In particular, heme maquettes have turned out to be good examples of how the redox activity of a cofactor can be modulated in a de novo polypeptide (for a review, see [37]). Like natural heme proteins where the redox potential spans over a range of 800 mV (from +400 mV for cytochrome $b_{559}$ to –400 mV for cytochrome $c_3$), the redox potential in the simple heme maquette could be varied by 435 mV [38]. The redox potential was modulated by altering the electron-donating/withdrawing nature of the peripheral macrocycle substituents of the heme and the protonation/deprotonation state of neighboring charged residues. In addition, structural distortion of the porphyrin macrocycle may also play a role in tuning the $E_m$. By analyzing more than 800 normal-coordinated heme groups from protein crystal structures, the effects of non-planar distortion of the heme were analyzed [39, 40]. The distortion force was found to be caused mainly by the formation of hydrogen bonds and by the steric interactions between the protein and the heme, with some role for the axial ligand orientation. Since the head group of metalloporphyrins is about two helix turns long, it can potentially interact with the peptide over a surface area of approximately 1.0 nm$^2$.

Several attempts to incorporate electron donors and acceptors into synthetic peptides to allow for light-induced ET have been made. A de novo–designed metalloprotein containing a ruthenium-bipyridine complex and a heme was constructed and was used to study the laser-induced electron transfer across the helix [41]. Similarly, flash photolysis and pulse radiolysis measurements on a designed metalloprotein in both the folded and unfolded states were used to investigate effects of helical conformation on electron transfer rates [42]. As a mimic for in vivo electron transfer, a dimeric 31-mer synthetic metalloprotein containing two ruthenium complexes was designed to accomplish electron transfer across the noncovalent peptide-peptide interface [43]. This design allowed Ogawa and coworkers to show a long-range intraprotein ET ($\sim 24$ Å) with a rate constant of 380 s$^{-1}$ across a noncovalent peptide-peptide interface [44]. The rates of photoinduced ET from a pyrenyl group to a nitrophenyl group attached to an $\alpha$-helical polypeptide was determined as the function of the number of spacer amino acids and was shown to be dependent on the edge-to-edge distance between the two chromophores [45]. It was also shown that a light-induced triplet ET between a Zn(II)-protoporphyrin bound to a histidine in the interior of a synthetic four-helix bundle protein and an anthraquinone molecule as an external electron acceptor occurred [46]. The protein was constructed by coupling four amphiphilic synthetic helical peptides via a linker to a cyclic decapeptide template.

A directional ET was reported between an N,N-dimethylaniline donor and a pyrene acceptor attached to two peptides that were different in the positions of the donor and acceptor with respect to the helix ends [47]. The directional ET was ascribed to the helix dipole moment; however, such effect was not observed when ET between polar metal complexes in a metallopeptide was studied [48]. The two metal complexes were a Ru polypyridyl donor and a pentammine Ru(III) acceptor covalently attached to a 30-residue polypeptide via a Cys and a His, respectively. The ET direction was reversed by switching the position of the Cys and His residues on the peptide. A redox-active tryptophan residue was created as a result of multi-step electron transfer on a de novo helix bundle peptide [49]. The electron donation of the tryptophan residue to the photo-oxidized pyrene effectively competed with the back electron transfer between pyrene$^{+\bullet}$ and methyl viologen$^{+\bullet}$. Both pyrene and methyl viologen moieties were bound to the peptide. By varying the distance between two ruthenium ions attached to a three-helix bundle motif, a strong distance dependence for the electron transfer rate was observed, indicating the structural integrity of the robust designed protein [50]. It was also shown that electron transfer occurred between a covalently attached flavin and a *bis*-His ligated heme in a light-activatable molecular maquette under continuous illumination [51].

### 7.3.2.1 Electron Acceptor
Two types of electron acceptors are used by natural RCs as the ultimate electron acceptors: quinone molecules and $Fe_4S_4$ clusters. Photosynthetically active quinones are highly hydrophobic, having a long hydrocarbon side chain for bind-

ing purposes and membrane solubility. These natural quinones are not covalently bound to the protein; however, the covalent attachment is a more practical approach to creating synthetic RCs. In addition to these hydrophobic natural quinones, synthetic analogues such as 2,6-dichloro-1,4-benzoquinone, p-benzoquinone, phenyl-p-benzoquinone, and 2,3-dimethoxy-5-methyl-1,4-benzoquinone are also commercially available and can be used in aqueous solution. They are routinely used with the purified RC in biochemical assays to compensate for the loss of quinone pool during purification steps. The reduced forms of quinones are semiquinone (singly reduced) and quinol, the doubly reduced protonated species. They are very efficient electron acceptors in their oxidized forms. The chemical reaction of some of these quinones with a –SH group of Cys residues in proteins has been reported [52].

Iron-sulfur clusters are iron atoms that are wholly or predominantly coordinated by sulfur atoms usually provided by Cys residues (Fig. 7.3). The $E_m$ of these clusters varies between proteins from +450 mV to −700 mV, which is believed to be controlled by solvent accessibility, hydrogen bonding, and the distribution of neighboring charge residues [53]. Different methods have been used to incorporate Fe-S clusters into proteins. For example, Gibney et al. successfully incorporated two distinct redox cofactors into a synthetic four-$a$-helix bundle, making a maquette containing four hemes and two $Fe_4S_4$ clusters [54]. Each dimeric peptide contains two helical regions of ~27 residues with a clostridial ferredoxin-binding domain for the cluster (CEGGCIACGAC) inserted into the glycine loop connecting the helices. Two engineered *bis*-histidines provided ligation sites for binding hemes. The self-assembly of the $Fe_4S_4$ cluster with the peptide was achieved by adding a solution of ferric chloride and sodium sulfide to the chemically reduced peptide under anaerobic conditions. The formation of the $Fe_4S_4$ cluster was confirmed by EPR spectroscopy and its re-

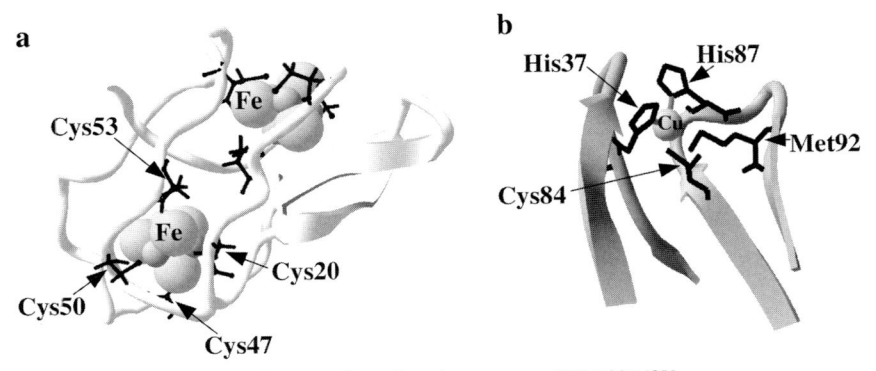

**Fig. 7.3** (a) PsaC protein of PSI RC from *Synechococcus* sp. PCC 7002 [81] and (b) spinach plastocyanin [82]. Ligating residues for Fe ions in $F_A$ and $F_B$ clusters and for Cu ions in plastocyanin are highlighted. Figures were generated by Swiss-PdbViewer from Protein Data Bank codes 1K0T and 1AG6.

duction potential was determined as −350 mV. Further studies on this design were made to determine the most important amino acids in the binding domain for the cluster formation [55, 56]. A similar approach was applied to form a single $Fe_4S_4$ cluster resembling the $F_X$ cluster of PSI by inserting the 10-residue CDGPGRGGTC into interhelical loops 1 and 3 of $α4$ protein, which was originally designed by DeGrado's group [57] to form a synthetic four-helix bundle protein [58]. The cluster formation in the holoprotein was confirmed by EPR and was shown to have a $E_m$ of −422 mV, which is more positive than that of the natural counterpart. Binding sites for a $Fe_4S_4$ cluster and a Ni(II) ion were also included in the design of a four-$α$-helical protein formed by two helix-loop-helix 63-mer peptides, as a mimic of the A-cluster of carbon monoxide dehydrogenase [59]. One cluster was incorporated into each 63-mer peptide when the peptide was reacted with a preformed cluster $[Fe_4S_4(SCH_2CH_2OH)_4]^{2-}$. A miniferredoxin protein was also created from a 31-residue peptide carrying a $Fe_4S_4$ cluster with a redox potential of −370 mV [60].

It is also shown that natural proteins can be modified to bind Fe-S clusters. A binding site was generated using the Dezymer program and was introduced into thioredoxin through site-directed mutagenesis [61, 62]. To create the $Fe_4S_4$ cluster site, four residues were mutated to add new Cys residues (Leu24Cys, Leu42Cys, Val55Cys, and Leu99Cys), while Cys residues of the native disulfide bridge were used to create a mononuclear cluster. The mononuclear cluster was redox-active capable of undergoing successive cycles of air oxidation/reduction by $β$-mercaptoethanol. The $Fe_4S_4$ cluster was assembled inside the protein by incubating the partially denatured thioredoxin with a synthetic $[Fe_4S_4(S-EtOH)_4]$ $(Me_4N)_2$ under anaerobic conditions.

#### 7.3.2.2 Electron Donor

Redox metals of Fe, Cu, and Mn act as electron donors in photosynthetic RCs. Even though these donors are positioned in the vicinity of $P^+$ for efficient ET, unlike other cofactors their interactions with the dimeric core of RC is somewhat different. In the case of the BRC and the PSI RC, the electron donor is brought into contact with the dimeric core of the RC using a separate protein (cytochrome or plastocyanin, respectively), while some of the Mn cluster-ligating residues are believed to be provided by D1 and D2 subunits. The Mn cluster is a part of the oxygen-evolving complex and, like the electron donor of the BRC and the PSI RC, is shielded from solvent by extrinsic proteins. The Mn cluster can be removed from PSII centers by biochemical treatments without the loss of photochemistry. Unlike the other two electron donors, this cluster, which contains four Mn ions, is involved in oxygen chemistry, and therefore its structural organization is rather complex. From the electron density map [4], it is suggested that Mn ions are arranged by protein ligands in a cubane-like $Mn_3CaO_4$ cluster connected to the fourth Mn in the extended region (Fig. 7.4). In this model, Mn ions that are linked by μ-oxo bridges are separated by 2.7 Å in the cubane-like structure and by 3.3 Å for the fourth Mn. In contrast to the

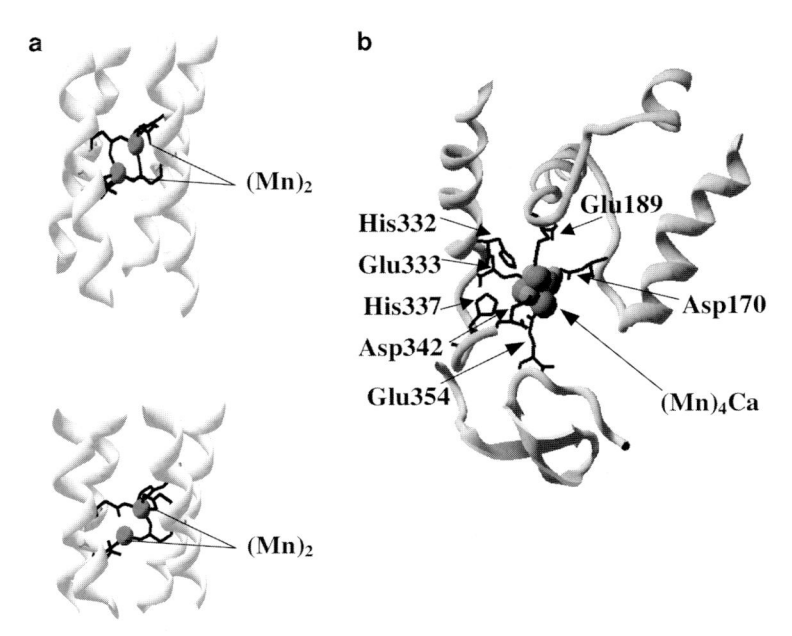

**Fig. 7.4** (a) Mn dimers in a de novo synthetic peptide and (b) PSII RC. Glu, Asp, and His residues that are proposed for forming the Mn cluster are highlighted. Figures were generated by Swiss-PdbViewer from Protein Data Bank codes 1LT1 and 1S5L.

Mn cluster, the interaction mode of Cu or heme with proteins is well understood. In plastocyanin, copper is ligated to the protein by two His nitrogen atoms, a Cys sulfur donor, and a Met sulfur atom at a longer distance in a distorted tetrahedral geometry (see Fig. 7.3). This type-1 copper exhibits an intense blue color ($\sim 600$ nm) in the oxidized form. In addition to Cu(II) coordination environment, which appears to play a role in giving an $E'$ in the range of $+200$ mV to $+1000$ mV in different proteins, other factors such as the protein second shell are also involved in adjusting the potential [63]. For example, a similar coordination pattern is used in rusticyanin and plastocyanin, but the redox potential of Cu(II) in the latter is much lower.

Several attempts have been made to design a binding site for Cu and Mn ions in natural and synthetic proteins. Histidines are the foremost ligands for metals, and as such, metals are known to bind nonspecifically to a histidine repeat incorporated at the N- or C-termini of proteins. Such repeats that can bind metals such as Cu(II), Ni(II), Zn(II), and Cd(II) have been used to immobilize proteins on surfaces. More specific binding is then encoded in their natural binding motifs. The coordination environment of metal can be determined by analyzing the structure of natural proteins [64]. The second shell for blue copper appears to include some solvent-exposed residues, while for Mn ions they are mostly buried. Interestingly, unlike Cu ions that are never ligated by acidic residues, the primary ligands for Mn ions are mostly Glu or Asp residues. In the case of copper, a tetrahedrally distorted square plane with a redox potential of +75 mV was created by including all four blue Cu(II) ligands into a 22-residue peptide [65]. The peptide, which contained the natural binding loop with the Cys, Met, and His ligands (YCSPHQGAGMVG K), was extended by a flexible Gly linker to add the fourth ligand. This design was based on the structure of Poplar plastocyanin in which the second His ligand is brought to the binding site by the protein fold. In the case of manganese, a di-Mn(II) center has been successfully incorporated into a four-$\alpha$-helical bundle protein, due-ferri-1, through engineering a binding site near the center of the bundle by replacing a Leu by an Ala residue [66]. The two Mn(II) ions ligated by Glu and His residues with a DMSO molecule, from the crystallization buffer, appear to form a mono-$\mu$-oxo bridge with the dimer. The Mn-Mn distances range from 3.6 Å to 3.9 Å. When the binding site was further expanded by introducing a Gly in the place of Leu, two Mn ions appeared to be coordinated to two terminal water molecules or bridged by a single water molecule (Fig. 7.4) [67].

### 7.3.2.3 Photocatalyst: Photoactive Peptides

BChl pigments are the photocatalyst in natural photosynthetic systems. Chlorophylls are chlorin tetrapyrroles with a fifth ring fused to one of pyrrole and Mg or Zn as their central metal. Even though all known eukaryotic photosynthetic organisms use Mg-Chl $a$ as their major pigment, some cyanobacteria are adapted to use other chlorophylls. For example, the major chlorophyll pigment in *Acaryochloris marina* and prochlorophytes are Chl $d$ and Chl $b$, respectively. Some species of the purple photosynthetic bacteria, which are found in highly acidic environments, use Zn-Chls instead [68]. This functional flexibility of the natural system in employing various chlorophylls even in the RC as the primary electron donors results from the fact that they all share the common chlorin ring. The central metal ligation appears to be a decisive factor for the natural system choosing between a Chl or a Pheo molecule [10]. This implies that the binding site is actually tailored for Chl, with the entire ring and the phytyl tail buried inside the protein making contact with residues in distant helices. This complicated binding mode makes the design of small proteins with such a binding site very challenging.

Binding studies of Chl molecules to peptides require the presence of detergent to increase the solubility of Chl in aqueous solutions when usually larger complexes resembling light-harvesting (LH) complexes are formed. Such artificial LH complexes were created from synthetic $\alpha$-helical peptides (14 residues long) containing a single His residue and BChl or Zn-BChl molecules. These peptides are water-soluble and exist dominantly as four-$\alpha$-helical structures; upon interaction with Chl molecules, however, they assemble into much larger complexes. The binding experiments were performed in the presence of $N$-octyl-$\beta$-D-glycopyranoside (OG), and the complex formation was verified by a large red-shift at the $Q_y$ absorption band (777 → 863 nm) of Chl [69]. This is because the interaction of the Chl ring with peptide residues or another Chl causes changes in spectral features such as linewidth or peak position. The optical spectrum of Chl has two major absorption bands: Soret in the blue or near-UV and $Q_y$ in the red and near-IR region. Based on the four-orbital model, four $\pi \rightarrow \pi^*$ electronic transitions ($a_{2u}$, $a_{1u}$, $e_{gx}$, and $e_{gy}$) are predicted to give rise to the UV-Vis absorption spectrum of metallochlorins [70, 71]. The Soret band components, $B_x$ and $B_y$, are very intense and are theoretically very close in energy as they have not been resolved at room temperatures. On the other hand, the visible bands, weak $Q_x$ and strong $Q_y$, are well separated. A pigment-peptide complex containing a Chl dimer was constructed by adding Chl dissolved in THF to the peptide containing a His residue [72]. The binding was completed after 22 h and was shown to be His-dependent, as the control peptide appears to form aggregates with Chls.

Segments of the polypeptides (up to 31 residues long) of the core LH complex were chemically synthetized to determine structural features required for making natural-like subunit complexes [73]. The segments, which are membrane-spanning peptides, carry a His residue for BChl binding, which was performed in the presence of 1% OG. Chl interaction with a peptide (16 residues long) containing a conserved motif in protein of LH complexes was also studied in a buffering system containing detergents above the critical micelle concentrations [74]. In this work, fluorescence spectroscopy was applied to determine the specific binding of Chl to the peptide. The specific binding to the peptide was achieved only when 1 mM DM (CMC 0.17 mM) was included in the buffering system. It was shown that 30 mM OG (CMC 25 mM) was not sufficiently nonpolar to prevent Chl aggregation with the peptide lacking ligation residues (His or Glu). The peptide was denatured prior to binding studies. The peptides used in these studies were membrane-spanning helices, which are different from those found on the surface of a globular protein. These membrane helices have a high proportion of amino acids and hydrophobic side chains located throughout the entire structure, which are likely to be partitioned into micelles especially when associated with Chls.

Studies of Chl binding can also be done with some relatively simple natural proteins to gain insight into the mode of Chl-protein interaction and to construct an RC in these proteins by introducing electron acceptors and donors. The water-soluble proteins that can bind one single Chl molecule per dimer have been shown to be useful for such studies [75, 76]. The water-soluble Chl-binding (WSCP) protein from cauliflower exists in tetrameric form only in the presence of Chl, indicating that the phytyl tail is shielded from the solvent when oligomerization occurs. Other Chl derivatives such as Zn-pheophorbide *a* or chlorophyllide *a* and *b* can bind to the WSCP protein but do not make oligomers. Binding appears to be driven by axial ligation since Pheo cannot bind to this protein. The artificial RC design can take advantage of these Chl derivatives, which can be made by enzymatic reactions or organic chemistry removing the phytyl tail while maintaining the chlorin ring. Some examples of these compounds are chlorin *e*6 (C*e*6), pyropheophorbide, and tetracarboxyphenylchlorin.

These compounds, like their parental molecule of Chl, are photoactive but offer more desirable physicochemical properties. They are more stable than Chls, but, more importantly, their state of aggregation can be controlled to allow their specific binding to engineered binding sites in a protein. In these cases, binding modes other than His ligation become feasible. For example, using a chemoselective method, a Zn-pheophorbide *b* derivative was covalently attached to a modified lysine residue on a synthetic four-helix bundle protein through an aldehyde group [77]. Similarly, it was also shown that light-induced electron transfer could occur between a chromophore bound in the interior of a synthetic four-helix bundle protein and exogenously added anthraquinone [46].

The first example of an artificial BRC was reported in Dutton's group [19]. The design was based on covalent attachment of two coproporphyrins at the N-termini of a four-$\alpha$-helix bundle forming a pair with four hemes placed inside the hydrophobic core of the bundle. The four-$\alpha$-helix bundle protein is a self-assembled dimer of helix-loop-helix structures each formed by two peptides (CGGG ELWKLHE ELLKKFE ELLKLHE RRLKKL-CoNH$_2$) linked through a disulfide bond in the loop region. The dimeric porphyrin appears to be positioned on the exterior of the helix bundle, allowing *bis*-His-ligated hemes to occupy the interior. To make these peptides photoactive, Zn(II) derivative porphyrin was used, or, as an alternate approach, a Ru(II) porphyrin was incorporated into one of the heme-binding pockets via His ligation. In the first design, the photoactive moiety is solvent-exposed, while in the second, one is placed in the interior. This shows a major advantage for the designed peptides, as the same basic design can be used for different modes of ligation. The importance of the distance between heme and photoactive electron donor was analyzed in a series of designs, and the creation of a successful charge separation was predicted by taking into account a second redox cofactor, such as quinone, in the system. The simplified RC versions capable of doing charge separation are then intended to include a photocatalyst and a properly positioned electron donor and electron acceptor. To further develop this design, we have recently reported a His-dependent binding of Zn-C*e*6 to such synthetic four-helix bundle proteins [78]. By preparing two variants of the peptide lacking either one or both of the His residues, we showed that each monomer can bind two Zn-C*e*6 molecules. Light-induced ET between Zn-C*e*6 and a quinone as the external electron acceptor was established and measured using time-resolved EPR spectroscopy.

## 7.4
**Perspective**

Protein design has now become a powerful tool to create synthetic peptides with catalytic activities by incorporating active pigments and redox cofactors. As this is a constructive approach, the biological complexity can be gradually introduced into the design to make functional complexes. The creation of these complexes not only provides valuable information on how the protein environment controls the redox activity, but also gives an opportunity to construct membrane assemblies as mimics of natural systems. Progress toward the construction of synthetic reaction centers is a step forward in creating artificial photosynthetic membranes.

## Acknowledgments

This work was supported by an ARC Discovery Grant (DP0450421). We thank P. L. Dutton for encouragement and support.

## References

1 Deisenhofer, J., Epp, O., Miki, K., Huber, R., Michel, H. (1985). Structure of the protein subunits in the photosynthetic reaction center of *Rhodopseudomonas viridis* at 3 Å resolution. *Nature* 318, 618–624.

2 Kamiya, N., Shen, J.-R. (2003). Crystal structure of oxygen-evolving photosystem II from *Thermosynechococcus vulcanus* at 3.7-Å resolution. *Proc. Natl. Acad. Sci. USA* 100, 98–103.

3 Zouni, A., Witt, H.-T., Kern, J., Fromme, P., Kraub, N., Saenger, W., Orth, P. (2001). Crystal structure of photosystem II from *Synechococcus elongatus* at 3.8 Å resolution. *Nature* 409, 739–743.

4 Ferreira, K. N., Iverson, T. M., Maghlaoui, K., Barber, J., Iwata, S. (2004). Architecture of the photosynthetic oxygen-evolving center. *Science*, in press.

5 Ben-Shem, A., Frolow, F., Nelson, N. (2003). Crystal structure of plant photosystem I. *Nature* 426.

6 Jordan, P., Fromme, P., Witt, H. T., Klukas, O., Saenger, W., Kraub, N. (2001). Three-dimensional structure of cyanobacterial photosystem I at 2.5 Å resolution. *Nature* 411, 909–917.

7 Page, C. C., Moser, C. C., Chen, X., Dutton, P. L. (1999). Natural engineering principles of electron tunnelling in biological oxidation-reduction. *Nature* 402, 47–52.

8 Razeghifard, M. R., Klughammer, C., Pace, R. J. (1997). Electron paramagnetic resonance kinetic studies of the S-states in spinach thylakoids. *Biochemistry* 36, 86–92.

9 Krabben, L., Schlodder, E., Jordan, R., Carbonera, D., Giacometti, G., Lee, H., Webber, A. N., Lubitz, W. (2000). Influence of the axial ligands on the spectral properties of P700 of photosystem I: A study of site-directed mutants. *Biochemistry* 39, 13012–13025.

10 Chirino, A. J., Lous, E. J., Huber, M., Allen, J. P., Schenck, C. C., Paddock, M. L., Feher, G., Rees, D. C. (1994). Crystallographic analyses of site-directed mutants of the photosynthetic reaction center from *rhodobacter sphaeroides*. *Biochemistry* 33, 4584–4593.

11 Lin, X., Murchison, H. A., Nagarajan, V., Parson, W. W., Allen, J. P., Williams, J. C. (1994). Specific alteration of the oxidation potential of the electron-donor in reaction centers from *Rhodobacter sphaeroides*. *Proc. Natl. Acad. Sci. USA* 91, 10265–10269.

12 Ivancich, A., Artz, K., Williams, J. C., Allen, J. P., Mattioli, T. A. (1998). Effects of hydrogen bonds on the redox potential and electronic structure of the bacterial primary electron donor. *Biochemistry* 37, 11812–11820.

13 Kálmán, L., LoBrutto, R., Allen, J. P., Williams, J. C. (1999). Modified reaction centers oxidize tyrosine in reactions that mirror photosystem II. *Nature* 402, 696–699.

14 Kálmán, L., LoBrutto, R., Allen, J. P., Williams, J. C. (2003). Manganese oxidation by modified reaction centers from *Rhodobacter sphaeroides*. *Biochemistry* 42, 11016–11022.

15 Haffa, A. L. M., Lin, S., Katilius, E., Williams, J. C., Taguchi, A. K. W., Allen, J. P., Woodbury, N. W. (2002). The dependence of the initial electron transfer rate on driving force in *Rhodobacter sphaeroides* reaction center. *J. Phys. Chem. B* 106, 7376–7384.

16 Johnson, E. T., Parson, W. W. (2002). Electrostatic interactions in a integral membrane protein. *Biochemistry* 41, 6483–6494.

17 Semenov, A. Y., Vassiliev, I. R., van der Est, A., Mamedov, M. D., Zybailov, B., Shen, G. Z., Stehlik, D., Diner, B. A., Chitnis, P. R., Golbeck, J. H. (2000). Recruitment of a foreign quinone into the A(1) site of photosystem I – Altered kinetics of electron transfer in phylloquinone biosynthetic pathway mutants studied by time-resolved optical, EPR, and electrometric techniques. *J. Biol. Chem.* 275, 23429–23438.

18 Hu, Q., Miyashita, H., Iwasaki, I., Kurano, N., Miyachi, S., Iwaki, M. & Itoh, S. (1998). A photosystem I reaction center driven by chlorophyll d in oxygenic photosynthesis. *Proc. Natl. Acad. Sci. USA* 95, 13319–13323.

19 Rabanal, F., Gibney, B. R., DeGrado, W. F., Moser, C. C., Dutton, P. L. (1996). Engineering photosynthesis: synthetic redox proteins. *Inorg. Chim. Acta* 243, 213–218.

20 Alegria, G., Dutton, P. L. (1991). Langmuir-Blodgett monolayer films of bacterial photosynthetic membranes and isolated reaction centers. *Biochim. Biophys. Acta* 1057, 239–257.

21 Packham, N. K., Mueller, P., Dutton, P. L. (1988). Photoelectric currents across planar bilayer membranes containing bacterial reaction centers: the response under conditions of multiple reaction-center turnovers. *Biochim. Biophys. Acta* 933, 70–84.

22 Tiede, D. M., Mueller, P., Dutton, P. L. (1982). Spectrophotometirc and voltage clamp characterization of monolayers of bacterial photosynthetic reaction centers. *Biochim. Biophys. Acta* 681, 191–201.

23 Hara, M., Majima, T., Ajiki, S., Sugino, H., Toyotama, H., Ueno, T., Asada, Y., Miyake, J. (1996). Multilayer preparation of bacterial photosynthetic membrane with a certain orientation immobilized on the solid surface by avidin-biotin interaction. *Bioelectrochem. Bioenerg.* 41, 127–129.

24 Zhao, J., Zou, Y., Liu, B., Xu, C., Kong, J. (2002). Differentiating the orientations of photosynthetic reaction centers on Au electrodes linked by different bifunctional reagents. *Biosen. Bioelec.* 17, 711–718.

25 Gust, D., Moore, T. A. (1989). Mimicking photosynthesis. *Science* 244, 35–41.

26 SteinbergYfrach, G., Liddell, P. A., Hung, S. C., Moore, A. L., Gust, D., Moore, T. A. (1997). Conversion of light energy to proton potential in liposomes by artificial photosynthetic reaction centers. *Nature* 385, 239–241.

27 Liddell, P. A., Kodis, G., de la Garza, L., Bahr, J. L., Moore, A. L., Moore, T. A., Gust, D. (2001). Photoinduced electron transfer in tetrathiafulvalene-porphyrin-fullerene molecular triads. *Helvetica Chimica Acta* 84, 2765–2783.

28 Imahori, H., Guldi, † D. M., Tamaki, K., Yoshida, Y., Luo, C., Sakata, Y., Fukuzumi, S. (2001). Charge separation in a novel artificial photosynthetic reaction center lives 380 ms. *J. Am. Chem. Soc.* 123, 6617–6628.

29 Hu, Y. Z., Takashima, H., Tsukiji, S., Shinkai, S., Nagamune, T., Oishi, S., Hamachi, I. (2000). Direct comparison of electron transfer properties of two distinct semisynthetic triads with non-protein based triad: Unambiguous experimental evidences on protein matrix effects. *Chem. Eur. J.* 6, 1907–1916.

30 Hu, Y. Z., Tsukiji, S., Shinkai, S., Oishi, S., Hamachi, I. (2000). Construction of artificial photosynthetic reaction centers on a protein surface: vectorial, multistep, and proton-coupled electron transfer for long-lived charge separation. *J. Am. Chem. Soc.* 122, 241–253.

31 Robertson, D., Farid, R., Moser, C., Urbauer, J., Mulholland, S., Pidikiti, R., Lear, J., Wand, A., DeGrado, W., Dutton, P. (1994). Design and synthesis of multi-Haem proteins. *Nature*, 425–431.

32 DeGrado, W. F., Summa, C. M., Pavone, V., Nastri, F., Lombardi, A. (1999). *De novo* design and structural character-

isation of proteins and metalloproteins. *Annu. Rev. Biochem.* 68, 779–819.

33  Skalicky, J. J., Gibney, B. R., Rabanal, F., Bieber-Urbauer, R. J., Dutton, P. L., Wand, A. J. (1999). Solution structure of a designed four-α-helix bundle maquette scaffold. *J. Am. Chem. Soc.* 121, 4941–4951.

34  Huang, S. S., Gibney, B. R., Stayrook, S. E., Dutton, P. L., Lewis, M. (2003). X-ray structure of a maquette scaffold. *J. Mol. Biol.* 326, 1219–1225.

35  Rosenblatt, M. M., Wang, J., Suslick, K. S. (2003). *De novo* designed cyclic-peptide heme complexes. *Proc. Natl. Acad. Sci. USA* 100, 13140–13145.

36  Gibney, B. R., Rabanal, F., Skalicky, J. J., Wand, A. J., Dutton, L. P. (1999). Iterative protein design. *J. Am. Chem. Soc.* 121, 4952–4960.

37  Gibney, B. R., Dutton, P. L. (2001). *De novo* design and synthesis of heme proteins. *Adv. Inorg. Chem.* 51, 409–455.

38  Shifman, J. M., Gibney, B. R., Sharp, R. E., Dutton, P. L. (2000). Heme redox potential control in *de novo* designed four-αα-helix bundle proteins. *Biochemistry* 39, 14813–14821.

39  Jentzen, W., Song, X.-Z., Shelnutt, J. A. (1997). Structural characterization of synthetic and protein-bound porphyrins in terms of the lowest-frequency normal coordinates of the macrocycle. *J. Phys. Chem. B* 101, 1684–1699.

40  Jentzen, W., Ma, J.-G., Shelnutt, J. A. (1998). Conservation of the conformation of the porphyrin macrocycle in hemoproteins. *Biophys. J.* 74, 753–763.

41  Rau, H. K., DeJonge, N., Haehnel, W. (1998). Molecular synthesis of *de novo*-designed metalloproteins for light-induced electron transfer. *Proc. Natl. Acad. Sci. USA* 95, 11526–11531.

42  Mutz, M. W., McLendon, G. L., Wishart, J. F., Gaillard, E. R., Corin, A. F. (1996). Conformational dependence of electron transfer across *de novo* designed metalloproteins. *Proc. Natl. Acad. Sci. USA* 93, 9521–9526.

43  Kozlov, G. V., Ogawa, M. Y. (1997). Electron transfer across a peptide-peptide interface within a designed metalloprotein. *J. Am. Chem. Soc.* 119, 8377–8378.

44  Kornilova, A., Wishart, J., Xiao, W., Lasey, R., Fedorova, A., Shin, Y., Ogawa, M. (2000). Design and characterization of a synthetic electron-transfer protein. *J. Am. Chem. Soc.* 122, 7999–8006.

45  Sisido, M., Hoshino, S., Kusano, H., Kuragaki, M., Makino, M., Sasaki, H., Smith, T. A., Ghiggino, K. P. (2001). Distance dependence of photoinduced electron transfer along alpha-helical polypeptides. *J. Phys. Chem. B* 105, 10407–10415.

46  Fahnenschmidt, M., Bittl, R., Schlodder, E., Haehnel, W., Lubitz, W. (2001). Characterisation of *de novo* synthesized four-helix bundle proteins with metalloporphyrin cofactors. *Phys. Chem. Chem. Phys.* 3, 4082–4090.

47  Fox, M. A., Galoppini, E. (1997). Electric field effects on electron transfer rates in dichromophoric peptides: the effect of helix unfolding. *J. Am. Chem. Soc.* 119, 5277–5285.

48  Fedorova, A., Chaudhari, A., Ogawa, M. (2003). Photoinduced electron-transfer along alpha-helical and coiled-coil metallopeptides. *J. Am. Chem. Soc.* 125, 357–362.

49  Jones, G., Vullev, V., Braswell, E. H., Zhu, D. (2000). Multistep photoinduced electron transfer in a *de novo* helix bundle: multimer self-assembly of peptide chains including a chromophore special pair. *J. Am. Chem. Soc.* 122, 388–389.

50  Mutz, M. W., Case, M. A., Wishart, J. F., Ghadiri, M. R., McLendon, G. L. (1999). *De novo* design of protein function: predictable structure-function relationship in synthetic redox proteins. *J. Am. Chem. Soc.* 121, 858–859.

51  Sharp, R. E., Moser, C. C., Rabanal, F., Dutton, P. L. (1998). Design, synthesis, and characterization of a photoactivable flavocytochrome molecular maquette. *Proc. Natl. Acad. Sci. USA* 95, 10465–10470.

52  Pfeiffer, E., Mezler, M. (1996). Interaction of *p*-benzoquinone and *p*-biphenoquinone with microtubule proteins *in vitro*. *Chemico-Biological Interactions* 102, 37–53.

53  Stephens, P. J., Jollie, D. R., Warshel, A. (1996). Protein control of redox poten-

tials of iron-sulfur proteins. *Chem. Rev.* 96, 2491–2513.

54 Gibney, B. R., Mulholland, S. E., Rabanal, F., Dutton, P. L. (1996). Ferredoxin and ferredoxin-heme maquettes. *Proc. Natl. Acad. Sci. USA* 93, 15041–15046.

55 Mulholland, S. E., Gibney, B. R., Rabanal, F., Dutton, P. L. (1999). Determination of nonligand amino acids critical to [4Fe-4S]2+/+ assembly in ferredoxin maquettes. *Biochemistry* 38, 10442–10448.

56 Mulholland, S. E., Gibney, B. R., Rabanal, F., Dutton, P. L. (1998). Characterization of the fundamental protein ligand requirements of [4Fe-4S]2+/+ clusters with sixteen amino acid maquettes. *J. Am. Chem. Soc.* 120, 10296–10302.

57 Regan, L., DeGrado, W. F. (1988). Characterization of a helical protein designed from first principles. *Science* 241, 976–978.

58 Scott, M. P., Biggins, J. (1997). Introduction of a [4Fe-4S (S-cys)4]+1+2 iron-sulfur center into a four-α-helix protein using design parameters from the domain of the Fx cluster in the photosystem I reaction center. *Protein Sci.* 6, 340–346.

59 Laplaza, C. E., Holm, R. H. (2001). Helix-loop-helix peptides as scaffolds for the construction of bridged metal assemblies in proteins: the spectroscopic A-cluster structure in carbon monoxide dehydrogenase. *J. Am. Chem. Soc.* 123, 10255–10264.

60 Sow, T. C., Pedersen, M. V., Christensen, H. E. M., Ooi, B. L. (1996). Total synthesis of a miniferredoxin. *Biochem. Biophys. Res. Com.* 223, 360–364.

61 Coldren, C. D., Hellinga, H. W., Caradonna, J. P. (1997). The rational design and construction of a cuboidal iron-sulfur protein. *Proc. Natl. Acad. Sci. USA* 94, 6635–6640.

62 Benson, D. E., Wisz, M. S., Liu, W. T., Hellinga, H. W. (1998). Construction of a novel redox protein by rational design: conversion of a disulfide bridge into a mononuclear iron-sulfur center. *Biochemistry* 37, 7070–7076.

63 Gray, H. B., Malmstrom, B. G., Williams, R. J. P. (2000). Copper coordination in blue proteins. *J. Biol. Inorg. Chem.* 5, 551–559.

64 Karlin, S., Zhu, Z.-Y., Karlin, K. D. (1997). The extended environment of mononuclear metal centers in protein structures. *Proc. Natl. Acad. Sci. USA* 94, 14225–14230.

65 Daugherty, R. G., Wasowicz, T., Gibney, B. R., DeRose, V. J. (2002). Design and spectroscopic characterization of peptide models for the plastocyanin copper-binding Loop. *Inorg. Chem.* 41, 2623–2632.

66 Costanzo, L. D., Wade, H., Geremia, S., Randaccio, L., Pavone, V., DeGrado, W. F., Lombardi, A. (2001). Toward the *de novo* design of a catalytically active helix bundle: a substrate-accessible carboxylate-bridged dinuclear metal center. *J. Am. Chem. Soc.* 123, 12749–12757.

67 DeGrado, W. F., Costanzo, L. D., Geremia, S., Lombardi, A., Pavone, V., Randaccio, L. (2003). Sliding helix and change of coordination geometry in a model di-MnII protein. *Angew. Chem. Int. Ed.* 42, 417–420.

68 Wakao, N., Yokoi, N., Isoyama, N., Hiraishi, A., Shimada, K., Kobayashi, M., Kise, H., Iwaki, M., Itoh, S., Takaichi, S., Sakurai, Y. (1996). Discovery of natural photosynthesis using Zn-containing bacteriochlorophyll in an aerobic bacterium *Acidiphilium rubrum*. *Plant Cell Physiol.* 37, 889–893.

69 Kashiwada, A., Nishino, N., Wang, Z.-Y., Nozawa, T., Kobayashi, M., Nango, M. (1999). Molecular assembly of bacteriochlorophyll *a* and its analogues by synthetic 4α-helix polypeptides. *Chem. Lett.* 2, 1301–1302.

70 Hartwich, G., Fiedor, L., Simonin, I., Cmiel, E., Schäfer, W., Noy, D., Scherz, A., Scheer, H. (1998). Metal-substituted bacteriochlorophylls. 1. Preparation and influence of metal and coordination on spectra. *J. Am. Chem. Soc.* 120, 3675–3683.

71 Singh, A., Huang, W.-Y., Egbujor, R., Johnson, L. W. (2001). Single site electronic spectroscopy of zinc and magnesium chlorin in *n*-Octane matrixes at 7 K. *J. Phys. Chem. A* 105, 5778–5784.

72 Dudkowiak, A., Kusumi, T., Nakamura, C., Miyake, J. (1999). Chlorophyll *a* aggregates stabilized by a synthesized peptide. *J. Photochem. Photobiol. B: Biol.* 129, 51–55.

73 Meadows, K. A., Parkes-Loach, P. S., Kehoe, J. W., Loach, P. A. (1998). Reconstitution of core light-harvesting complexes of photosynthetic bacteria using chemically synthesized polypeptides. 1. Minimal requirements for subunit formation. *Biochemistry* 37, 3411–3417.

74 Eggink, L. L., Hoober, J. K. (2000). Chlorophyll binding to peptide maquettes containing a retention motif. *J. Biol. Chem.* 275, 9087–9090.

75 Schmidt, K., Fufezan, C., Krieger-Liszkay, A., Satoh, H., Paulsen, H. (2003). Recombinant water-soluble chlorophyll protein from *Brassica oleracea* Var. botrys binds various chlorophyll derivatives. *Biochemistry* 42, 7427–7433.

76 Satoh, H., Nakayama, K., Okada, M. (1998). Molecular cloning and functional expression of a water-soluble chlorophyll protein, a putative carrier of chlorophyll molecules in cauliflower. *J. Biol. Chem.* 273, 30568–30575.

77 Rau, H. K., Snigula, H., Struck, A., Robert, B., Scheer, H., Haehnel, W. (2001). Design, synthesis and properties of synthetic chlorophyll proteins. *Eur. J. Biochem.* 268, 3284–3295.

78 Razeghifard, M. R., Wydrzynski, T. J. (2003). Binding of Zn-chlorin to a synthetic four-helix bundle peptide through histidine ligation. *Biochemistry* 42, 1024–1030.

79 Guex, N., Peitsch, M. C. (1997). SWISS-MODEL and the Swiss-PdbViewer: An environment for comparative protein modeling. *Electrophoresis* 18, 2714–2723.

80 Deisenhofer, J., Epp, O., Sinning, I., Michel, H. (1995). Crystallographic refinement at 2.3 Å resolution and refined model of the photosynthetic reaction center from *Rhodopseudomonas viridis*. *J. Mol. Biol.* 246, 429–457.

81 Antonkine, M. L., Liu, G., Bentrop, D., Bryant, D. A., Bertini, I., Luchinat, C., Golbeck, J. H., Stehlik, D. (2002). Solution structure of the unbound, oxidized photosystem I subunit PsaC, containing [4Fe-4S] clusters FA and FB: a conformational change occurs upon binding to photosystem I. *J Biol. Inorg. Chem.* 7, 461–472.

82 Xue, Y., Okvist, M., Hansson, O., Young, S. (1998). Crystal structure of spinach plastocyanin at 1.7 Å resolution. *Protein Sci.* 7, 2099–2105.

# 8
# Novel Geometry Polynorbornane Scaffolds for Chromophore Linkage and Spacing

*Ronald N. Warrener, Davor Margetic, David A. Mann, Zhi-Long Chen, and Douglas N. Butler*

## 8.1
## Introduction

Studies of natural photosynthesis (NP) have continued to provide the informational and inspirational basis for the development of artificial systems for photon capture required for clean energy production, e.g., artificial photosynthesis (AP). In particular, X-ray structural data, such as that for the bacterial photosynthetic reaction center [1], have stimulated much activity in the development of artificial photosynthetic systems in the last decade. The synthesis of donor-linker-acceptor molecules joined by conformationally flexible alkyl chains to mimic the light-harvesting capacity of NP have played an important role in the early understanding of photoinduced charge separation, electron transfer, and energy transfer processes that are at the heart of AP systems [2]. Incorporation of a rigid frame into such molecules to form donor-spacer-acceptor molecules has allowed the study of more precise effects involving the distance between the chromophores as well as their fixed geometric interrelationships. Our own interest in this field has been two-pronged, dealing first with the development of new synthetic methodology to prepare chromophore-spacer-chromophore models (abbreviated as **CR-n$\sigma$-CR**, where CR stands for the generic chromophore and $n$ is the number of $\sigma$-bonds in the rigid frame separating the chromophores) [3] and subsequently with their use in photophysical studies of electron transfer [4].

At the synthetic level, the development of the ACE coupling reaction (acronym: **A**=alkene; **CE**=cyclobutene epoxide) [5–9] has played an integral role in the formation of polynorbornane scaffolds. The ACE reaction, first reported from our laboratory in 1997, employed the in situ generation of 1,3-dipoles, e.g., **2**, by thermal ring-opening of ester-activated cyclobutene epoxides **1** and their trapping with alkene dipolarophiles to form cycloadducts **3** (Scheme 8.1). Carboxylic acids and carboxamides can replace the ester-substituents, the former providing access to water-soluble products [5j].

*Artificial Photosynthesis: From Basic Biology to Industrial Application*
Edited by Anthony F. Collings and Christa Critchley
Copyright © 2005 WILEY-VCH Verlag GmbH & Co. KGaA, Weinheim
ISBN: 3-527-31090-8

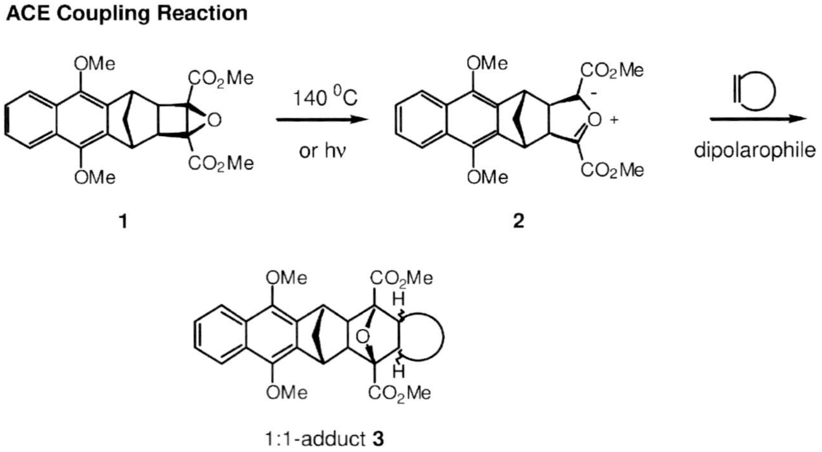

## ACE Coupling Reaction

**Scheme 8.1** The ACE reaction involving the cycloaddition of an alkene with the 1,3-dipole **2** formed by ring opening of cyclobutene epoxide **1**

The ACE reactions reported to date have predominantly used symmetrical BLOCK components in the coupling process, thereby leading to adducts that retained a $C_s$-symmetry plane. This has meant that the individual chromophores in the resultant **CR-spacer-CR** were either tilted away from coplanar (type I) or orthogonally related (type II) (Scheme 8.2). Other research groups have used alternative scaffolds to space-separate chromophores, but again most are of type 1 and were characterized by having $C_s$ symmetry [10].

The stereoselectivity of the ACE coupling reaction can vary, depending on the type of alkene employed, e.g., cyclic alkenes such as acenaphthylene **4** gave a mixture of the *exo-* and *endo*-isomers **5a,b** (Scheme 8.2a), whereas reaction of 1,4-benzoquinone **6** with CE-**1** gave exclusively the *exo*-adduct **7** (Scheme 8.2b). Further, alicyclic alkenes such as norbornadiene **8** (Scheme 8.2c) and the fused cyclobutene diester **11** (Scheme 8.2d) were both active dipolarophiles and produced fused products **9** and **12**, respectively, with complete stereoselectivity. Depending on the stereochemistry required in the dyad, it is possible to use these reaction stereoselectivities to provide a guide to achieve specific architectural outcomes.

Norbornanes were also found to undergo cycloaddition with complete *exo,exo*-diastereoselectivity, e.g., formation of 2:1 adduct **10** on further reaction of **9** with CE-**1** (Scheme 8.2c). Functionalized norbornanes have played a prominent role as the 2-BLOCK in our ACE coupling synthetic strategies. In this way, it was possible to transfer local stereochemistry or chirality present in the BLOCK reagent with integrity to the coupled product (see below).

The stereoselectivity of these reactions governed the topology of the polynorbornane frame, and by using $C_s$-symmetrical reagents, one carrying the donor (**D**) and the other the acceptor (**A**), it was possible to form rigid-framed

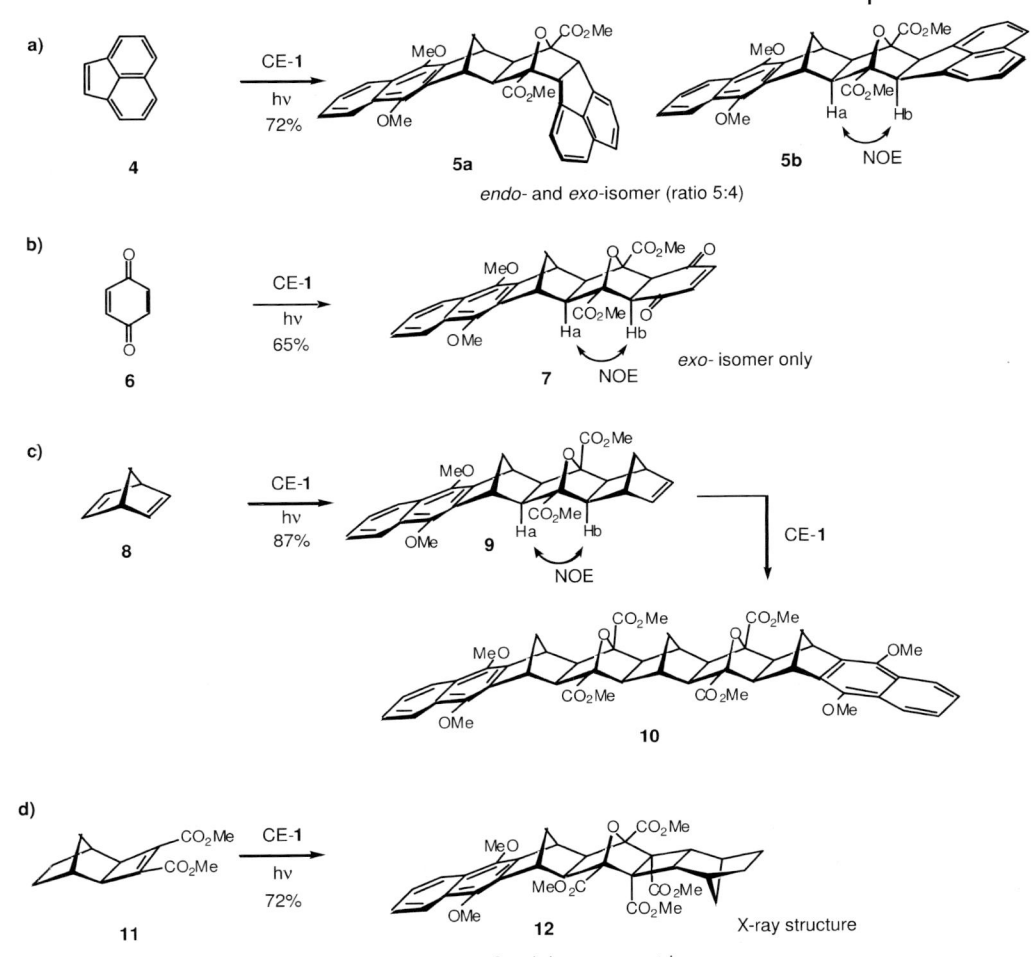

**Scheme 8.2** Representative ACE coupling reactions of cyclobutene epoxide **1** with cyclic alkenes with capacity to form type I dyads [5k]

D-spacer-A dyads of type I and type II and preserve the $C_s$-symmetry regime (Fig. 8.1). This strategy has been applied to the synthesis of dyads such as **POR-n-POR** (**POR**=porphyrin) [3, 5f], **POR-nσ-DMN** dyads (**DMN**=1,4-dimethoxy naphthalene) [7], and **MPN-n-MPN** dyads (**MPN**=metal-complexed phenanthroline) [8].

The question we addressed in the present study was how to develop an ancillary method of general application that could be used to form type III scaffolds in which one of the chromophores was no longer symmetrically positioned about the σ-plane of the scaffold. One recent solution by our group was afforded by the surprising reaction of CE-1 at the $\Delta^{1,2}$ π-bond of anthracene **13** (naphthalene and many other polycyclic aromatic hydrocarbons reacted similarly) to form

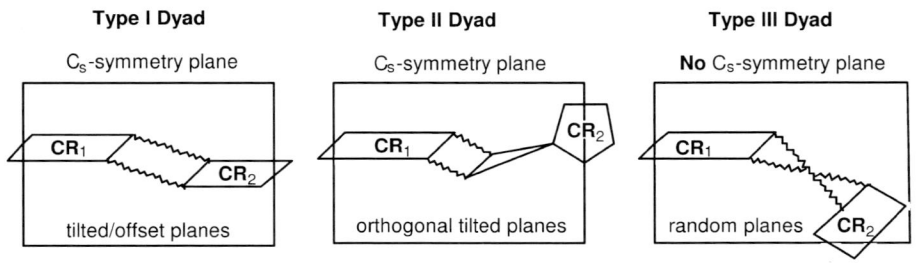

# 8 Novel Geometry Polynorbornane Scaffolds for Chromophore Linkage and Spacing

**Type I Dyad**
C_s-symmetry plane
CR_1 / CR_2
tilted/offset planes

**Type II Dyad**
C_s-symmetry plane
CR_1 / CR_2
orthogonal tilted planes

**Type III Dyad**
**No** C_s-symmetry plane
CR_1 / CR_2
random planes
(can also be spiro-linked, see text)

CR=unspecified chromophore

**Fig. 8.1** Schematic representation of the three types of rigid-frame (scaffold), chromophore-spacer-chromophore dyads: type I (parallel, tilted, or offset CR planes; $C_s$ symmetry), type II (orthogonal CR planes, $C_s$ symmetry), and type III (no symmetry).

adduct **14**, in which the anthracene chromophore was restored by dehydrogenation with DDQ to form *unsym*-**DMN**-4(4)σ-**ANTH** dyad **15** (Scheme 8.3) (see Section 8.5 for discussion on dyad nomenclature) [11]. This result is a variant of the type 1 protocol and falls into the type III category only because the site of attachment to the anthracene is offset; the reaction with acenaphthylene is at a symmetrically positioned π-bond and the product is of type 1 (Scheme 8.2a).

**13** → CE-1, 140 °C, 62% → **14** (exo- and endo- isomers, endo-favoured 5.5:1) → DDQ → **15** (reference 10)

**Scheme 8.3** ACE route to *unsym*-**DMN**-4σ-**ANTH** dyad **15** using anthracene as a dipolarophile

In the present study we report the use of carbonyl dipolarophiles with cyclobutene epoxides to form adducts which, by the unsymmetrical nature of the CO cycloaddend, must yield unsymmetrically aligned type III dyads. This new reaction provides access to systems with new interchromophoric alignments of a type not previously described.

This new reaction has been given the acronym COCE (**CO**=carbonyl; **CE**= cyclobutene epoxide) as a shorthand way to describe the cycloaddition of a carbonyl group to cyclobutene epoxide-derived dipoles such as **2**. In this way the relationship with the ACE reaction is preserved (see above).

## 8.2
Results and Discussion

### 8.2.1
Reaction at Carbonyl Groups to Form Unsymmetrical Type III Dyads

The reactions of CE-1 with benzaldehyde **16** produced a single 1:1 adduct. NMR spectral data were in accord with structure **17** in which the phenyl substituent was positioned in the *exo* position. The stereochemistry was confirmed by NOE measurements showing that the newly introduced proton Hd derived from the aldehyde was *syn*-related space-wise to protons Ha, Hb on the rack. Such an interrelationship confirmed that all three protons had the same *endo* geometry and fully defined the stereochemistry of **17**. The other NOE observed between Hd and aryl protons Hc was in keeping with the structure but not stereo-defining (Scheme 8.4).

A similar reaction using naphthalene-2-aldehyde **18** as the CO component also gave a single adduct **19** in which the naphthalene ring was again *exo*-fused, thereby confirming the *exo* stereoselectivity of the aldehyde reaction.

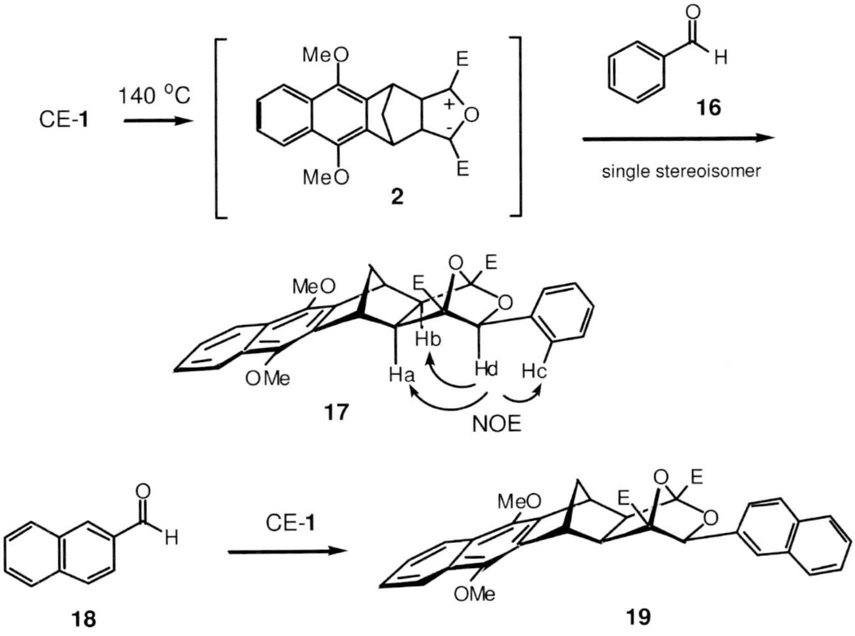

*unsym,sigma*-**DMN**-4(5),1σ-**NAPH** dyad

**Scheme 8.4** The COCE reaction of benzaldehyde **16** and 2-naphthaldehyde **18** with the 1,3-dipole **2** formed by the thermal ring opening of the ester-activated cyclobutene epoxide **1**

Clearly this approach offers a stereochemically defined way to form *unsym,sigma*-**DMN**-5(6)σ-**CR** dyads (**CR**=unspecified chromophore) in which one of the chromophores is attached to the rigid scaffold by a single σ-bond.

These results were the first examples of a carbonyl group acting as a 2π-reagent with the dipolar species generated from ring opening of a cyclobutene epoxide, although carbonyl groups have been reported to act as dienophiles [12] and dipolarophiles [13] in other [4π+2π] reactions.

The COCE reaction was not limited to aldehydes, as some ketones have been found to react, e.g., aromatic and heterocyclic compounds containing a carbonyl group as part of a ring; however, aliphatic ketones (acetone, cyclohexanone) or non-cyclic aromatic ketones (benzophenone) do not participate. The carbonyl groups of amides, esters, and imides are inactive, while the cyclic carbonyl group of camphor is also unreactive, presumably owing to steric crowding.

Reaction of fluorenone **20** with CE-1 formed the *unsym,spiro*-**FLU**-4(5),1σ-**DMN** dyad **21** (**FLU**=fluorene) (Scheme 8.5 a). Symmetry played an interesting role in this cycloaddition, and while both reagents have $C_s$ symmetry, the transition state geometry, and hence the structure of the product, was unsymmetrical. In the product, the local $C_s$ symmetry of the spiro substituents originating from the carbonyl reagent ensured that no *exo,endo* stereoisomers were possible.

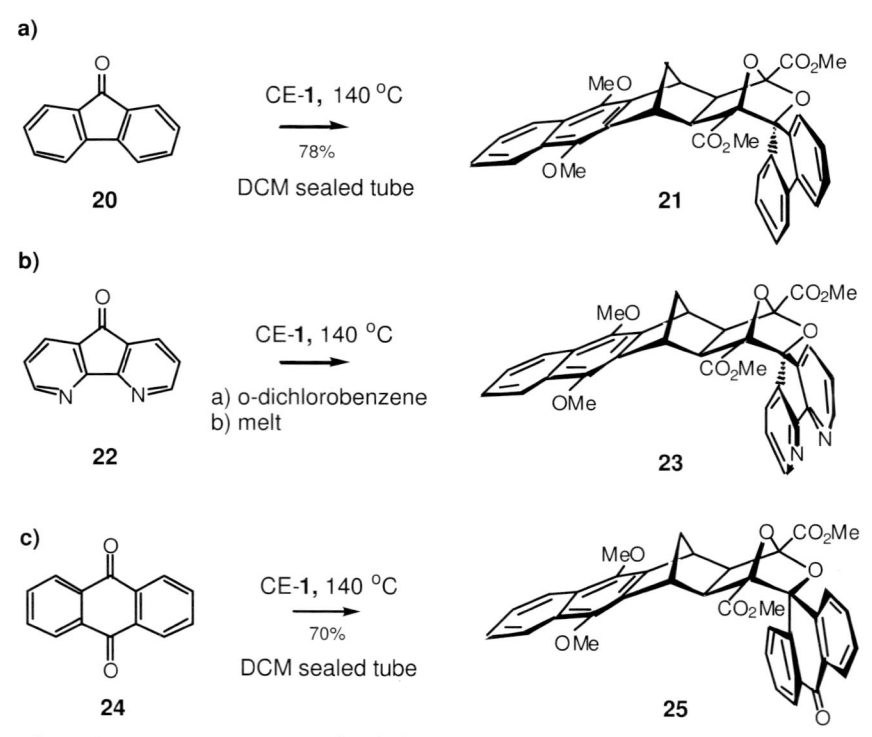

**Scheme 8.5** New COCE reactions of cyclic ketones or quinones with CE-1

This means of producing a rigid **CR-nσ-CR** dyad containing an aromatic chromophore fused to the central scaffold in a geometrically fixed, off-centered orientation was used for the preparation of the *unsym,spiro*-**DAF**-4(5),1σ-**DMN** dyad **23** (DAF = 4,5-diazafluorene). In this case, the 4,5-diazafluorenone **22** delivered the bidentate ligand moiety and the cyclobutene epoxide reagent CE-1 was the source of the 1,4-dimethoxynaphthalene chromophore (Scheme 8.5b).

In view of the fact that anthracene reacted as a dipolarophile with CE-1 to form rigid scaffold **14** (Scheme 8.3), the question that arose when employing 9,10-anthraquinone **24** was whether it would act as an aromatic C=C or a C=O dipolarophile. Heating 9,10-anthraquinone **24** with CE-1 at 140 °C in methylene chloride in a sealed tube led to the formation of a single product that was shown by mass spectrometry to be a 1:1 adduct ($m/z$ = 618.1900) (Scheme 8.5c). This product, assigned structure **25**, was unsymmetrical and displayed four methoxyl resonances ($\delta$ 3.39, 3.58, 3.83, 3.99) in the $^1$H NMR spectrum corresponding to the methoxy and ester methyl groups, and a count of the aromatic protons (12H) demonstrated that reaction had not occurred at the aromatic ring. The structural assignment was further supported by the presence of three $^{13}$C-NMR carbonyl resonances (two ester carbonyls plus one carbonyl at $\delta$ 164.9, 165.3, and 184.2, respectively) rather than four if cycloaddition had occurred at a C=C bond.

From the viewpoint of type III dyad construction, the reaction of coumarin **26** with CE-1 turned out to be a win-win situation since attack at either π-center formed an unsymmetrical product. In fact, a single adduct **27** was produced by site-selective attack at the C=C bond rather than at the lactone C=O (Scheme 8.6). The *endo* stereochemistry assigned to **27** was supported by the upfield shift observed in the $^1$H NMR spectrum of Ha relative to its cross-frame partner Hb ($\delta$ 2.18, 2.46, $J$ = 7.1 Hz) and was attributed to the ring current of the proximate

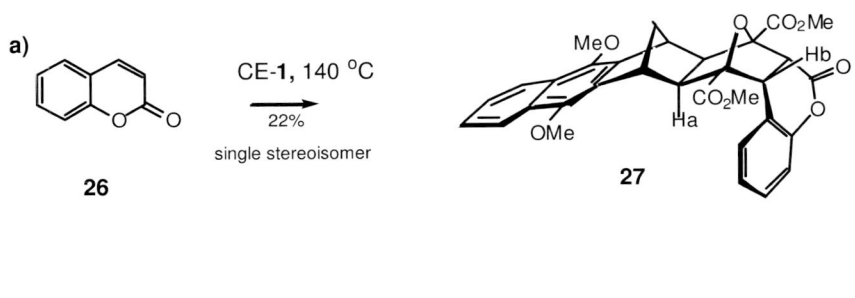

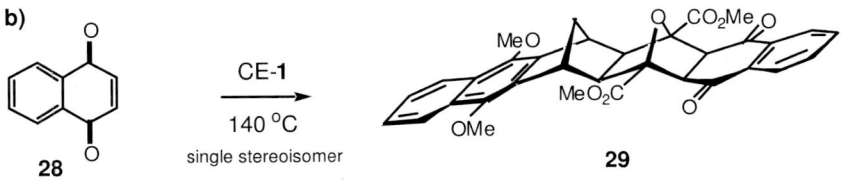

**Scheme 8.6** Site selectivity of C=C and C=O containing 2π-reagents

benzenoid ring in that geometry and the lack of an NOE between Ha and Hb in the NOESY spectrum.

Reaction of CE-1 with 1,4-naphthoquinone **28** exhibited the same site selectively at the C=C $\pi$-bond to form adduct **29** (Scheme 8.6b) and in this respect followed the pattern exhibited by 1,4-benzoquinone (Scheme 8.1b).

### 8.2.2
### Extended-frame Dyads

Starting with the known molrac **30** [14], it was possible to form the cyclobutene-1,2-diester **31** using the Ru-catalyzed cycloaddition of dimethyl acetylenedicarboxylate reported by Mitsudo et al. [15]. Formation of the stretched cyclobutene epoxide **32** was achieved by nucleophilic epoxidation (tBuOOH, tBuO$^-$) of **31** as described by us earlier [5]. Reaction of this stretched cyclobutene epoxide **32** with 9-fluorenone **20** occurred at 140 °C to form the extended-frame type III dyad **33** (Scheme 8.7). This approach has the potential to be applied to higher

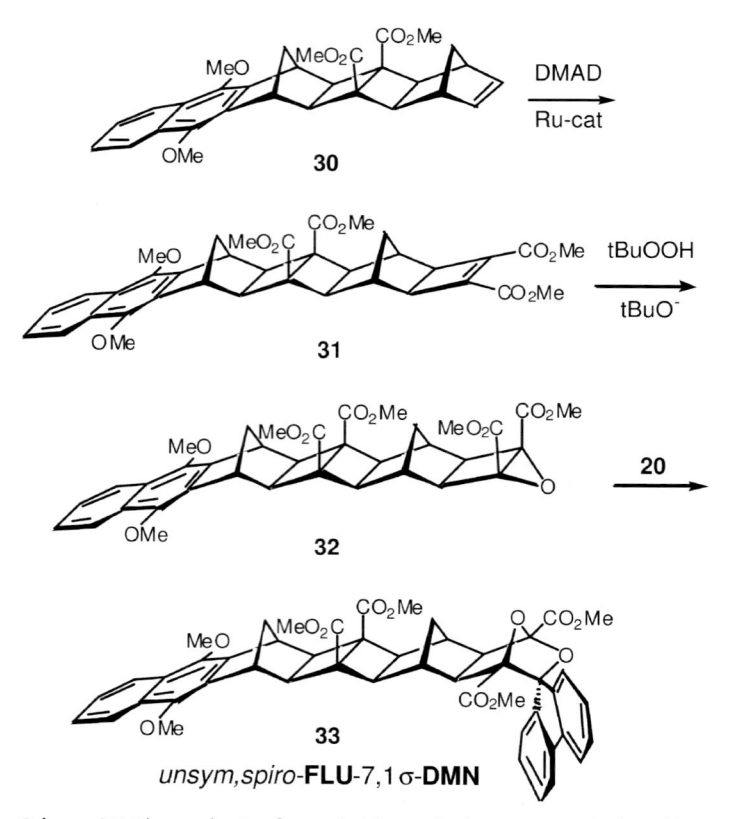

**Scheme 8.7** The synthesis of extended-frame dyads using stretched CE blocks

norbornylogues to produce giant type III dyads and allow their photophysics to be compared with the type I analogues reported by the Paddon-Row group [10].

## 8.3
## Preliminary Results

### 8.3.1
### The Use of Multicarbonyl Reagents for Dyad Formation

It has now been found that $\alpha$-diones are key CO reagents in the COCE reaction, provided that the CO groups are incorporated in a ring. Thus, phenanthroquinone **34**, but not benzil, reacted with CE-1 to produce a single 1:1 adduct, assigned the stereochemistry depicted in structure **36** on the basis of NOE measurement (Scheme 8.8a). This same approach served as an alternative way to introduce bidentate ligand capacity into the scaffold by reaction of CE-1 with 1,10-phenanthroline-5,6-dione **35** (Scheme 8.3a). The stereochemistry of the derived 1:1 adduct **37** was again based on NOE measurement between protons Ha, Hb on the underside of the scaffold with the aromatic protons Hc, Hd on the spiro-phenanthroline ring, a relationship available only to that geometry.

This dione variant of the COCE reaction has been used to make porphyrin-containing dyads. Thus, heating porphyrin dione **38** [16] with CE-1 formed a stereoisomeric mixture of 1:1 adducts **39** (Scheme 8.8b). No indication of 2:1 adducts was observed, and this has been attributed to the steric crowding around the remaining CO group in **39** (supported by molecular modeling). This reaction was a rare example where the stereoselectivity of the reaction was not specific and roughly equal amounts of each isomer were produced. The COCE reaction of porphyrin dione **38** with the 3,6-di(2-pyridyl)pyridazine (dpp)-containing cyclobutene epoxide **40** also produced a stereoisomeric mixture of adducts, indicating that the stereoselectivity is governed by the CO reagent (Scheme 8.8c). A direct consequence of the use of the ponderous dipolarophile **38** in COCE reactions was the fact that porphyrin dyads could now be designed with two different geometrical alignments. This outcome has design advantages, as the spiro-fused chromophore can be delivered by the CE reagent directly spiro-fused onto the modified porphyrin ring.

The reactions of anhydrous ninhydrin **42** represent a situation where there are three contiguous carbonyl groups and they fall into a central one with two equivalent flanking COs directly conjugated with the phenylene ring. Reaction with CE-1 occurred with the red-colored trione **42** or its hydrate to form a mixture of two 1:1 adducts **43** and **44** (Scheme 8.9). The minor adduct **44** was still colored and reacted with phenylene diamine **45** to form quinoxalines of type **46**; the major isomer **43** remained unchanged under these conditions. This difference in reactivity was used to assign structures to these isomers. No evidence for further reaction between CE-1 and **43** was observed, and again this was attributed to steric screening by the spiro-fused scaffold to approach to either CO

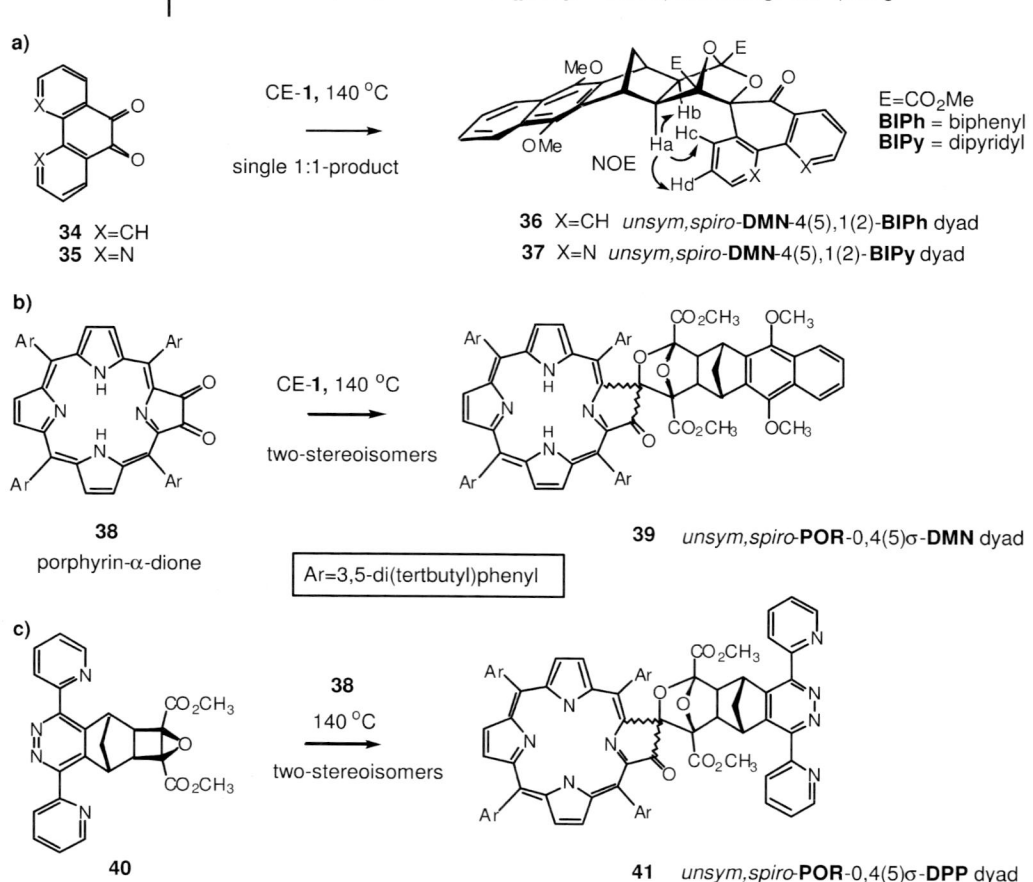

**Scheme 8.8** COCE reactions involving α-diones

group irrespective of the facial selectivity. Importantly, the two-step sequence involving COCE cycloaddition (step 1a) followed by keto-amine condensation (step 2a) provided a stereochemically precise way to link chromophores onto a rigid scaffold and where the entire carbon skeleton of ninhydrin is incorporated into the final product.

A second approach involving ninhydrin as the central core can also lead to spiro-linked bi-chromophoric systems (route b, highlighted in the box of Scheme 8.9). The reaction sequence is essentially order-reversed: initial reaction involves the keto-amine condensation between ninhydrin **42** and a primary amine **47** capable of carrying a chromophore (chromophore amine) to form site-selectively the iminodione **48** by attack at the central CO group. This derivative is itself active in the COCE reaction, with the cyclobutene epoxide attacking one of the equivalent conjugated CO groups rather than the central imine. In this way the DMN-containing product **49** can be formed from reaction with CE-1.

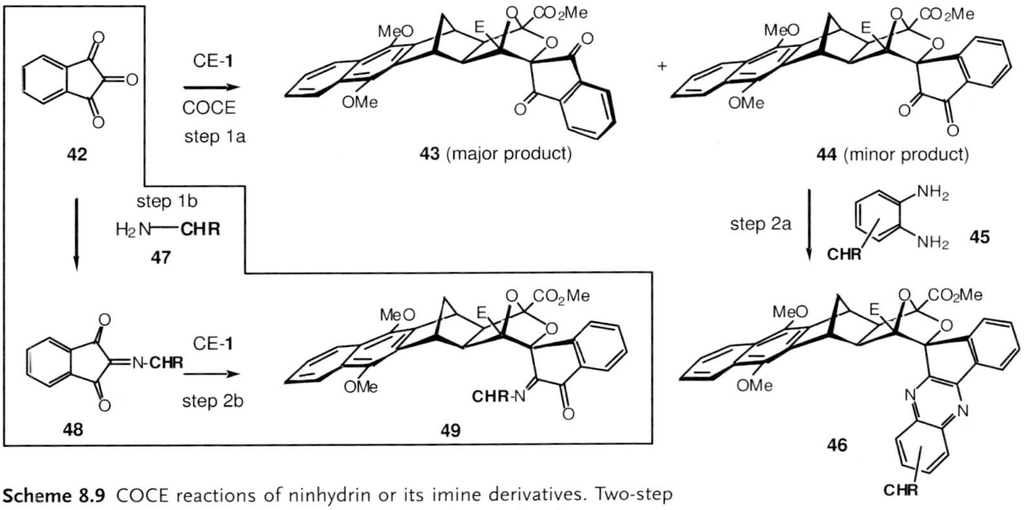

**Scheme 8.9** COCE reactions of ninhydrin or its imine derivatives. Two-step reaction sequences for preparing dyads of unusual spiro-fused geometry. CHR = unspecified chromophore.

These approaches to scaffold-separated *bis*-chromophores are entirely new and have much potential for producing new architectures for the scaffolds and the attendant geometrical interrelationship between the chromophores.

## 8.4
## Conclusions

The work carried out in this study has concentrated on the use of the **DMN** chromophore embodied in the cyclobutene epoxide reagent CE-1 as the donor component in rigid dyad formation. The use of carbonyl reagents as dipolarophiles in the COCE reaction, reported here for the first time, has considerably broadened the scope of BLOCK coupling protocols, especially as it has allowed access to dyads with new geometries (type III dyads). The simplicity of this BLOCK synthesis hopefully will encourage others to develop new dyads using the ACE and COCE methodologies and to advance, inter alia, the cause of artificial photosynthesis. The preliminary data employing $\alpha$-diones and the cyclic trione ninhydrin have considerably widened the scope of the COCE reaction and shown their potential in dyad formation.

## 8.5
## Dyad Nomenclature

We have identified six classes of scaffold-linked dyads, each containing two different chromophores (Fig. 8.2). Classes I–III are symmetrical, whereas classes IV–VI are unsymmetrical. The nomenclature used to describe these dyads uses letter abbreviations for each chromophore (bold capitals) and indicates their separation using a count of the sigma bonds ($n\sigma$) in the scaffold linking them. This number is inserted between the codes for the different chromophores in the geometrical alignment of the chromophores. In the simplest case (class I), both chromophores are fused to the scaffold (two-bond linkage) and the molecule

**CLASS I**

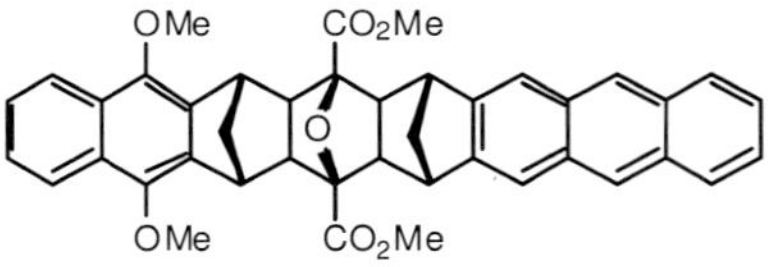

*sym*-**DMN**-6σ-**ANTH** dyad
**50**

**CLASS II**

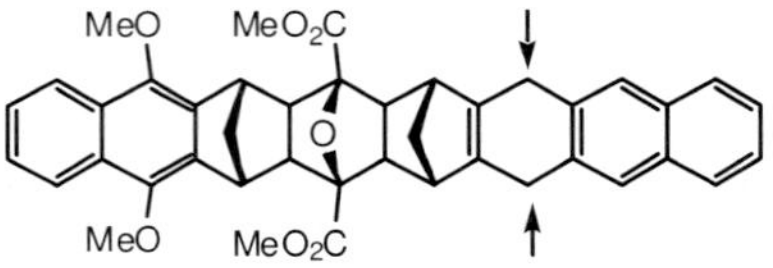

*sym,hinged*-**DMN**-7,1σ-**NAPH** dyad
**51**

**CLASS III**

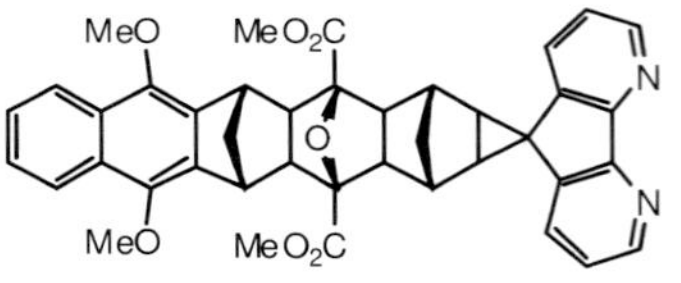

*sym,spiro*-**BDMN**-7,1σ-**DAF** dyad
**52**

**CLASS IV**

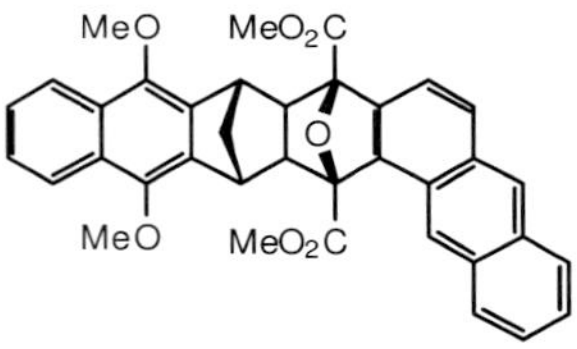

*unsym*-**DMN**-4σ-**ANTH** dyad
**53**

**CLASS V***

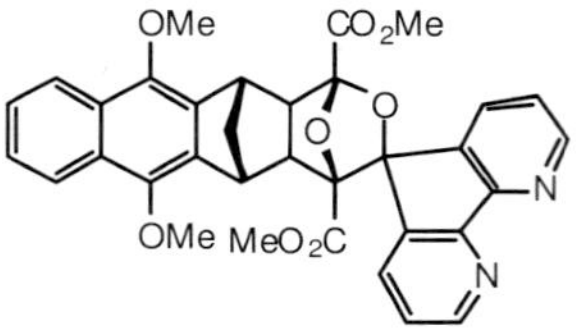

*unsym,spiro*-**DMN**-4(5),1σ-**DAF** dyad
**23**

**CLASS VI***

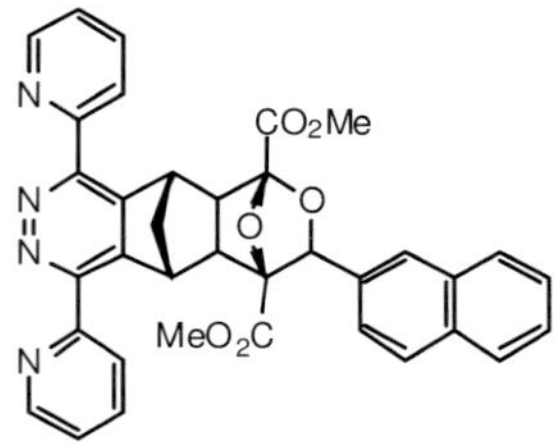

*unsym,sigma*-**DPP**-4(5),1σ-*exo*-**NAPH** dyad
**54**

**DMN** = 1,4-dimethoxynaphthalene; **ANTH** = anthracene; **DAF** = 4,5-diazafluorene; **NAPH** = naphthalene; **DPP** = 3,6-di(2-pyridyl)pyridazine

* reported for the first time in this publication

**Fig. 8.2** The six classes of scaffold-linked dyads, each containing two different chromophores. For a detailed description, see text.

has $C_s$ symmetry. Compound **50** is the example for this class of dyad (Fig. 8.2) and has 1,4-dimethoxynaphthalene (**DMN**) and anthracene (**ANTH**) chromophores separated by six sigma bonds in the alicyclic scaffold. The symmetry is designated by a prefix (*sym* or *unsym*); this symmetry differentiates classes I–III from classes IV–VI. Thus, the name for **50** is *sym*-**DMN**-6σ-**ANTH**.

Class II is a special subtype of class I and differs from the fully rigid scaffold system to one that contains a conformationally flexible ring capable of acting as a hinge. The hinge system is indicated as an italicized *hinge* prefix in the name, and its position is defined by using an a,b prefix directly in front of the σ-bond linker. Thus, in dyad **51** the 8σ-bond linker is designated 7,1σ to show that the hinge point is seven σ-bonds from the dimethoxynaphthalene (**DMN**) chromophore and one σ-bond from the naphthalene (**NAPH**) chromophore. The *sym* prefix defines the overall symmetry of the molecule and precedes secondary definer, in this case, *hinged*. The attachment points to each chromophore are automatically defined by the overall symmetry designator. Thus, the name for dyad **52** is *sym,hinged*-**DMN**-7,1σ-**NAPH**.

Class III dyads have one of the chromophores attached to the scaffold via spiro linkage. The same nomenclature system used to indicate the hinge point in class II dyads is used here to indicate the position of the spiro attachment. Thus, dyad **52** contains **DMN** and diazafluorene (**DAF**) chromophores; they are separated by eight σ-bonds, and the spiro linkage occurs seven σ-bonds from the **DMN** chromophore and one σ-bond from the **DAF**. The overall symmetry of the system is designated by the *sym* prefix, and the full name for **52** is *sym,-spiro*-**DMN**-7,1σ-**DAF**.

Class IV dyads are the first of the unsymmetrical systems. Compound **53** involves a **DMN** and an anthracene (**ANTH**) dyad that are separated by four σ-bonds, and the full name is *unsym*-**DMN**-4σ-**ANTH**. This class corresponds to the unsymmetrical version of class I systems and is fully rigid, with each chromophore linked by two sigma bonds to ensure rigidity.

Class V dyads are unsymmetrical and spiro-containing. Again they have a symmetrical counterpart (class III) and all the same rules apply. The unsymmetrical positioning of the spiro group means that there are linking arms of different length; the shorter linker is used as the primary indicator and the longer one designated in parenthesis; again the spiro linkage is specified by dual numbering. Thus, compound **23** is fully defined as *unsym,spiro*-**DMN**-4(5),1σ-**DAF**.

The class VI dyads have one of the chromophores linked by a single σ-bond and are capable of conformational rotation about that bond. An italicized sigma is used as a prefix to designate this classification, which, because of its origin from a COCE reaction onto an aldehyde, must also be unsymmetrical. As well, the σ-bonded chromophore can have either *exo* or *endo* stereochemistry and is so designated immediately in front of the qualified chromophore. Accordingly, the dyad **54** acquires the name *unsym,sigma*-**DPP**-4(5),1σ-*exo*-**NAPH**, thereby defining the position, bonding, and stereochemistry of the naphthalene to the **DPP** scaffold.

## Acknowledgments

It is a pleasure to acknowledge the Australian Research Council (ARC) for funding this research, Central Queensland University for continuing to host the Centre for Molecular Architecture (1992–2003), and the dedication of the participants working on this project. RNW also thanks the organizers of the Boden Research Conference on Artificial Photosynthesis for the opportunity to present a preliminary report on this work (Sydney, January 2003).

## Appendix – Experimental

**Formation of Dyad 17:** Reaction of Benzaldehyde **16** with Cyclobutene Epoxide **1**

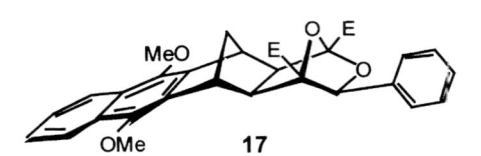

A solution of epoxide **1** (410 mg, 1.00 mmol) and benzaldehyde **16** (117 mg, 1.1 mmol) in 1,1,2,2-tetrachloroethane (20 mL) was heated at 140–147 °C under $N_2$ atmosphere for 6 h. Reaction progress was monitored by TLC. The solvent was removed in vacuo and residue was subjected to column chromatography (petroleum ether-ethyl acetate 4:1) and recrystallized from petroleum ether-ethyl acetate (1:1) to afford product **17** as a colorless solid (430 mg, 83.3%), m.p. 172–174 °C. $^1$H NMR (CDCl$_3$), $\delta$/ppm: 1.52 (1H, d, $J=10.0$ Hz), 2.71 (1H, d, $J=6.5$ Hz), 2.72 (1H, d, $J=10.0$ Hz), 2.86 (1H, d, $J=6.5$ Hz), 3.59 (3H, s), 3.71 (1H, s), 3.87 (1H, s), 4.02 (6H, s), 4.05 (3H, s), 4.86 (1H, s), 7.29 (3H, m), 7.34 (2H, m), 7.46 (2H, m), 8.09 (2H, m); $^{13}$C NMR (CDCl$_3$), $\delta$/ppm: 42.2, 43.2, 53.0, 53.3, 53.9, 55.8, 62.2, 62.3, 85.1, 92.2, 106.4, 122.8, 126.3, 127.8, 128.8, 128.8, 129.5, 134.3, 134.4, 137.1, 145.0, 145.2, 165.5, 166.9; MS ($m/z$, EI): 516 (M$^+$, 100%); HRMS ($m/z$): calculated for $C_{30}H_{28}O_8$ 516.1784 found: 516.1793.

**Formation of Dyad 19:** Reaction of 2-Naphthaldehyde **18** with Cyclobutene Epoxide **1**

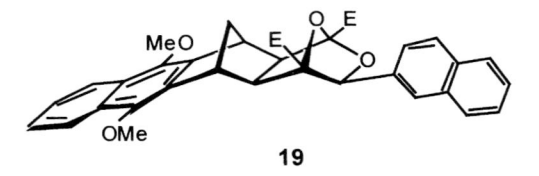

A solution of epoxide **1** (50 mg, 0.122 mmol) and 2-naphthaldehyde **18** (100 mg, 0.555 mmol) in dichloromethane (1 mL) was heated at 140 °C for 2 h. Solvent was removed in vacuo to afford yellow oily residue. Radial chromatography (starting with petroleum ether-ethyl acetate 10:1, then solvent polarity was gradually increased to 10% ethyl acetate) afforded adduct **19** as a yellow-colored solid (46 mg, 64%), m.p. 182–184 °C. $^1$H (CDCl$_3$), $\delta$/ppm: 1.55 (1H, td, $J$=9.9 Hz, $J$=1.4 Hz), 2.76 (1H, td, $J$=9.9 Hz, $J$=1.4 Hz), 2.78 (1H, d, $J$=6.2 Hz), 2.92 (1H, d, $J$=6.2 Hz), 3.54 (3H, s), 3.74 (1H, s), 3.91 (1H, s), 4.05 (6H, s), 4.11 (3H, s), 5.05 (1H, s), 7.47–7.52 (5H, m), 7.81–7.83 (4H, m), 8.11 (2H, d, $J$=6.2 Hz, $J$=3.3 Hz); $^{13}$C (CDCl$_3$), $\delta$/ppm: 42.1, 42.2, 42.8, 42.9, 52.8, 53.2, 53.6 (q, $J$=16.6 Hz), 55.6, 61.8, 61.9, 85.0, 92.0, 106.3, 122.5, 122.6, 125.0, 126.0 (2C), 126.5, 126.8, 127.3, 128.0, 128.4, 128.5, 128.6, 128.7, 133.2, 133.9, 134.0, 134.1, 134.3, 144.8, 149.0, 165.3, 166.6; HRMS (m/z): calculated for C$_{34}$H$_{30}$O$_8$: 566.1941 found: 566.1935.$_{28}$O$_8$ 516.1784 found: 516.1793.

**Formation of Dyad 21**: Reaction of Fluorenone **20** with Cyclobutene Epoxide **1**

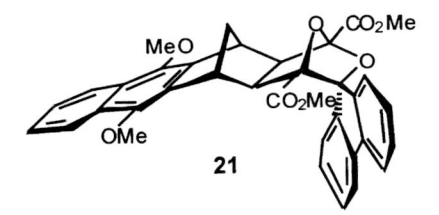

A solution of epoxide **1** (100 mg, 0.244 mmol) and fluorenone **20** (200 mg, 1.212 mmol) in dichloromethane (1 mL) was heated in a sealed glass tube at 140 °C for 2 h. Evaporation of the solvent in vacuo yielded solid material that was purified by radial chromatography (petroleum ether-ethyl acetate 20:1, then the solvent polarity gradually increased to ethyl acetate) to afford the product **21** (113 mg, 78%,), a colorless solid, m.p. 208–210 °C. $^1$H NMR (CDCl$_3$), $\delta$/ppm: 1.59 (1H, d, $J$=9.8 Hz), 2.94 (1H, d, $J$=9.8 Hz), 3.17 (1H, d, $J$=6.4 Hz), 3.26 (1H, d, $J$=6.4 Hz), 3.39 (3H, s), 3.58 (1H, s), 3.83 (3H, s), 3.99 (3H, s), 4.06 (3H, s), 4.14 (3H, s), 7.11–7.36 (5H, m), 7.51–7.61 (4H, m), 7.87 (1H, d, $J$=7.7 Hz), 8.13–8.18 (2H, m); $^{13}$C NMR (CDCl$_3$), $\delta$/ppm: 42.5, 42.6, 43.1, 48.9, 52.5, 53.6, 56.4, 61.9, 62.2, 93.1, 95.2, 106.3, 119.7, 120.6, 122.1, 122.8, 125.7, 126.1 (2C), 126.7, 127.6, 128.3, 128.6, 128.9, 130.7, 130.8, 133.9, 134.4, 139.8, 140.4, 142.4, 144.3, 144.7, 145.1, 169.5, 169.7.

**Formation of Dyad 23:** Reaction of 4,5-Diazafluorenone **22** with Cyclobutene Epoxide **1**

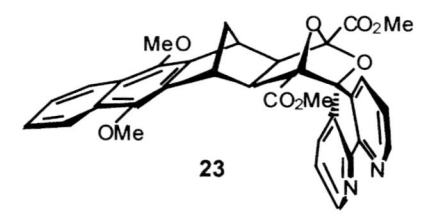

A solution of epoxide **1** (100 mg, 0.246 mmol) and 4,5-diazafluoren-9-one **22** (30 mg, 0.246 mmol) in dichloromethane (1 mL) was heated at 140 °C for 2 h. Solvent was removed in vacuo to afford a yellow residue. Radial chromatography (starting with petroleum ether-ethyl acetate 5:1, then solvent polarity was gradually increased to ethyl acetate) afforded adduct **23** as a yellow-colored solid (39 mg, 27%), m.p. 243–244 °C. $^1$H (CDCl$_3$), $\delta$/ppm: 1.60 (1H, d, $J$=10.2 Hz), 2.88 (1H, td, $J$=10.2 Hz, $J$=1.6 Hz), 3.05 (1H, d, $J$=6.6 Hz), 3.15 (1H, d, $J$=6.6 Hz), 3.41 (3H, s), 3.59 (1H, s), 3.84 (3H, s), 3.98 (1H, s), 4.08 (3H, s), 4.13 (3H, s), 7.14 (1H, dd, $J$=7.9 Hz, $J$=4.9 Hz), 7.27 (1H, $J$=7.9 Hz, $J$=4.9 Hz), 7.51–7.59 (3H, m), 8.09–8.11 (1H, m), 8.14–8.19 (2H, m), 8.68 (2H, td, $J$=1.6 Hz, $J$=4.9 Hz); $^{13}$C (CDCl$_3$), $\delta$/ppm: 42.6 (d, $J$=14.9 Hz), 42.7, 43.1, 49.5, 53.0, 53.9 (q, $J$=18.4 Hz), 56.4, 62.0 (q, $J$=15.3 Hz), 62.6 (q, $J$=15.5 Hz), 93.1, 96.6, 105.4, 122.3, 122.7, 123.1, 124.1, 126.3 (2C), 128.6, 128.7, 133.1, 133.5, 133.9, 134.0, 135.9, 139.5, 144.8, 145.2, 152.3, 152.5, 157.7, 160.0, 164.6, 165.2; HRMS ($m/z$): calculated for C$_{34}$H$_{28}$N$_2$O$_8$: 592.1846 found: 592.1843.

**Formation of Dyad 25:** Reaction of Anthraquinone **24** with Cyclobutene Epoxide **1**

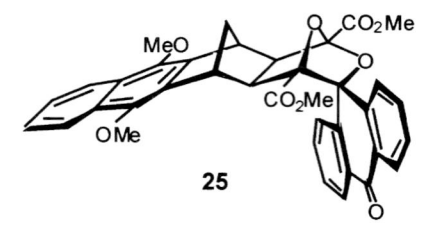

A solution of epoxide **1** (100 mg, 0.244 mmol) and 9,10-anthraquinone **24** (200 mg, 1.087 mmol) in dichloromethane (1 mL) was heated in a sealed glass tube at 140 °C for 2 h. Evaporation of the solvent in vacuo yielded solid material that was subjected to radial chromatography (petroleum ether-ethyl acetate 20:1, then the solvent polarity gradually increased to pure ethyl acetate) to afford the yellow-colored adduct **25** (101 mg, 70%), m.p. 244–247 °C. $^1$H NMR (CDCl$_3$), $\delta$/

ppm: 2.34 (1H, td, $J=9.9$ Hz, $J=1.6$ Hz), 1.46 (1H, td, $J=9.9$ Hz, $J=1.6$ Hz), 2.36 (1H, d, $J=6.6$ Hz), 3.10 (1H, d, $J=6.6$ Hz), 3.39 (1H, s), 3.52 (3H, s), 3.62 (3H, s), 3.93 (1H, s), 4.07 (3H, s), 4.10 (3H, s), 7.41–7.50 (3H, m), 7.63–7.68 (2H, m), 7.79 (1H, dd, $J=5.8$ Hz, $J=3.3$ Hz), 7.89 (1H, dd, $J=7.7$ Hz. $J=0.9$ Hz), 8.00–8.03 (1H, m), 8.11–8.15 (2H, m), 8.31 (1H, dd, $J=5.8$ Hz, $J=3.3$ Hz); $^{13}$C NMR (CDCl$_3$), $\delta$/ppm: 42.7, 42.8, 42.9, 48.8, 52.9, 53.9, 54.0, 55.7, 61.3, 62.2, 85.9, 94.6, 107.2, 122.6, 125.6, 126.0, 126.1, 127.1, 128.2, 128.5, 128.6, 129.3, 129.4, 131.6, 132.6, 133.0, 133.6, 133.7, 133.9, 134.5, 138.6, 141.6, 144.7, 144.8, 164.9, 165.3, 184.2. HRMS (m/z): calculated for C$_{35}$H$_{30}$O$_9$: 594.1889 found: 594.1895.

**Formation of Adduct 27:** Reaction of Coumarin **26** with Cyclobutene Epoxide **1**

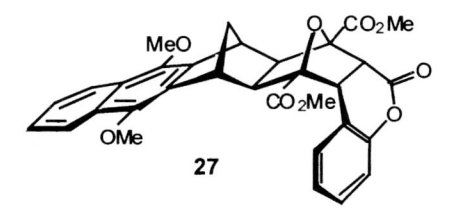

A solution of epoxide **1** (60 mg, 0.150 mmol) and coumarin **26** (120 mg, 0.822 mmol) in dichloromethane (1 mL) was heated in a sealed glass tube at 140 °C for 2 h. Evaporation of the solvent in vacuo yielded a colorless solid that was subjected to radial chromatography (petroleum ether-ethyl acetate 20:1, then the solvent polarity gradually increased to pure ethyl acetate) to afford the colorless product **27** (16 mg, 21.6%), m.p. 272–273 °C. $^1$H NMR (CDCl$_3$), $\delta$/ppm: 1.44 (1H, d, $J=10.2$ Hz), 2.18 (1H, d, $J=7.1$ Hz), 2.46 (1H, d, $J=7.1$ Hz), 2.63 (1H, td, $J=10.2$ Hz, $J=1.8$ Hz), 3.70 (3H, s), 3.72 (1H, s), 3.79 (1H, d, $J=12.6$ Hz), 3.92 (1H, d, $J=12.6$ Hz), 4.03 (3H, s), 4.05 (6H, s), 4.14 (1H, s), 6.91 (1H, dd, $J=8.2$ Hz, $J=1.1$ Hz), 7.13 (1H, dt, $J=7.7$ Hz, $J=1.1$ Hz), 7.18 (1H, dt, $J=7.7$ Hz, $J=1.1$ Hz), 7.33 (1H, d, $J=7.1$ Hz), 7.42–7.45 (2H, m), 7.97–8.00 (1H, m), 8.07–8.11 (1H, m); $^{13}$C NMR (CDCl$_3$), $\delta$/ppm: 42.4, 43.2, 43.9, 47.3, 49.2, 50.3, 50.9, 53.3, 53.5, 61.5, 61.6, 89.8, 92.1, 116.4, 118.2, 122.3, 122.7, 125.3, 125.9 (2C), 128.3, 128.5, 130.0, 130.1, 133.6, 134.2, 144.4, 144.7, 151.3, 164.1, 167.8, 169.3,

HRMS (m/z): calculated for C$_{32}$H$_{28}$O$_9$: 556.1733 found: 556.1729.

## Formation of Stretched Cyclobutene Epoxide 32

Reaction of Alkene 30 with DMAD

A suspension of alkene 30[14] (250 mg, 0.514 mmol), DMAD (2.00 g, 16.9 mmol), and $RuH_2CO(PPh_3)_3$ catalyst (200 mg, 0.222 mmol) in benzene (20 mL) was refluxed for 3 days under a nitrogen atmosphere. Solution was filtered through a short silica column, solvent was removed in vacuo, and excess of DMAD was distilled off in high vacuo. Brown-colored residue was treated with methanol and precipitate was collected by filtration to afford pure product 31 as a colorless solid (210 mg, 65%), m.p. 182–183 °C. $^1H$ (CDCl$_3$), $\delta$/ppm: 1.28 (1H, d, $J=12.2$ Hz), 1.63 (1H, d, $J=10.4$ Hz), 2.07 (1H, d, $J=12.2$ Hz), 2.31 (2H, s), 2.34 (2H, s), 2.40 (1H, d, $J=10.4$ Hz), 2.57 (2H, s), 2.60 (2H, s), 3.73 (2H, s), 3.78 (6H, s), 3.83 (6H, s), 3.99 (6H, s), 7.46 (2H, dd, $J=6.4$ Hz, $J=3.3$ Hz), 8.09 (2H, dd, $J=6.4$ Hz, $J=3.3$ Hz); $^{13}C$ (CDCl$_3$), $\delta$/ppm: 25.3, 36.6, 41.5, 43.4, 45.2, 50.1, 51.2, 51.7, 52.3, 62.2, 54.4, 122.2, 125.8, 128.5, 133.6, 141.9, 145.1, 161.1, 165.5, 170.5; HRMS ($m/z$): calculated for $C_{36}H_{36}O_{10}$: 628.2309 found 628.2307.

Reaction of Cyclobutene-1,2-diester 31 with tBuOOH

To a stirred solution of tetraester 31 (1.320 g, 2.102 mmol), in dry THF (15 mL), tBuOOH (3.3 M in toluene, 1.01 mL, 3.153 mmol) and tBuOOK (70 mg, 0.631 mmol) were added under a nitrogen atmosphere at 0 °C and the reaction mixture was stirred at room temperature for 3 h. Reaction was quenched with 20% sodium sulfite and diluted with DCM, extracted with DCM (3×). Organic extracts were collected and washed with sodium sulfite, water (2×), and dried (MgSO$_4$). Solvent was removed in vacuo and product was recrystallized from methanol to afford the colorless solid 32 (1.012 g, 75%), m.p. 144–146 °C. $^1H$ (CDCl$_3$), $\delta$/ppm: 1.47 (1H, d, $J=12.0$ Hz), 1.62 (1H, d, $J=10.5$ Hz), 2.22 (2H, s), 2.30 (2H, s), 2.34 (1H, d, $J=10.5$ Hz), 2.46 (2H, s), 2.69 (1H, d, $J=12.0$ Hz), 2.76 (2H, s), 3.71 (2H, s), 3.78 (6H, s), 3.84 (6H, s), 4.05 (6H, s), 7.44 (2H, dd, $J=6.5$ Hz, $J=3.2$ Hz), 8.07 (2H, dd, $J=6.5$ Hz, $J=3.2$ Hz); $^{13}C$ (CDCl$_3$), $\delta$/ppm: 28.4, 38.6, 41.4, 42.8, 48.6, 50.3, 50.5, 51.8 (d, $J=11.6$ Hz), 53.8 (d, $J=13.6$ Hz), 55.0, 62.2 (d, $J=11.2$ Hz), 64.3, 122.5, 125.8, 128.5, 133.5, 134.3, 165.8, 170.4; HRMS ($m/z$): calculated for $C_{36}H_{36}O_{11}$: 644.2258 found 644.2259.

**Formation of Dyad 33:** Reaction of 9-Fluorenone **20** with Cyclobutene Epoxide **32**

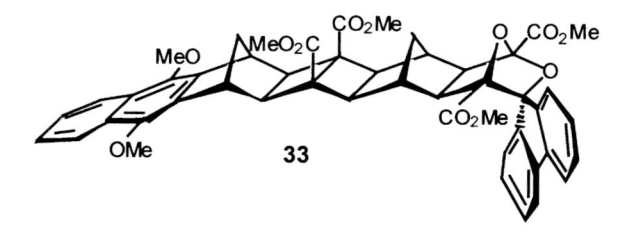

A solution of epoxide **32** (48 mg, 0.075 mmol) and 9-fluorenone **20** (100 mg, 0.156 mmol) in dichloromethane (2 mL) was heated in a sealed glass tube at 140 °C for 2 h. Solvent was removed in vacuo and the yellow residue was subjected to radial chromatography (starting with petroleum ether-ethyl acetate 5:1, then solvent polarity was gradually increased to ethyl acetate) to afford adduct **33** as a colorless solid (27 mg, 45%), m.p. 231–232 °C. $^1$H (CDCl$_3$), $\delta$/ppm: 1.63 (1H, d, $J$=10.7 Hz), 1.96 (1H, d, $J$=12.0 Hz), 2.03 (1H, s), 2.08 (1H, s), 2.38 (1H, d, $J$=10.7 Hz), 2.39 (1H, d, $J$=12.0 Hz), 2.46 (1H, d, $J$=5.6 Hz), 2.56 (1H, d, $J$=5.6 Hz), 2.58 (1H, d, $J$=5.5 Hz), 2.60 (1H, d, $J$=5.5 Hz), 2.76 (1H, d, $J$=6.7 Hz), 2.85 (1H, d, $J$=6.7 Hz), 3.24 (3H, s), 3.68 (1H, s), 3.75 (4H, s), 3.85 (3H, s), 3.93 (3H, s), 3.96 (3H, s), 3.99 (3H, s), 7.20 (2H, t, $J$=7.8 Hz), 7.35 (1H, t, $J$=7.5 Hz), 7.37 (1H, t, $J$=7.5 Hz), 7.42–7.49 (3H, m), 7.48 (1H, d, $J$=7.5 Hz), 7.57 (1H, d, $J$=7.5 Hz), 7.74 (1H, d, $J$=7.5 Hz), 8.08 (2H, m); $^{13}$C (CDCl$_3$), $\delta$/ppm: 28.6, 39.7 (d, $J$=17.9 Hz), 41.4, 43.3 (2C), 47.7, 50.5 (2C), 51.0, 51.3, 51.7 (d, $J$=11.2 Hz), 51.8 (d, $J$=15.8 Hz), 52.3 (d, $J$=14.3 Hz), 53.4 (d, $J$=18.2 Hz), 54.3, 54.4, 54.5, 62.1 (d, $J$=16.5 Hz), 62.2 (d, $J$=20.4 Hz), 92.9, 93.9, 106.7, 119.4, 1207. 122.5 (2C), 125.6, 125.8 (2C), 126.5, 127.4, 128.5, 128.7, 130.3 (2C), 133.3, 133.4, 139.5, 140.7, 142.5, 144.1, 145.1, 145.2, 163.8, 164.9, 170.3, 170.5; HRMS-ES (m/z): calculated for C$_{49}$H$_{44}$N$_4$O$_{11}$: 824.2833, calculated for C$_{49}$H$_{44}$N$_4$O$_{11}$+Na$^+$: 847.2730 found 847.2731.

## References

1 Deisenhofer, J.; Epp, O.; Miki, K.; Huber, R.; Michel, H. *Nature* **1985**, *318*, 618. Deisenhofer, J.; Michel, H. *EMBO J.* **1989**, *8*, 2149. Huber, R. *Angew. Chem., Int. Ed. Engl.* **1989**, *101*, 849; *28*, 848.

2 Gust, D.; Moore, T. A.; Moore, A. L. *Acc. Chem. Res.* **1993**, *26*, 198 and references therein. Wasielewski, M. R. *Chem. Rev.* **1992**, *92*, 435 and references therein. Kurreck, H.; Huber, M. *Angew. Chem. Int. Ed. Engl.* **1995** *107*, 929; *34*, 849. Osuka, A.; Nakajima, S.; Okada, T.; Taniguchi, S.; Nozaki, K.; Ohno, T.; Yamazaki, I.; Nishimura, Y.; Matage, N. *Angew. Chem., Int. Ed. Engl.* **1996**, *108*, 98; *35*, 92. Imahori, H.; Yamada, K.; Hasegawa, M.; Taniguchi, S.; Okada, T.; Sakata, Y. *Angew. Chem. Int. Ed. Engl.* **1997**, *109*, 2740; *36*, 2626. Wiederrecht, G. P.; Neimczyk, M. P.; Svec, W. A.; Wasielewski, M. R. *J. Am. Chem. Soc.* **1996**, *118*, 81 and references therein. Dixon, I. M.; Collin J.-P.; Sauvage, J.-P.; Flamigni, L. *Inorganic Chem.* **2001**, *40*, 5507–5517.

Flamigni, L.; Marconi, G.; Dixon, I. M.; Collin, J.-P.; Sauvage, J.-P. *J. Phys. Chem. B* **2002**, *106*, 6663–6671.

3 Warrener, R. N.; Sun, H.; Johnston, M. R. *Aust. J. Chem.* **2003**

4 (a) Warrener, R. N.; Ferreira, A. B. B.; Schultz, A. C.; Butler, D. N.; Keene, F. R.; Kelso, L. S. *Angew. Chem. Int. Ed. Engl.* **1996**, *35*, 2485–2487. (b) Kelso, L. S.; Smith, T. A.; Schultz, A. C.; Junk, P. C.; Warrener, R. N.; Ghiggino, K. P.; Keene, F. R. *J. Chem. Soc., Dalton Trans.*, **2000**, 2599–2606. (c) Flamigni, L.; Johnston, M. R.; Giribabu, L. *Chem.-A European J.* **2002**, *8*, 3938–3947. (d) Flamigni, L.; Talarico, A. M.; Barigelletti, F.; Johnston, M. R. *Photochem. Photobiol. Sci.* **2002**, *1*, 190–197.

5 (a) Warrener, R. N.; Schultz, A. C.; Butler, D. N.; Wang, S.; Mahadevan, I. B.; Russell, R. A. *Chem. Commun.*, **1997**, 1023–1024. (b) Warrener, R. N.; Butler, D. N.; Russell, R. A. *Synlett*, **1998**, 566–573. (c) Margetic, D.; Johnston, M. R.; Tiekink, E. R. T.; Warrener, R. N. *Tetrahedron Lett.*, **1998**, *39*, 5277–5280. (d) Schultz, A. C.; Johnston, M. R.; Warrener, R. N.; Gunter, M. J. Article 077, *Electronic Conference on Heterocyclic Chemistry '98*, **1998**, H. S. Rzepa and O. Kappe (Eds), Imperial College Press, ISBN 981-02-3594-1 (http://www.ch.ic.ac.uk/ectoc/echet98/pub/077/index.htm). (e) Warrener, R. N.; Margetic, D.; Sun, G.; Amarasekara, A. S.; Foley, P.; Butler, D. N.; Russell, R. A. *Tetrahedron Lett.*, **1999**, *40*, 4111–4114. (f) Warrener, R. N.; Schultz, A. C.; Johnston, M. R.; Gunter, M. J. *J. Org. Chem.*, **1999**, *64*, 4218–4219. (g) Warrener, R. N.; Margetic, D.; Amarasekara, A. S.; Butler, D. N.; Mahadevan, I. B.; Russell, R. A. *Org. Lett.*, **1999**, *1*, 199–202. (h) Warrener, R. N.; Margetic, D.; Amarasekara, A. S.; Russell, R. A. *Org. Lett.*, **1999**, *2*, 203–206. (i) Butler, D. N.; Hammond, M. L. A.; Johnston, M. R.; Sun, G.; Malpass, J. R.; Fawcett, J.; Warrener, R. N. *Org. Lett.*, **2000**, *2*, 721–724. (j) Warrener R. N.; Butler, D. N.; Margetic, D.; Pfeffer, F. M.; Russell, R. A. *Tetrahedron Lett.*, **2000**, *41*, 4671–4675. (k) Margetic, D.; Russell, R. A.; Warrener, R. N. *Org. Lett.*, **2000**, *2*, 4003–4006. (l) Warrener, R. N.; Margetic, D.; Foley, P. J.; Butler, D. N.; Winling, A.; Beales, K. A.; Russell, R. A. *Tetrahedron*, **2001**, *57*, 571–582. (m) Johnston, M. R.; Latter, M. J.; Warrener, R.aN. *Aust. J. Chem.* **2001**, *54*, 633–636. (n) Johnston, M. R.; Gunter, M. J.; Warrener, R. N. *Tetrahedron* **2002**, *58*, 3445–3451. (o) Johnston, M. R.; Latter, M. L.; Warrener, R. N. *Aust. J. Chem.*, **2002**, *55*, 633–636.

6 Gunter, M. J.; Tang, H.; Warrener, R. N. *J. Porphyrins Phthalocyans*, **2002**, *6*, 673–684.

7 Warrener, R. N.; Johnston, M. R.; Gunter, M. J. *Synlett*, **1998**, 593–595.

8 Schultz, A. C.; Kelso, L. S.; Johnston, M. R.; Warrener, R. N.; Keene, F. R. *Inorg. Chem.* **1999**, *38*, 4906–4909.

9 Warrener, R. N. *Eur. J. Org. Chem.* **2000**, 3363–3380.

10 (a) Napper, A. M.; Head, N. J.; Oliver, A. M.; Shephard, M. J., Paddon-Row, M. N. *J. Amer. Chem. Soc.*, **2002**, *124*, 10171–10181. (b) Hviid, L.; Verhoven, J. W.; Brouwer, A. M.; Paddon-Row, M. N. *Photochem. Photobiol. Sciences*, **2004**, *3*, 246–251.

11 Margetic, D.; Warrener, R. N.; Butler, D. N. 7th Electronic Conference on Synthetic Organic Chemistry, Article A009 (ECSOC-7, www.mdpi.net/ecsoc-7/index.htm).

12 (a) Jurczak, J.; Golebiowski, A.; Bauer, T. *Synthesis* **1985**, 928–929. (b) Achmatowicz, O., Jr.; Szymoniak, J. *Tetrahedron* **1982**, *38*, 1299–1302. (c) Sulan, Y.; Roberson, M.; Reichel, F.; Hazell, R. G.; Jorgensen, K. A. *J. Org. Chem*, **1999**, *64*, 6677–6687. (d) Bennett, D. M.; Okamoto, I.; Danheiser, R. L. *Org. Lett.* **1999**, *1*, 641–644. (e) Achmatowicz, O.; Zamojski, A. *Roczniki Chem.* **1961**, *35*, 799–812.

13 (a) Chen, I.-L.; Chen, Y.-L.; Tzeng, C.-C. *Helv. Chim. Acta* **2002**, *85*, 2214–2221. (b) Reissig, H. U. *Tetrahedron Lett.* **1981**, *22*, 2981–2984.

14 Paddon-Row, M. N.; Cotsaris, E.; Patney, H. K. *Tetrahedron*, **1986**, *42*, 1779–1788.

15 Mitsudo, T.; Kokuryo, K.; Shinsugi, T.; Nakagawa, Y.; Watanabe, Y.; Takegami, Y. *J. Org. Chem.*, **1979**, *44*, 4492–4496.

16 Crossley, M. J.; McDonald, *J. Chem. Soc., Perkin Trans 1*, **1999**, 2429–2431.

# Part III
# Feeding the Grid from the Sun

# 9
# Very High-efficiency in Silico Photovoltaics

*Martin A. Green*

## 9.1
### Introduction

Over the last seven years, the rapid growth of silicon photovoltaic sales has seen the associated industry evolve from a small base to a multibillion dollar a year activity. Most present product is based on the use of silicon wafers, similar to those used in microelectronics, with a typical module shown in Fig. 9.1.

This "first-generation" product has enormous strengths, having demonstrated exceptional durability (25-year warranties are common), reasonable energy conversion efficiency, high production yield, and steadily decreasing costs. Another strength is that this approach allows manufacturers to benefit from developments in the still much larger microelectronics industry. Scrap silicon from this industry provides low-cost source material, particularly during periods when microelectronics activities are in recession, and items such as the furnaces, screen-printers, and contacting pastes used in cell production were originally developed for microelectronics.

However, the wafer-based approach does have weaknesses stemming from its material intensiveness. The thickness of the silicon wafers used in the cells

**Fig. 9.1** Silicon wafer-based photovoltaic module (photo courtesy of BP Solar).

(presently 0.3 mm) is determined by the need for high mechanical yields during production rather than by operational issues. The cells require quite elaborate encapsulation to provide the chemical and mechanical protection to ensure a 20- to 30-year operational field life. High-quality, low-iron toughened glass is used as a superstrate. UV-resistant, high-quality transparent polymeric layers are used to enshroud the cells, holding them in place as well as accommodating the differential in thermal expansion between silicon and the glass superstrate.

Although costs are steadily reducing as a result of higher production volumes, improving energy conversion efficiency, and the use of larger and thinner wafers, the material intensiveness of the approach means associated costs will always be reasonably high compared to other electricity-generation options. The modularity and deployability of photovoltaics will ensure large niche markets for such a premium product, e.g., in residential rooftop and building-integrated grid-connected systems, as well as in remote-area power supplies. However, it is unlikely that costs of the wafer-based approach can be sufficiently low for bulk power generation, taking into account the low capacity factors (10–25%) resulting from the cyclic nature of solar energy availability.

To reach the low costs required for bulk power generation, the author believes that a transition to a "second-generation" thin-film photovoltaic technology is essential. In the thin-film approach, the photovoltaically active material is deposited onto a large-area supporting substrate or superstrate. Apart from reduced material costs, this increases the area of each production unit from that of a 10- or 15-cm square wafer to that of a much larger glass sheet, as well as allowing a monolithic construction where cell interconnection occurs during the deposition stages. As indicated in Fig. 9.2, the low material cost is traded for slightly lower efficiency but gives a cost per watt, the key market differentiator, that could be less than half that of the wafer-based approach.

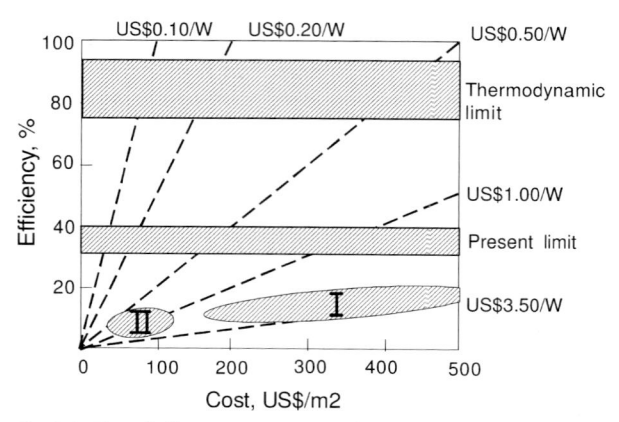

**Fig. 9.2** Plot of efficiency versus cost for first-generation wafer-based technology (area labeled "I") and for a mature thin-film technology (area labeled "II"). The straight lines show the corresponding dollar-per-watt cost.

Although the advantages of a thin-film approach are fairly obvious, it has been difficult for emerging thin-film technologies to make much impact upon the established wafer product. There are several reasons for this. One has been stability and/or durability issues with the early key thin-film contenders that have raised questions about their comparative robustness. Another is the relatively high capital cost of thin-film manufacturing facilities compared with that of a plant that purchases silicon wafers and sells processed cells to a module assembler. The thin-film approaches are less suited to trading capital cost for operational costs in this way. More recently, the development of thin-film technologies based on silicon in various forms is likely to better address these issues, as is subsequently discussed.

A feature apparent from Fig. 9.2 is the dollar-per-watt cost leverage provided by high-energy conversion efficiency. This leads to the conclusion that if photovoltaics is to reach its full potential, it must evolve ultimately to a "third generation" of very high-efficiency thin-film photovoltaic modules. Suggested approaches for improving energy conversion that could underpin this third generation are discussed in the final sections of this chapter.

## 9.2
## Silicon Wafer Approach

Over 90% of recent photovoltaic sales involve products based on silicon wafers, using the traditional monocrystalline wafers (as used in microelectronics), multicrystalline wafers cut from a large ingot (several hundred kilograms in weight) prepared by direction solidification, or wafers prepared directly as ribbon or sheets [1].

Fig. 9.3 shows typical commercial module energy conversion efficiencies, extracted from a recent survey of silicon module manufacturers [2]. Although there is considerable overlap, the most efficient modules are based on monocrystalline silicon, followed by multicrystalline wafers, in turn followed by the ribbon/sheet approaches. There has been considerable progress in the efficiency of commercial product over recent years as the market has become more competitive. Not only does efficiency provide a more prestigious product and the cost leverage as noted in connection with Fig. 9.2, but also in the residential rooftop market, high efficiency allows a higher rating system to be installed in the often limited available space. In the German market, this is important under the guaranteed feed-in tariff scheme in place in order to maximize the income-earning potential of the system. In the Japanese market, high efficiency allows purchasers to come closer to their apparent goal of providing all the electricity for their home from their solar system (only 9%, however, reach this target [3]).

The process used to convert the starting wafers into cells is generally based on a simple metal paste screen-printing approach to contact both the wafer and a diffused top surface layer [4]. More sophisticated processing approaches are now claiming an increasing share of production, notably through efforts at BP

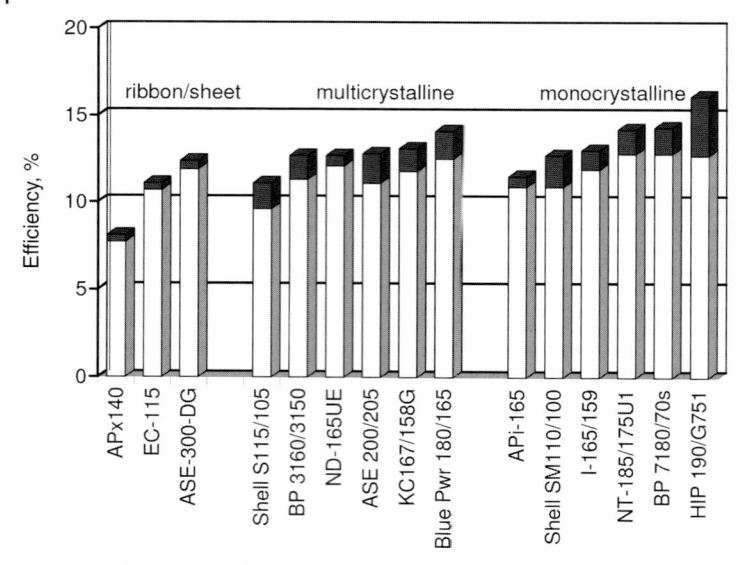

**Fig. 9.3** Wafer-based module efficiencies (total area). The band shows the range expected for modules meeting specifications.

Solar and Sanyo [5]. These allow improved efficiency and the use of thinner wafers, where difficulties arise with screen-printing.

As indicated in Fig. 9.4, deduced from several sources [6–9], costs of this technology have been steadily decreasing as the accumulated production volume increases, following an 80% "learning curve" (i.e., costs reduce to 80% of previous

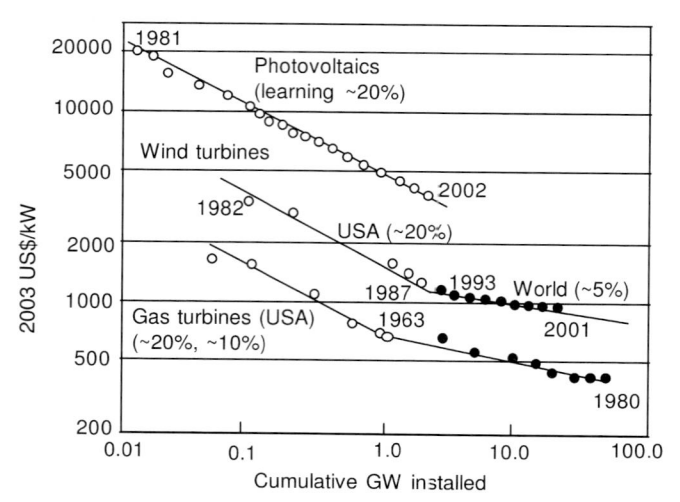

**Fig. 9.4** Cost in 2004 (U.S.$) versus accumulated production volume for photovoltaics compared with both wind generators and gas turbines.

for each doubling of this volume). However, experience with related products, such as wind and gas turbines, shows that this behavior is unlikely to continue forever but is rather a feature of a fledgling industry as it struggles to reach maturity. Lower learning rates for the wafer-based approach are expected in the not-too-distant future, with costs settling down to values at the high end of those for other energy generation options.

Recent experience in Japan, in particular, suggests that there could still be enormous markets for photovoltaics even if positioned at this high-cost periphery of the market. Japanese photovoltaic sales are booming, even though the government rebate for residential rooftop systems has dropped from 50% in 1994 to a token value of about 10% in 2004 [10]. However, as noted, thin-film technology has the potential for much lower cost and could tap much larger markets if it is able to match the wafer approach in some key performance areas.

## 9.3
## Thin-film Approaches

A potentially much lower cost approach to photovoltaics is to deposit the active material as a thin layer on a supporting superstrate or substrate, as suggested in Fig. 9.5. Despite their potential, thin films have had a rather checkered commercial history in photovoltaics. Early attempts to commercialize cells based on the chalcogenide semiconductor, CdS, failed because the cells were intrinsically unstable [11]. More success was obtained with cells based on hydrogenated alloys of amorphous silicon (a-Si:H), where a market was found for small consumer products such as pocket calculators and digital watches. However, the material proved disappointingly unstable outdoors. The only solution to date to address this problem has been to design around it, at the cost of greatly increased complexity, by using stacks of two or even three cells on top of one another to reduce the required thickness of each [12].

The disappointing performance of a-Si:H opened the door to two more chalcogenide-based compound semiconductors, CdTe and $CuInSe_2$ (CIS). $CdCl_2$

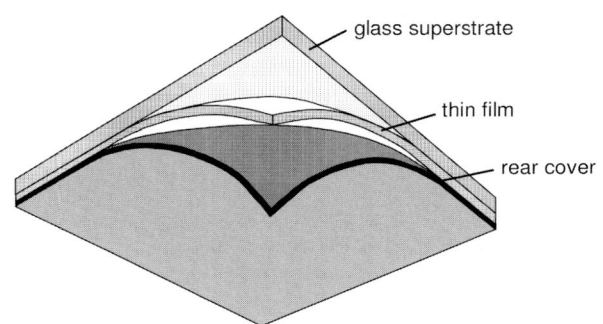

**Fig. 9.5** Thin-film solar cell.

treatments after deposition can greatly increase grain size in the former, allowing deposition by very simple techniques such as screen-printing or electrodeposition [13]. However, early stability problems with the rear contact seem to remain in the background, with special encapsulation to prevent moisture ingress deemed essential [14]. However, a more fundamental problem seems to be the lack of market acceptance of clearly toxic photovoltaic products such as this.

The case for CIS is more promising. This material and its alloys appear to be moisture-sensitive in all compositions [15], with special encapsulation required to pass standard qualification tests [14, 16, 17]. The use of a CdS junction layer again raises the issue of material toxicity. Although replacements have been found, they are reported to be even more moisture-sensitive [14]. Even if these problems are surmounted, known reserves of indium would limit the contribution to world energy demands this technology could make to about 30 times that being made by photovoltaics in 2004 [18] (or comparable to that being made by wind generators at that time).

The situation would appear fairly bleak were it not for recent progress with silicon thin-films on two fronts. Microcrystalline silicon ($\mu$c-Si), a mixed phase of amorphous and very fine-grain polycrystalline silicon (Fig. 9.6), has long been used as a contact in a-Si:H technology. In 1991, interesting photovoltaic properties of this material, in its own right, were demonstrated [20], stimulating programs to develop this material for use as the lower cell in a two-cell stack with an upper a-Si:H device [21] (Fig. 9.7).

The top a-Si:H device has the higher bandgap and responds to light at the blue end of the spectrum, while the lower $\mu$c-Si cell responds to the red end. Since the two cells are connected in series, performance is poor unless the current generated by each is matched. This has been a source of difficulty with this technology, since the a-Si:H cell has to be thin for good stability but thick to provide this current matching. Thus, although high performance is often re-

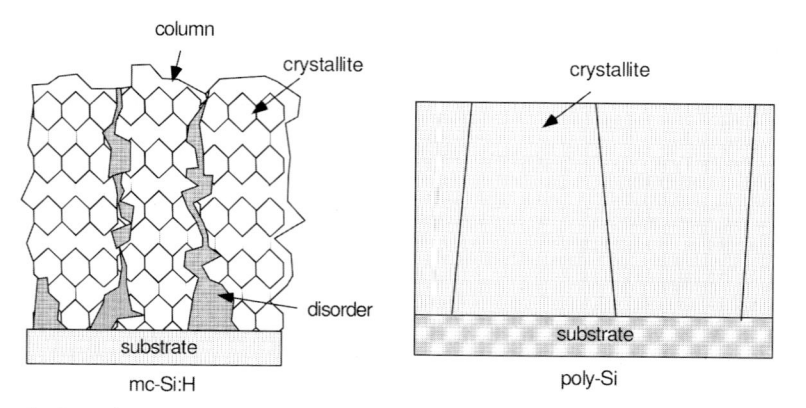

**Fig. 9.6** Schematic comparing (left) mixed-phase microcrystalline silicon with (right) polycrystalline silicon (after [19]).

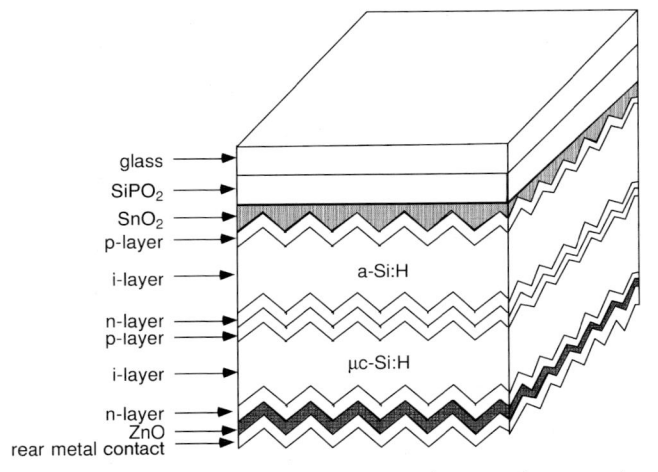

**Fig. 9.7** Tandem cell structure based on a combination of an amorphous silicon and microcrystalline silicon cell.

ported for these devices [21], the best performance is obtained from the most unstable devices.

Compared to the performance of other thin-film products on the market in 2004 [2] (Fig. 9.8), this "hybrid" approach already offers among the highest efficiencies. However, there are indications that stability still remains an issue. Although prod-

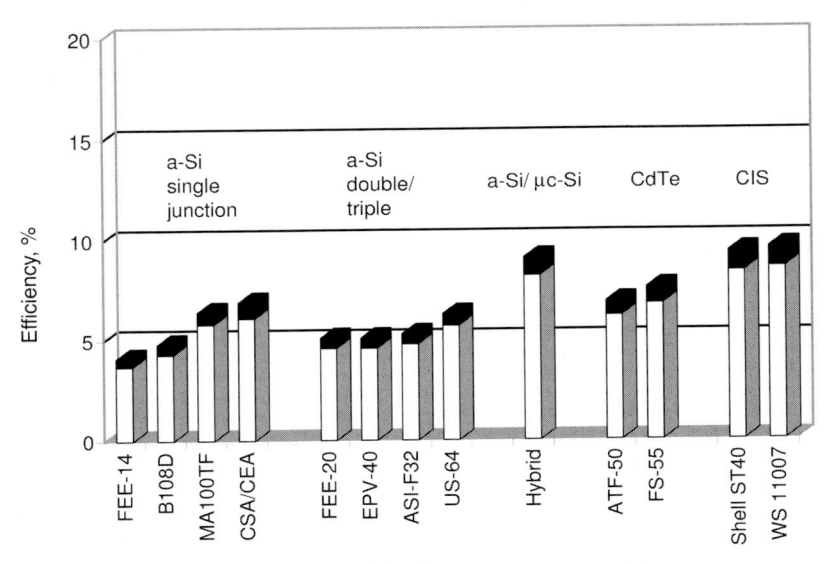

**Fig. 9.8** Energy conversion efficiency of thin-film modules commercially available in 2004. Also shown is the range expected for products falling within specifications.

ucts have not been generally available outside Japan for independent testing, one module that did find its way to a European test center was close to specifications prior to installation (8.7% initial efficiency) but degraded 19% (to 7.0% efficiency) in the first three months in the field [22]. On the basis of the previous discussion, this suggests that the a-Si:H layer was too thick, although a thinner layer would have given lower initial efficiency. However, even this module was clearly superior to earlier a-Si:H based products. A triple junction cell, the premium of such an a-Si:H product, started with an efficiency of 6.4% in the same testing, but degraded 22% (to 5.0%) over the same three-month period [22].

In response to such perceived difficulties with the key traditional thin-film contenders, the author's group began a program in 1988 [23] to develop thin-film technology from scratch based on silicon with a quality similar to that used in wafer modules. This might seem an obvious step, but this silicon is quite a weak absorber of light, causing one scientist to estimate the prospect for developing a 10% thin-film cell technology based on this approach as "one in a million" [24]. However, experience in our high-efficiency wafer cell program had encouraged us to believe that such silicon cells could be made very much thinner than commonly supposed by using "light trapping" [25].

In 1995, a "spin-off" company, Pacific Solar, took over these activities, commencing pilot-line fabrication of thin-film (poly-)crystalline silicon on glass (CSG) modules in 1998. Module, rather than cell, efficiency rapidly approaching 10% has now been demonstrated [26], defying the million-to-one odds previously mentioned.

One feature of this technology is its simplicity (Fig. 9.9) [26]. After texturing the glass layer (not shown), a layer of silicon nitride followed by the silicon layers are deposited in the same deposition chamber at low temperature. After a high-temperature step to crystallize the deposited amorphous silicon, it is patterned by laser and a polymeric insulating layer is applied to its rear. This is patterned by ink-jet printing to allow contacting to the different sides of the junction. A metal contact layer is finally deposited and patterned by laser or ink-jet printing [26].

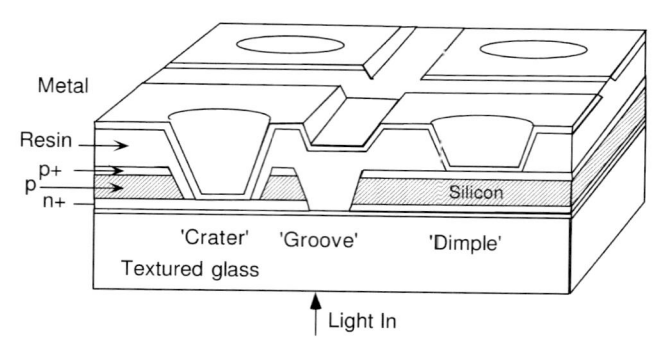

**Fig. 9.9** Schematic of the (poly-)crystalline silicon on glass (CSG) cell structure and interconnection scheme.

## 9.3 Thin-film Approaches

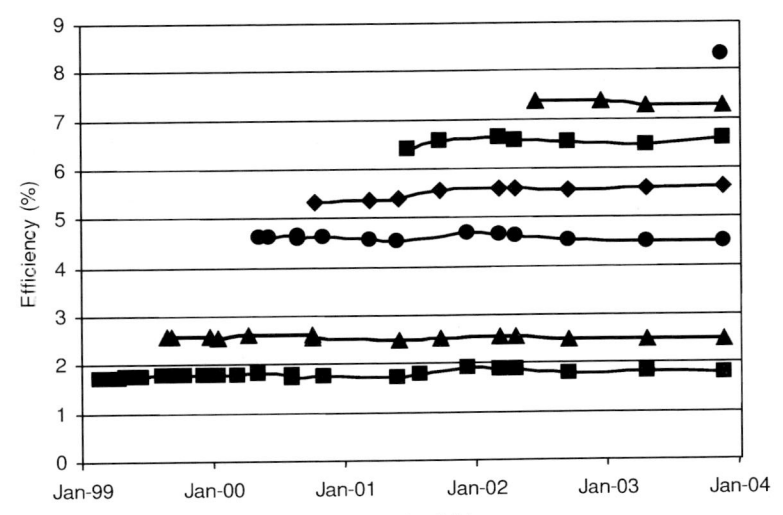

**Fig. 9.10** Performance of CSG modules in the field.

Another feature of this technology has been its excellent stability and durability, which seem likely to surpass even that of the very reliable wafer-based standard. Fig. 9.10 shows the performance of modules installed in the field at various points in time, as well as the evolution of the efficiency of modules available for field-testing. Even more impressive has been the performance of CSG modules under accelerated life testing [27]. Commercial thin-film modules generally are subjected to an IEC61646 qualification test. The three toughest tests are the damp-heat, humidity-freeze, and temperature-cycling tests. In the standard test, a separate module is subjected to each of these tests. In the accelerated test developed by Pacific Solar [27], the same module is subjected to all three tests. If it survives these, it is subjected to another cycle until performance drops to 80% of its initial value.

As shown in Fig. 9.11, wafer-based modules generally survive two to three cycles of such testing, while thin-film CIS modules have difficulty surviving even one. However, the performance of CSG modules under this test has been out-

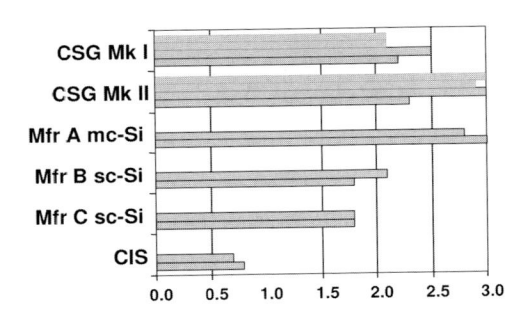

**Fig. 9.11** Accelerated testing of a range of modules based on the damp-heat, humidity-freeze, and temperature-cycling qualification tests of IEC61646. Shown is the number of cycles passed by different module types.

standing, suggesting that, with attention paid to this aspect, they can be far more durable than even wafer-based modules. This is probably not unexpected given that there are no wafers to break, interconnects to fatigue, solder joints to age, or encapsulants between the cell and glass to discolor or delaminate, some of the main failure modes of wafer-based modules [26].

In a market where investors wishing to invest in photovoltaics can get a good return by a low-risk investment in established wafer-based technology, there are challenges in attracting investment in a more capital-intensive thin-film technology, even if it has much stronger long-term potential. This is the challenge facing the market entry of thin-films at the present time.

## 9.4
## Third-generation Technologies

Figs. 9.3 and 9.8 document the quite modest energy conversion efficiencies demonstrated by present first- and second-generation products. These relatively modest values might suggest that there may be a thermodynamic reason that the conversion of dispersed sunlight into high-grade electrical energy is low. An investigation below, however, shows that no such argument can be sustained.

Fig. 9.12 shows the energy and entropy flows in an idealised solar converter [28]. By balancing first energy and then entropy fluxes and then eliminating the common term in $\dot{Q}$, the heat rejection rate to the ambient, the efficiency can be expressed as:

$$\eta = \dot{W}/\dot{E}_S = (1 - T_A \dot{S}_S / \dot{E}_S) - (1 - T_A \dot{S}_C / \dot{E}_C) \dot{E}_C / \dot{E}_S - T_A \dot{S}_G / \dot{E}_S \quad (1)$$

where $\dot{E}_S$ and $\dot{E}_C$ are energy fluxes from the sun and cell and $\dot{S}_S$ and $\dot{S}_C$ are corresponding entropy fluxes. $T_S$ and $T_A$ are the sun ambient temperatures. In the

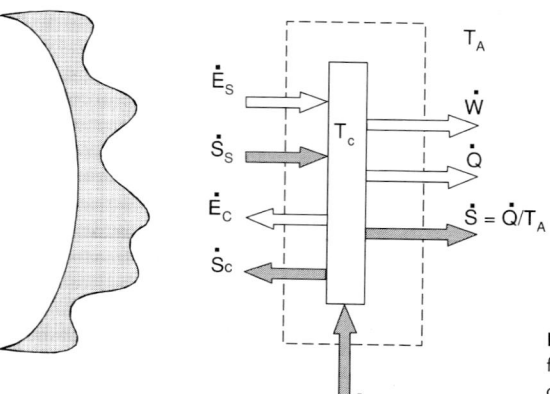

**Fig. 9.12** Energy and entropy fluxes involved in the conversion of sunlight to electricity or other useful work.

limit of zero entropy generation during the conversion process ($\dot{S}_G = 0$), eliminating terms that are negligibly small, the limit becomes:

$$\eta_{\lim} \approx 1 - T_A \dot{S}_S / \dot{E}_S \tag{2}$$

If the sun is modeled as a black body (at 6000 K), the entropy-to-energy flux ratio is well known and the limiting efficiency becomes:

$$\eta_{\lim} \approx 1 - \frac{4}{3} T_A / T_S \tag{3}$$

This limit is slightly lower than the Carnot efficiency, which can apply only in the limit of infinitesimal work output. It has a numerical value of 93.3% for an ambient (sink) temperature of 300 K.

This limit applies only to conversion of direct sunlight and would apply to a system that accepted light only from the direction of the sun's disc. For a system such as a normal photovoltaic system that responds almost equally well to sunlight from any direction, the limit that applies is the same as that calculated if the sun's energy were distributed over the hemisphere. This increases its entropy per unit energy about four times [29], reducing the limit given by Eq. (2) to 73.7%.

This result shows that there is no fundamental reason why photovoltaic energy conversion efficiency has to be as low as that demonstrated by present commercial products. Rather, this low efficiency is due to the specific conversion process used by this product that gives similarly fundamental conversion efficiency limits of 40.8% and 31.0% for direct and omnidirectional light, respectively, less than half that thermodynamically possible.

Fig. 9.13 shows the cost leverage possible if it were possible to approach the higher efficiency limits more closely, while retaining areal production costs intermediate between thin films and wafer-based products. Dollar-per-watt costs could be reduced by a further factor of two, making photovoltaics one of the lowest-cost approaches yet suggested for electricity generation.

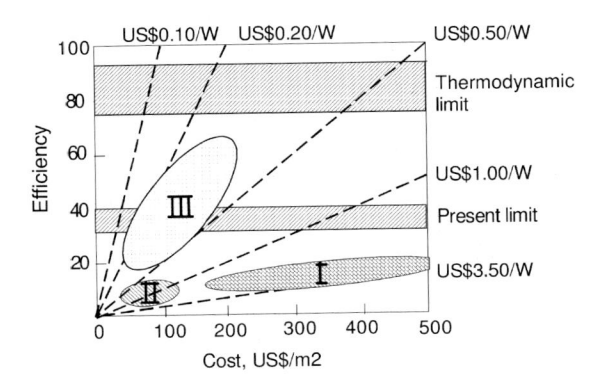

**Fig. 9.13** Efficiency versus areal cost as in Fig. 9.2, but including a possible target area (labeled III) for third-generation photovoltaics.

What would be the key features of such a third-generation technology? One requirement is that it be based on thin films to retain the associated cost and production advantages. It should also have a high efficiency, an appreciably higher efficiency than is possible from standard cells using similar materials. It has to be based on abundant and nontoxic materials, and it must have the potential to be even more stable and durable than wafer-based products.

Have any approaches been suggested to be capable of demonstrating these third-generation ideals? Several speculative ideas that seem particularly promising are discussed below, but the stacked a-Si:H/μc-Si cells discussed in the previous section (Fig. 9.7) come close to fulfilling all the previous third-generation criteria. The fundamental efficiency limit for such a two-cell stack is 42.9% under omnidirectional light, compared with 31% for a single cell, a 38% relative performance gain. In comparison, the manufacturers of the "hybrid" a-Si:H/μc-Si module in Fig. 9.8 expect a 32% relative gain over the same company's single-junction a-Si:H module (labeled "CSA/CEA").

Generalizing to an $n$-cell tandem stack as shown in Fig. 9.14, where each cell has a slightly lower bandgap than its neighbor on top, the limiting efficiency under omnidirectional light increases to 68.2% as $n$ approaches infinity, quite close to the thermodynamic limit of 73.7%. It turns out that 68.2% efficiency is the limit for a normal time-symmetric conversion system, with the higher value possible only by the use of time-asymmetric components such as circulators [30].

For a tandem stack involving a large number of cells, a generic approach to cell fabrication would be ideal to keep costs under control. One such approach is suggested in Fig. 9.15, where parameters of a quantum well or quantum dot superlattice are controlled in a way to give steadily decreasing bandgap in the same materials system. This gives scope for depositing such stacks in a single deposition chamber by controlling the flow of the source gases.

Given that the (poly-)crystalline silicon on glass (CSG) approach is one of the most promising thin-film approaches and that it admirably meets all third-generation criteria except efficiency, such a generic approach to providing ongoing CSG efficiency improvements is particularly relevant. Our group is exploring

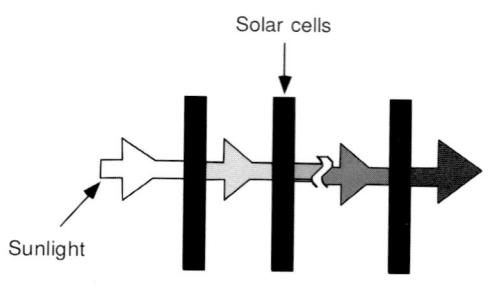

**Fig. 9.14** Tandem cell stack involving $n$ cells.

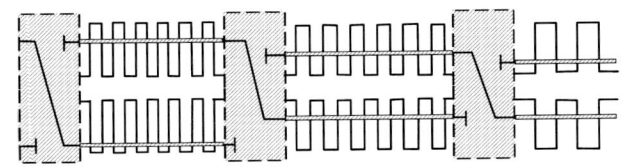

**Fig. 9.15** Generic approach to a tandem stack using bandgap control by quantum confinement in superlattices of quantum wells or quantum dots.

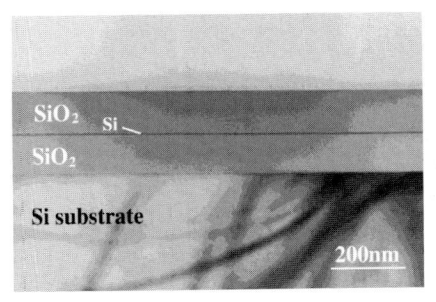

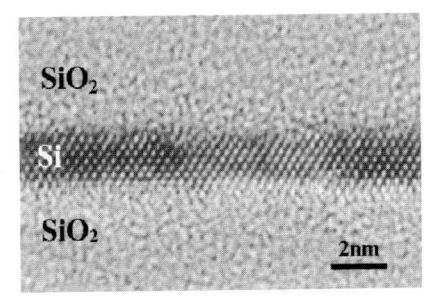

**Fig. 9.16** Quantum well based on a thin layer of silicon formed from a commercial SOI (silicon on insulator) wafer shown at two different magnifications.

the potential of quantum dot and quantum well superlattices compatible with this approach, where the dot or well is crystalline silicon while the barrier region is amorphous silicon oxide, nitride, or carbide. The first stage in this work was to demonstrate increased bandgap by the associated quantum confinement, using structures such as that in Fig. 9.16.

This stage has been successful, with photoluminescence measurements confirming the required bandgap control (Fig. 9.17). Bandgap as high as 1.7 eV was measured for layers of about 1-nm thickness. This is precisely the required bandgap for a two-cell stack using the CSG approach. Quantum dots of an intermediate size should allow a bandgap of 2.0 eV to be demonstrated, as required for the top cell of a three-cell stack, or even the value of 2.4 eV for a six-cell stack.

The photoluminescence intensity increased strongly for the intermediate thickness layers, demonstrating the expected strengthening of optical processes due to confinement. This means that each cell may need to be only 100 nm or so in thickness, minimizing both material and carrier mobility requirements. Present work within our group involves a systematic evaluation of these issues.

The stacked-cell approach accommodates the wide spectral content of sunlight by directing different spectral components to cells of an appropriate bandgap for efficient conversion. An alternative high-efficiency strategy is to try to manipulate the incoming spectrum itself, by photon frequency up- and down-con-

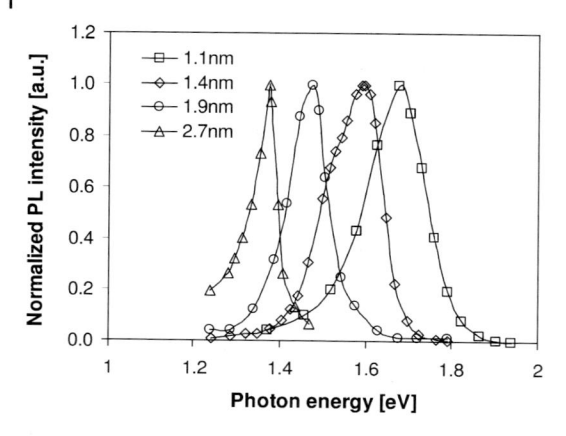

**Fig. 9.17** Luminescent intensity versus wavelength for silicon quantum wells of varying thickness.

version, so that it is well matched to what a single target cell can convert efficiently.

Fig. 9.18 shows up- and down-converters together with a target cell. The limiting performance of each type of conversion was analyzed recently, with two surprising conclusions [31, 32]. One is that a down-converter, which ideally absorbs incoming high-energy photons and converts each to two lower-energy photons, can be placed on the front of a cell without seriously impeding the transport of solar photons across it [31]. The second is that the efficiency of an up-converter, placed on the rear of the cell to convert two low-energy photons to a single useable photon, can be improved if energy relaxation occurs during the two-step transition involved [32]. Although efficient up- or down-converters are not generally available, these encouraging findings have prompted experimental work in this area, resulting in a very small, but measureable, increase in performance in a silicon test cell by the use of a rear down-converter [33]. An attraction of this

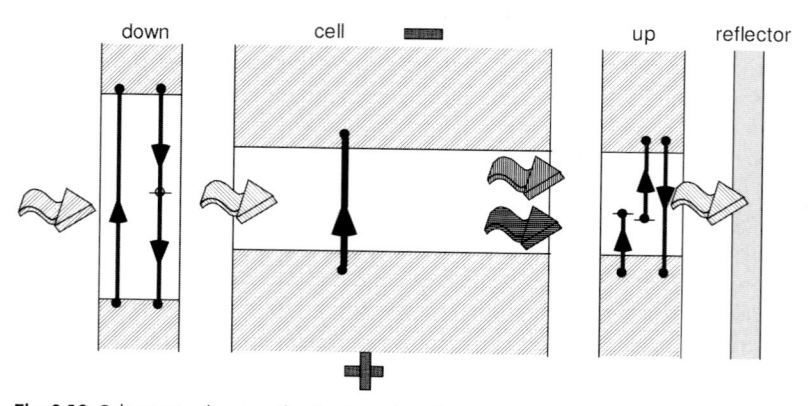

**Fig. 9.18** Schematic showing the front-surface down-conversion and the rear-surface up-conversion of light, resulting in more useable photons reaching the target cell.

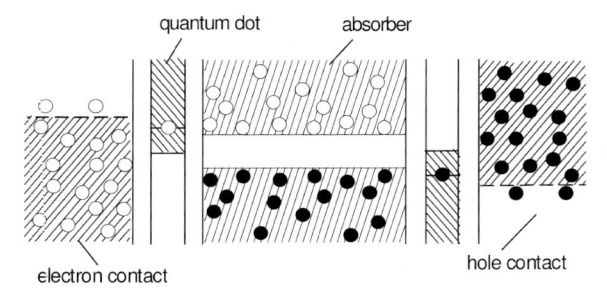

**Fig. 9.19** Schematic of a hot carrier cell showing hot carriers in the absorber and the monoenergetic transfer of electrons between the absorber conduction and valence bands and their respective contacts.

approach is that no electrical contact to the up-converters is required, since they are purely optical elements.

Perhaps the most elegant but also the most challenging third-generation approach yet suggested is the hot carrier cell of Fig. 9.19. In this approach, the normal lattice thermalization of photogenerated carriers in the absorber is suppressed in some as yet unidentified way, allowing "hot" carrier populations to evolve in the absorber. This would eliminate one of the major loss processes in conventional cells. Finding ways of preventing this lattice thermalization is a major challenge, but thin devices with strong optical processes, superlattices, and phononic engineering may have some potential [28]. As if this challenge were not enough, the interface between the absorber and the outside world, as represented by the device contacts, also has to have some quite specific features. To prevent cooling of carriers in the absorber by those in the contacts, the required transfer of carriers between these regions has to occur at a single energy. Resonant tunneling through quantum dots or the equivalent is the idea we are exploring to achieve this [28]. The motivation for these efforts is that the hot carrier cell, although a "simple" device with two terminal contacts, offers the same efficiency potential, in principle, as an infinite stack of tandem cells.

## 9.5
## Conclusions

The photovoltaic market is presently dominated by first-generation wafer-based products, while second-generation thin-film products are struggling to gain even a small market share. The recently developed (poly-)crystalline silicon on glass (CSG) technology would seem to overcome many of the limitations of the more established thin-film candidates and provides an in silico path to low costs in the longer term.

However, the low present energy conversion efficiency of commercial products, bounded by basic thermodynamics to a value of 31%, is not a consequence of any fundamental characteristic of solar energy, such as its diffuse nature. A general analysis suggests that efficiency as high as 74% is possible in principle. In the very long term, photovoltaics will have to access at least some of this performance differential if it is to reach its full commercial potential. This suggests a further transition to a third generation of high-efficiency thin-film products with other key attributes such as the use of abundant, nontoxic, stable, and durable materials.

Possible third-generation candidates are "all-silicon" tandem cell stacks where quantum confinement is used to provide the required control of bandgap; the use of up- and down-converters to manipulate the incoming spectrum; and hot carrier cells, which offer severe challenges in implementation but relatively simple final structures.

## References

1 F. Ferrazza, Crystalline silicon: Manufacture and properties, in *Practical Handbook of Photovoltaics*, T. Markvart and L. Castene (eds.), Elsevier, Oxford, 2003, 137.
2 S. von Aichberger, Strong east wind: market survey on photovoltaic modules, *Photon International*, February, 2004, 46–55.
3 K. Otani, T. Takashima, K. Sakuta, T. Yamaguchi, Performance analysis of hundred Japanese residential grid-connected photovoltaic systems based on five years' experience, Proceedings, *Int. Solar Eng. Society Conf.*, Stockholm, June, 2003.
4 M. A. Green, Silicon *Solar Cells: Advanced Principles and Practice* (Bridge Printery, Sydney, 1995).
5 J. Bernreuter, Higher, higher, the highest, *Photon International*, 5, 48–52, May, 2003.
6 V. Parente, J. Goldenberg, R. Zilles, *Prog. in Photovoltaics*, 2002, 10, 571–574.
7 A. McDonald, L. Schrattenholzer, Learning rates for energy technologies, *Energy Policy 20*, 255–261, 2001.
8 P. R. MacGregor, C. E. Maslak, H. G. Stoll, The market outlook for integrated gasification combined cycle technology, General Electric Company, New York, 1991.
9 L. Neij, Experience curves for wind turbines energy, *The International Journal* 24, 5, 1999.
10 W. P. Hirshman, Japanese PV program secure through FY 2005 as subsidies halved, *Photon International*, 28, 2004.
11 Martin A. Green, *SOLAR CELLS: Operating Principles, Technology and System Applications* (Prentice-Hall, New Jersey, 1982).
12 J. Yang, A. Banerjee, T. Glatfelter, K. Hoffman, X. Xu, S. Guha, *Conf. Record, $1^{st}$ World Conference on Photovoltaic Energy Conversion*, 380–385, 1994
13 R. H. Bube, *Photovoltaic Materials*, Imp. College Press, London, 1998.
14 E. Oszan, E. Dunlop, Workshop on Stability and Yield Issues in Module Production, *Symposium B, Thin-Film Chalcogenide Photovoltaic Materials*, European Materials Research Society, Strasbourg, June, 2002.
15 J. Malmstrom, J. Wennergerg, L. Stolt, A study of the influence of the Ga content on the long-term stability of Cu(In,Ga)Se$_2$ thin-film solar cells, *Symposium B, Thin-Film Chalcogenide Photovoltaic Materials*, European Materials Research Society, Strasbourg, June, 2002.
16 V. Probst, W. Stetter, J. Palm, R. Toelle, S. Visbeck, H. Calwer, T. Niesen, H. Vogt, O. Hernandez, M. Wendl, F.H. Karg, CIS module pilot processing from fundamental investigations to advanced performance, *Conf. Record, 3rd World Conference of Photovoltaic Energy Conversion*, Osaka, May, 2003.

17 M. Powalla, M. Cemernjak, J. Eberhardt, F. Kessler, R. Kniese, H. D. Mohring, Large-area CIGS modules: Pilot line production and new developments, *Technical Digest, International PVSEC-14*, Bangkok, January, 2004, 513.

18 B. J. Andersson, S. Jacobsson, Monitoring and assessing technology choice: the case of solar cells, *Energy Policy*, 28, 1037–1049, 2000.

19 W. Fuhs, Crystalline silicon thin-film technology for photovoltaics, *PVNET Workshop Proceedings, RTD Strategies for PV*, Ispra, 172–175, May, 2002.

20 C. Wang, G. Lucovsky, Intrinsic microcrystalline silicon deposited by remote PECVD: a new thin-film photovoltaic material, Conf. Record, *21st IEEE Photovoltaic Specialists Conference*, Kissimimee, 1614–1618, 1990.

21 K. Yamamoto, A. Nakajima, M. Yoshimi, T. Sawada, S. Fukuda, K. Hayashi, T. Suezaki, M. Ichikawa, Y. Koi, M. Goto, H. Takata, Y. Tawada, High efficiency thin-film silicon solar cell and module, *Conf. Record, 29th IEEE Photovoltaic Specialists Conf.*, New Orleans, May, 2002.

22 LEEE News, *Newsletter of the Laboratory of Energy, Ecology and Economy*, 3, 1, 2002.

23 M. A. Green, Solar cell research and development in Australia, *Solar Cells* 26, 1–11, 1989.

24 C. E. Backus (ed.), Solar cells, *IEEE*, New York, 1976, 404.

25 M. A. Green, Surface texturing and patterning in solar cells, in *Advances in Solar Energy*, 8, M. Prince (ed.), American Solar Energy Society, Boulder, 1993, 231–269.

26 M. A. Green, P. A. Basore, N. Chang, D. Clugston, R. Egan, R. Evans, J. Ho, D. Hogg, S. Jarnason, M. Keevers, P. Lasswell, J. O'Sullivan, U. Schuberg, A. Turner, S. R., Wenham, T. Young, Crystalline silicon on glass (CSG) thin-film solar cell modules, *Solar Energy, Special Issue on Thin Film Solar Cells*, 2004.

27 P. A. Basore, Large area deposition for crystalline silicon on glass modules, *Conf. Record, 3rd World Conference of Photovoltaic Energy Conversion*, Osaka, May, 2003.

28 M. A. Green, *Third Generation Photovoltaics: Advanced Solar Electricity Generation*, Springer, Berlin, 2003.

29 P. T. Landsberg, G. Tonge, Thermodynamic energy conversion efficiencies, *J. Appl. Phys.*, 51, R1–R20, 33, 1980.

30 H. Ries, Complete and reversible absorption of radiation, *Applied Phys. B* 32, 153, 1983.

31 T. Trupke, M. A. Green, P. Würfel, Improving solar cell efficiencies by down-conversion of high-energy photons, *J. Appl. Phys.*, 92, 1668–1674, 2002.

32 T. Trupke, M. A. Green, P. Würfel, Improving solar cell efficiencies by up-conversion of high-energy photons, *J. Appl. Phys.*, 92, 4117–4122, 2002.

33 T. Trupke, A. Shalav, P. Würfel, M. A. Green, Efficiency enhancement of solar cells by luminescent up-conversion of sunlight, *Tech. Digest, International PVSEC-14*, Bangkok, 753, 2004.

# 10
# Mimicking Bacterial Photosynthesis

*Devens Gust, Thomas A. Moore, and Ana L. Moore*

## 10.1
### Introduction

Photosynthesis, the process by which organisms harvest sunlight and convert it into useful electrochemical energy, is a prime example of naturally occurring nanotechnology. Complex supramolecular architectures have evolved to carry out the following steps in the conversion process:

1. Absorption of sunlight by light-harvesting antenna systems.
2. Transfer of the absorbed singlet excitation energy to the photochemical reaction center.
3. Primary charge separation, whereby light energy is converted by the reaction center into electrochemical potential energy.
4. Energy transduction, in which the electrochemical energy produced by the reaction center is converted to chemical energy that can be used and stored by the organism.

Photosynthesis fills humanity's food requirements; provides fuel in the form of coal, petroleum, natural gas, and firewood; and is the source of cellulosic building materials, paper, etc. It is also responsible for the oxygen in the atmosphere and removes potentially harmful carbon dioxide from the air. Given the importance and durability of this natural solar conversion "technology," it is not surprising that the idea of artificial photosynthesis has been current for about 100 years [1]. Mimicry of the natural process promises to provide new approaches to solar energy conversion, principles and materials for molecular-scale optoelectronics, and new information about how the natural process occurs. In this chapter, we will briefly describe natural photosynthesis as it occurs in bacteria and then exemplify some approaches to its biomimicry. The emphasis will be upon work performed in our own laboratories, although many researchers around the world have contributed substantially to the current state of knowledge.

*Artificial Photosynthesis: From Basic Biology to Industrial Application*
Edited by Anthony F. Collings and Christa Critchley
Copyright © 2005 WILEY-VCH Verlag GmbH & Co. KGaA, Weinheim
ISBN: 3-527-31090-8

## 10.2
## Natural Photosynthesis

Photosynthetic purple bacteria, some of the least complex organisms that carry out the process, employ a relatively straightforward set of elementary energy-conversion steps that are cyclic in terms of the materials involved (Fig. 10.1). These steps do not require water oxidation, a process that thus far eludes our understanding, and need no external donor or acceptor species to supply or remove electrons. Thus, they are attractive candidates for biomimicry. The various components of the natural system are proteins, with associated chromophores and other cofactors, which span a lipid bilayer membrane. Sunlight is absorbed by light-harvesting antenna arrays containing hundreds of bacteriochlorophyll and carotenoid chromophores. These colored materials absorb light in the visible and near-infrared regions. Photosynthesis is constrained to work in this spectral region for two reasons. First of all, relatively little ultraviolet light strikes the earth's surface, and therefore light in that spectral region cannot contribute much to overall energy conversion. Infrared light, on the other hand, is abundant, but its photons contain too little energy to perform the necessary chemical reactions.

The antennas of each organism are tuned to absorb light according to the solar spectrum available in the surrounding habitat. Because environments

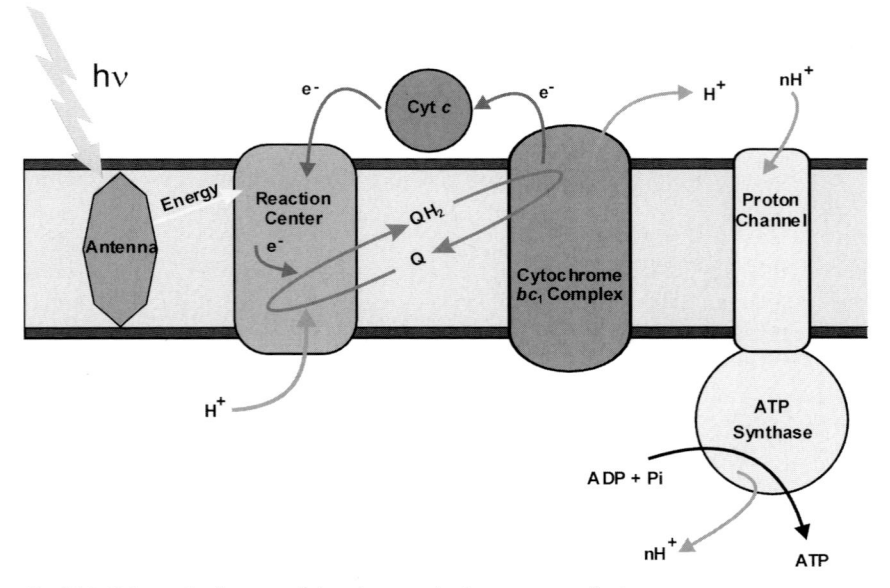

**Fig. 10.1** Schematic diagram of the photosynthetic apparatus of a bacterium. The horizontal bars define a lipid bilayer membrane that contains the proteinaceous components of the solar conversion system, which are represented as geometrical shapes.

(deserts, oceans, rain forests, etc.) differ considerably in the quantity and quality of sunlight available, antenna systems show marked variability across the range of photosynthetic organisms. Once absorbed by the antenna, light energy is preserved in the absorbing chromophore as singlet electronic excitation energy. This excitation energy is transported spatially to the chromophores of the photosynthetic reaction center via singlet-singlet energy transfer processes.

In the reaction center, the excitation energy localizes on a "special pair" of closely associated bacteriochlorophyll molecules. Within about 3 ps, an electron is transferred from the special pair to a nearby bacteriopheophytin (a chlorophyll lacking the central magnesium atom) via a monomeric accessory bacteriochlorophyll. From there, an electron moves to a quinone electron acceptor, and finally to a second quinone. At this stage, a significant amount of the photon energy has been captured as transmembrane charge separation (electrochemical energy). The net result of this process is a large spatial separation between the electron on the quinone and the positive charge (hole) on the special pair, which slows charge recombination and allows sufficient time for the organism to use the stored energy. Transmembrane electron transport via multiple steps is necessary because transfer across the membrane in a single step is too slow to compete with the photophysical processes that rapidly drain the energy from the special pair excited state and convert it into useless heat. After two sequential photoinduced electron transfer events, the final quinone acceptor has been reduced to a hydroquinone and leaves the reaction center.

The bacterium employs the energy of the reduced quinone to power a transmembrane hydrogen ion pump. The cytochrome $bc_1$ complex oxidizes the hydroquinone back to a quinone, using the electrons produced to reduce the special pair bacteriochlorophylls back to their original state and the energy released to transport protons across the membrane. Since the membrane is not permeable to hydrogen ions, a proton concentration gradient (proton-motive force, or pmf) develops across the membrane. Some of the light energy is thus converted into chemical potential energy stored in the proton gradient. Energy stored in such gradients is central to bioenergetics. Non-photosynthetic organisms establish a proton-motive force via their mitochondria by oxidizing sugars or other (ultimately photosynthetic) products.

Another enzyme, ATP synthase, spans the bacterial membrane and allows protons to travel through it in the direction of the proton gradient established by the cytochrome $bc_1$ complex. The energy released by this exergonic process is used to convert adenosine diphosphate (ADP) into energy-rich adenosine triphosphate (ATP), which is used to power essentially all energy-requiring processes in the organism. Thus, the overall result of bacterial photosynthesis is conversion of light energy into chemical energy stored in the bonds of ATP. The process is cyclic with respect to all of the conversion components.

Green plant photosynthesis is similar in many ways to bacterial photosynthesis. However, in addition to ATP, chemical reducing equivalents are generated that can be used to convert carbon dioxide into energy-rich carbohydrates. These electrons are obtained by oxidation of water, molecular oxygen being the byproduct. Two different photosynthetic reaction centers operating in series are necessary to generate the chemical potential energy necessary for the overall process.

## 10.3
## Artificial Photosynthesis

Many approaches to artificial photosynthesis are being pursued. Some of these employ the basic process of photoinduced electron transfer but use "non-biological" components such as semiconductors of various sorts or transition-metal complexes. In this chapter, we will focus on artificial photosynthetic systems that employ organic chromophores and are in this way closer to the natural process. A large number of molecules that mimic one or more aspects of this process have been studied, and reviews in the literature should be consulted for details [2–16].

### 10.3.1
### Artificial Antenna Systems

The basic photophysical process underlying antenna function is singlet-singlet energy transfer:

$$^1D + A \rightarrow D + {}^1A \tag{1}$$

Absorption of light by a donor chromophore $D$ generates an electronically excited singlet state $^1D$. This state transfers its excitation energy to an acceptor chromophore $A$, relaxing back to the ground state and generating an electronically excited acceptor $^1A$. The resulting acceptor state is the same as would result from direct light absorption by $A$. In an antenna, light is transferred from the initially absorbing chromophore, via a possibly large number of intermediate antenna chromophores, and thence on to the reaction-center chromophores. The process requires electronic interaction of the donor and acceptor that is strong enough so that energy transfer is much faster than the usual photophysical decay processes that depopulate the excited singlet states. The most common mechanism for singlet-singlet energy transfer is the Förster, or dipole-dipole, mechanism [17], which requires relatively close spatial proximity of the donor and acceptor; no overlap of electronic orbitals is necessary.

Although artificial antenna arrays based on self-assembly are under study, the majority of the artificial systems use covalent bonds to join the chromophores and ensure the necessary spatial proximity. There are many examples in the literature. We will illustrate the field with pentad **1**, which is shown in Fig. 10.2 [18]. The molecule consists of four zinc porphyrin chromophores covalently linked to a free-base porphyrin (in which the zinc atom is replaced by two hydrogen atoms). The zinc porphyrins act as an antenna. They collect light and

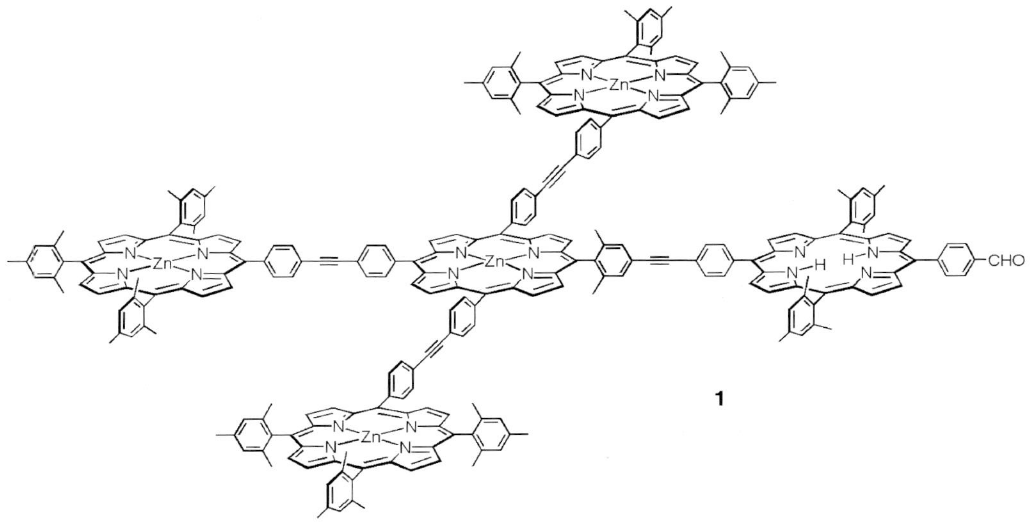

**Fig. 10.2** Porphyrin-based artificial photosynthetic antenna array **1**.

transport the excitation energy to the free-base porphyrin. This energy transfer is thermodynamically favored, as the energy of the first excited singlet state of the free-base porphyrin is 0.15 eV lower than those of the zinc porphyrins.

The photophysical events following absorption of light by molecule **1** have been studied using laser-based, time-resolved spectroscopic techniques. Investigations show that absorption of light by a peripheral zinc porphyrin is followed by relaxation of that chromophore to its first excited singlet state. Normally, this state would decay to the ground electronic state in $2.4 \times 10^{-9}$ s by the usual photophysical pathways of fluorescence, internal conversion, and intersystem crossing. In the pentad, there is a much faster decay pathway. This is singlet excitation energy transfer to the central zinc porphyrin. The time constant for this essentially isoenergetic process is $5.0 \times 10^{-11}$ s. Excitation energy on the central zinc porphyrin migrates in turn to the free-base porphyrin with a time constant of $3.2 \times 10^{-11}$ s. As these two time constants are much smaller than that for decay of the zinc porphyrin excited states by other photophysical processes, nearly every photon absorbed by the zinc antenna porphyrins is delivered as excitation energy to the free-base porphyrin. The free-base porphyrin excited state has a lifetime of $1.1 \times 10^{-8}$ s, ultimately decaying by the photophysical processes mentioned above.

In **1**, the zinc porphyrin array acts as an efficient antenna for the free-base energy sink. A variety of other approaches to forming artificial antenna systems are being investigated by various research groups. The antenna systems of photosynthetic organisms usually contain large numbers (tens, hundreds, or even more) of chromophores. Thus, it seems reasonable to expect that very large synthetic antennas will most easily be prepared by polymerization or self-assembly techniques.

Carotenoid polyenes play several very important roles in photosynthesis, including energy harvesting as antenna chromophores. Carotenoids absorb strongly in spectral regions where chlorophylls are not particularly good absorbers, and hence are excellent supplemental antennas. Singlet energy transfer from carotenoids to chlorophylls occurs with quantum yields approaching 1.0 in some photosynthetic systems. Thus, construction of synthetic systems with carotenoid antennas would seem to be an obvious goal.

Upon closer examination, the use of carotenoids as antennas for cyclic tetrapyrroles requires surmounting several serious problems. The light absorption transition from the carotenoid ground state ($S_0$) to the first excited singlet state ($S_1$) is formally symmetry forbidden in symmetric carotenoids. Thus, the extinction coefficient for the absorptive transition is too small to measure, and the fluorescence quantum yield from $S_1$ is negligible. The lifetime of $S_1$ is typically about $1 \times 10^{-11}$ s. Thus, the $S_1$ state is both a poor light absorber and a poor energy donor via the Förster dipole-dipole mechanism. By contrast, the transition to the $S_2$ state, typically in the 480-nm region, is highly allowed, and carotenoids show strong absorption in that spectral region. As is the case with the upper excited states of most molecules, the $S_2$ state is short-lived (ca. $1.5 \times 10^{-13}$ s), and this is the main obstacle to obtaining a high yield of singlet energy transfer from that state. Taken together, these factors suggest that in order to obtain a reasonable yield of singlet energy transfer from a carotenoid to a porphyrin or chlorophyll, the two chromophores would have to be in very close physical proximity. Such association would favor rapid transfer by the Förster mechanism, and possibly by an electronic orbital overlap (Dexter-type) mechanism.

Suitably designed, covalently linked carotenoid-porphyrin systems were found to demonstrate a degree of singlet energy transfer over 20 years ago [19–21]. More recently, energy transfer efficiencies approaching 100% have been achieved. An example is triad **2**, which consists of a phthalocyanine linked covalently to two carotenoids via the central silicon atom (Fig. 10.3) [22]. The molecule, dissolved in toluene solution, was investigated using several spectroscopic methods. Fig. 10.4 shows the absorption spectrum of **2** as a solid line. The absorbance in the 400–500 nm region is due mainly to the carotenoids, whereas the remainder of the absorption spectrum is attributed to the phthalocyanine. The figure also shows the corrected fluorescence excitation spectrum as a dotted line. This spectrum was obtained by monitoring the phthalocyanine fluorescence intensity at 790 nm while the excitation wavelength was varied from 300–750 nm. The excitation spectrum was normalized to the absorption spectrum at 693 nm, where only the phthalocyanine absorbs light. Note that the fluorescence excitation spectrum and the absorption spectrum are nearly coincident. This indicates that essentially all of the light absorbed by the carotene gives rise to fluorescence from the phthalocyanine, showing that the efficiency of singlet energy transfer from the carotene to the phthalocyanine is >95%.

Time-resolved absorption and emission studies of **2** allowed investigation of the kinetic details of the energy transfer process. It was found that energy transfer from the $S_2$ state of the carotenoid occurs with an efficiency of 35%. The re-

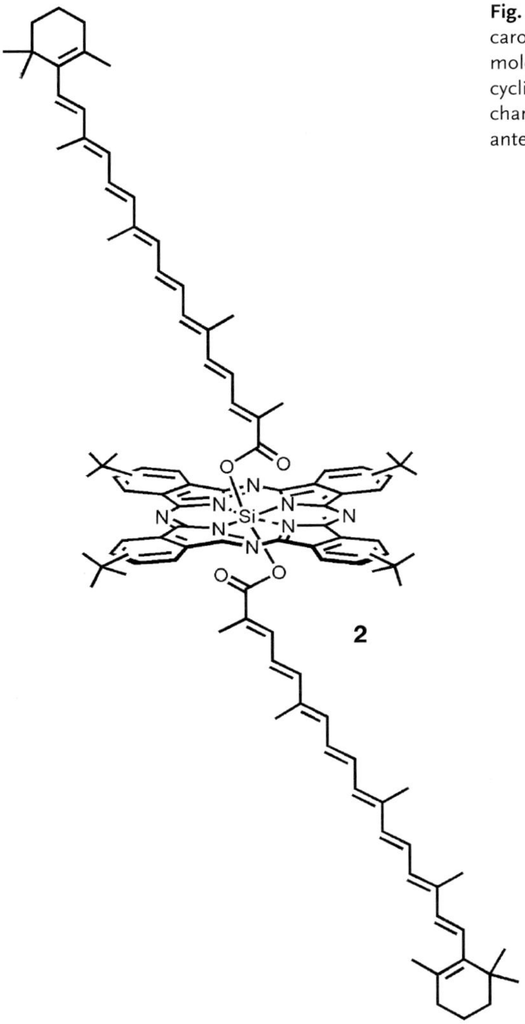

**Fig. 10.3** A phthalocyanine bearing two carotenoid polyenes. This synthetic molecule, **2**, mimics the carotenoid-to-cyclic tetrapyrrole energy transfer characteristic of photosynthetic antenna systems based on carotenes.

mainder of the $S_2$ state relaxes to the $S_1$ state, which also donates energy to the phthalocyanine. Overall, the efficiency of energy transfer, based on the kinetics of the excited states, is calculated to be ca. 93%, in excellent agreement with the experiment shown in Fig. 10.4. Recent work in this and related systems suggests that there may be excited states lying in energy between the two excited states described above, and that these may play a role in the overall energy transfer process. Triad **2** illustrates that even in synthetic systems, carotenoids can be excellent antennas for energy-conversion purposes if the electronic interactions between the carotene and porphyrin or other energy acceptor are carefully controlled.

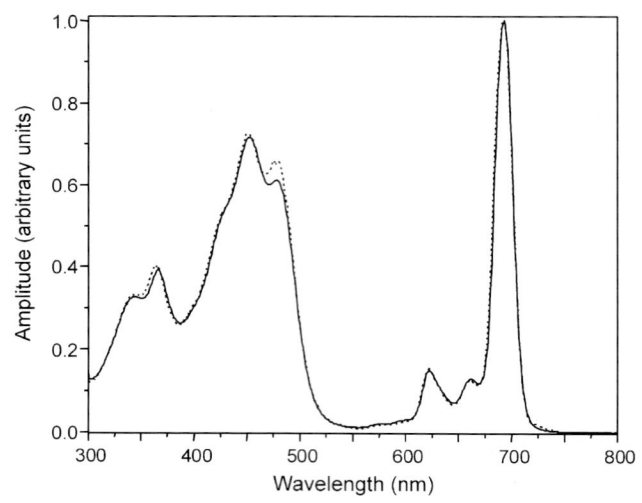

**Fig. 10.4** Spectral results for antenna mimic **2**. The solid line is the absorption spectrum of **2** in toluene solution. The dotted line is the corrected excitation spectrum for phthalocyanine emission at 790 nm. The excitation spectrum nearly tracks the absorption spectrum, even in the 450–500 nm region where most of the absorption is due to the carotenoid polyenes. This signifies essentially complete singlet energy transfer from the carotenes to the phthalocyanine.

## 10.3.2
### Artificial Reaction Centers

The fundamental process underlying photosynthetic reaction center function is photoinduced electron transfer:

$$^1D + A \rightarrow D^{\bullet+} + A^{\bullet-} \tag{2}$$

As with the antennas, the donor species is again an excited chromophore, but it relaxes by transferring an electron (rather than excitation energy) to a nearby acceptor, generating a charge-separated state that preserves some of the excitation energy as electrochemical potential. Thus, the redox properties of the donor and acceptor are important, in addition to their photophysical properties. Excited states of molecules are both better electron donors and better electron acceptors than the corresponding ground states. This is because the excited state has a vacancy in the orbital corresponding to the highest occupied molecular orbital of the ground state, and an electron in the higher-energy orbital corresponding to the lowest unoccupied molecular orbital of the ground state. The high-energy electron can be readily donated to an appropriate acceptor, or the vacancy can be filled by electron donation from a nearby moiety. An electron transfer process requires a suitable energetic driving force ($-\Delta G^0$ for the reaction). In practice, it is important to choose the energies of the donor chromophore and electron acceptor species so that the energy of the charge-separated state is lower than that of the excited state, but not so low that energy is wasted.

In addition to a suitable driving force, sufficient electronic coupling between the initial and final states is required for rapid photoinduced electron transfer. Suitable electronic orbital overlap can be achieved either by direct overlap of the donor and acceptor orbitals ("through space") or by participation of the bonds of a chemical linker joining the donor and the acceptor ("superexchange"). In most of the electron transfer reactions discussed in this review, transfer occurs in the nonadiabatic regime. Within this framework, the work of Marcus, Hush, Levitch, and others provides a good theoretical basis for understanding the electron transfer process [23–26]. Equation (3) has been developed to describe electron transfer rate constants ($k_{et}$) under these conditions.

$$k_{et} = \sqrt{(\pi/\hbar^2 \lambda k_B T)} |V|^2 \exp\left[-(\Delta G^0 + \lambda)^2 / 4\lambda k_B T\right] \tag{3}$$

The pre-exponential factor includes the electronic matrix element $V$ that describes the coupling of the reactant state with that of the product. Element $V$ is a function of the electronic overlap of the donor and acceptor orbitals. In addition to Planck's constant $\hbar$, Boltzmann's constant $k_B$, and the absolute temperature $T$, the pre-exponential factor also includes the reorganization energy for the reaction, $\lambda$. The reorganization energy is associated with the nuclear motions involved in transforming the molecule from the initial to the final state. It is convenient to express $\lambda$ as the sum of a solvent-independent term $\lambda_i$, which originates from internal molecular structural differences between the reactant and product, and $\lambda_s$, the solvent reorganization energy, which is due to differences in the orientation and polarization of solvent molecules around the (usually neutral) ground state and the (usually zwitterionic) charge-separated state. The exponential term, the Franck-Condon factor, includes the standard free-energy change for the reaction, $\Delta G^0$, as well as $\lambda$.

Electron transfer as described by Eq. (3) may occur in three regimes. In the "normal" region, increasing the thermodynamic driving force for electron transfer (as $\Delta G^0$ becomes more negative) leads to more rapid electron transfer, as would be expected for the usual chemical reaction. However, when $-\Delta G^0$ equals $\lambda$, the rate of electron transfer is maximized as a function of driving force, and electron transfer becomes activationless. In the third regime, when $-\Delta G^0 > \lambda$, increasing the driving force leads to a *decrease* in the electron transfer rate constant. This is the "inverted" region of Eq. (3). Semi-classical refinements of Eq. (3) are considered more appropriate for describing electron transfer rates in the "inverted" region [27, 28].

Based on the above, the simplest "artificial reaction center" will consist of an electron donor (or acceptor) chromophore electronically coupled to an electron acceptor (or donor). Ideally, this chromophore will absorb light strongly in the visible region of the spectrum, so as to function using natural sunlight. Thus, many otherwise-useful UV-absorbing organic chromophores are excluded, and visible-absorbing molecules such as porphyrins or other chlorophyll relatives are usually chosen. The first covalently linked molecules of this type designed as photosynthesis mimics consisted of a synthetic porphyrin bonded to a quinone electron acceptor [29, 30]. Since that time, hundreds of other systems have been prepared [8, 31].

The function of such molecules will be illustrated using porphyrin (P) fullerene ($C_{60}$) dyad **3** (Fig. 10.5), where the $C_{60}$ moiety plays the role of the electron acceptor [32]. Excitation of this molecule at wavelengths where the porphyrin absorbs light yields the porphyrin first excited singlet state, $^1P\text{-}C_{60}$. By using transient spectroscopic techniques, the fate of this state on very short timescales may be investigated. Such studies in the solvent 2-methyltetrahydrofuran at ambient temperatures show that after excitation, the porphyrin transfers an electron to the fullerene with a time constant of $2.5\times10^{-11}$ s, giving a $P^{\bullet+}\text{-}C_{60}^{\bullet-}$ charge-separated state. The state is generated with a quantum yield of unity, indicating that essentially every photon absorbed gives rise to charge separation. The porphyrin first excited singlet state is 1.9 eV above the ground state, and $P^{\bullet+}\text{-}C_{60}^{\bullet-}$ is at 1.67 eV. Thus, charge separation is achieved with high yield, and the resulting state stores a significant fraction of the excited state energy as chemical potential.

In spite of these facts, **3**, and indeed essentially all two-part donor-acceptor systems, is not a good mimic of photosynthetic reaction centers. This is because there is a strong tendency for the electron to return to the porphyrin radical cation, regenerating the ground state of the molecule and wasting the stored energy as heat. In **3**, this charge recombination process occurs in $3\times10^{-9}$ s. Rapid charge recombination was a major stumbling block in the development of artificial reaction centers for some time.

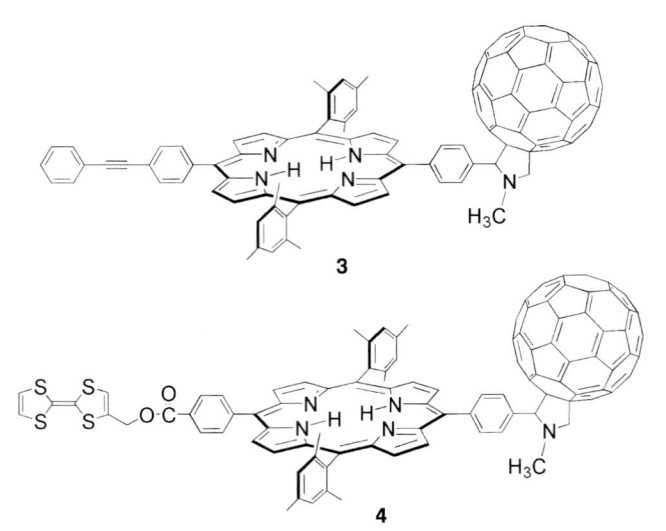

**Fig. 10.5** Structures of artificial photosynthetic reaction centers. Molecule **3** is a typical donor-acceptor dyad. The triad **4** features a tetrathiafulvalene moiety as a secondary electron donor.

The lifetimes of charge-separated states may be extended by using a multi-step electron transfer strategy, as is employed by natural reaction centers. If the electron and "hole" (radical cation) can be further separated before charge recombination occurs, the electronic coupling, and therefore the rate constant for recombination, may be drastically reduced. This was first done in artificial systems in the early 1980s [33–36], and the method has been used extensively since that time by our group and others. It will be illustrated with tetrathiafulvalene (TTF) porphyrin fullerene molecular triad **4** (Fig. 10.5) [32].

Triad **4** is similar to dyad **3**, but a TTF secondary electron donor has been added. Excitation of the porphyrin moiety of this molecule gives TTF-$^1$P-$C_{60}$, which undergoes photoinduced electron transfer as in **3** to yield TTF-P$^{\bullet+}$-$C_{60}^{\bullet-}$. The time constant for electron transfer in 2-methyltetrahydrofuran is $2.5 \times 10^{-11}$ s. Charge shift from the TTF secondary electron donor to the porphyrin radical cation, with a time constant of $2.5 \times 10^{-10}$ s, is faster than charge recombination (time constant = $3.0 \times 10^{-9}$ s), and the result is a TTF$^{\bullet+}$-P-$C_{60}^{\bullet-}$ charge-separated state, formed with an overall yield of 92%. In TTF$^{\bullet+}$-P-$C_{60}^{\bullet-}$, the radical ions are insulated from one another due to the presence of the intervening porphyrin, and this reduction in electronic coupling decreases the rate of charge recombination, relative to that in dyad **3**. The lifetime of TTF$^{\bullet+}$-P-$C_{60}^{\bullet-}$ is $6.6 \times 10^{-7}$ s. The increase in lifetime by a factor of 220 comes at the expense of stored energy (in **4**, the final state has an energy $\sim 1.0$ eV above the ground state), but the final state persists long enough that the stored energy can be accessed via subsequent diffusional or interfacial electron transfer processes.

Many artificial photosynthetic reaction center molecules based on these general principles have been prepared and studied. Some of these are very complex, with multiple donor and acceptor units, and some have charge-separation lifetimes on or near the millisecond timescale in solution at ambient temperatures [8]. One example is pentad **5** (Fig. 10.6), which illustrates how more complex molecular systems introduce new possibilities for the control of electron and energy transfer [37, 38].

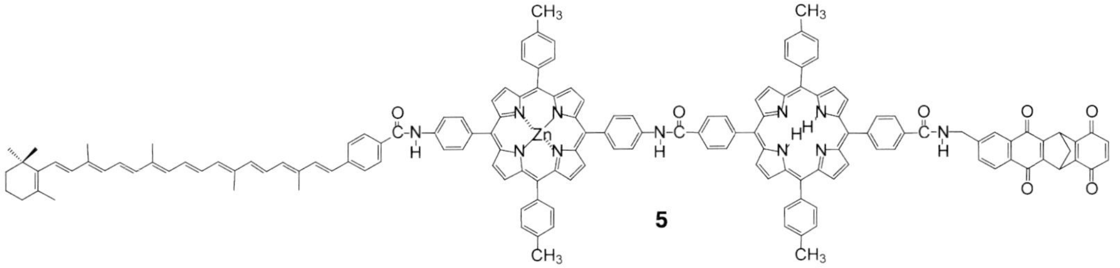

**Fig. 10.6** Pentad artificial reaction center **5**, containing a covalently linked zinc porphyrin-free-base porphyrin dyad. The zinc porphyrin bears a carotenoid secondary electron donor, and the free base a naphthoquinone electron acceptor, which in turn is linked to a benzoquinone.

The multi-step electron transfer strategy used by triad **4** and natural reaction centers is sequential. Photoinduced electron transfer produces an initial charge-separated state, and a series of subsequent dark reactions moves the positive and negative charges apart and thus enhances the charge-separation lifetime. In pentad **5**, parallel multi-step electron transfer pathways are also observed. At the heart of the pentad is a zinc porphyrin ($P_{Zn}$) covalently linked to a free-base porphyrin (P). The free-base porphyrin bears a naphthoquinone electron acceptor (NQ), which in turn is attached to a benzoquinone (Q). The zinc porphyrin is linked to a carotenoid (C), which serves as an electron donor rather than as an antenna moiety.

The pentad is designed so that photoinduced electron transfer from the free-base porphyrin to the attached naphthoquinone is followed by a cascade of electron transfer pathways that converge upon a final $C^{\bullet+}\text{-}P_{Zn}\text{-}P\text{-}NQ\text{-}Q^{\bullet-}$ charge-separated state (Fig. 10.7). The process begins with excitation of the free-base porphyrin moiety to give the first excited singlet state $C\text{-}P_{Zn}\text{-}^1P\text{-}NQ\text{-}Q$. This can occur either via direct absorption of light or by singlet-singlet energy transfer from the attached zinc porphyrin (step 1 in Fig. 10.7), which can act as an antenna. The time constant for the energy transfer process is $4.4 \times 10^{-11}$ s in chloroform solution, as determined from global analysis of the time-resolved fluorescence behavior of the molecule at 14 wavelengths. The energy transfer quantum yield is about 90%.

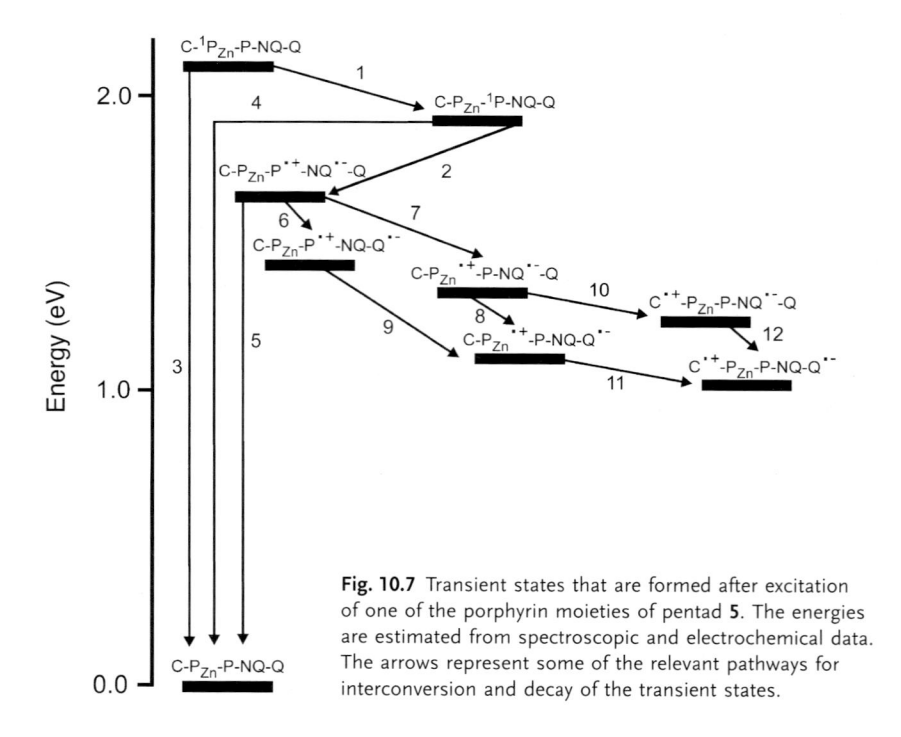

**Fig. 10.7** Transient states that are formed after excitation of one of the porphyrin moieties of pentad **5**. The energies are estimated from spectroscopic and electrochemical data. The arrows represent some of the relevant pathways for interconversion and decay of the transient states.

The free-base porphyrin first excited singlet state decays in part by electron transfer to the attached naphthoquinone (step 2 in Fig. 10.7) to produce C-$P_{Zn}$-$P^{\bullet+}$-$NQ^{\bullet-}$-Q. The time constant is $1.4 \times 10^{-9}$ s, and the quantum yield is 0.85. This initial charge-separated state can in principle decay to the ground state through charge recombination (step 5). However, two electron transfer steps operating in parallel compete with recombination. One of these (step 6) involves electron migration from the naphthoquinone radical anion to the attached benzoquinone, which is a better electron acceptor. This gives the new charge-separated state C-$P_{Zn}$-$P^{\bullet+}$-NQ-$Q^{\bullet-}$. Alternatively, electron donation from the zinc porphyrin to the free-base porphyrin radical cation of C-$P_{Zn}$-$P^{\bullet+}$-$NQ^{\bullet-}$-Q (step 7) can also occur to give C-$P_{Zn}^{\bullet+}$-P-$NQ^{\bullet-}$-Q. This reaction has a reasonable thermodynamic driving force, as the zinc stabilizes the positive charge on the macrocycle. Although the two new intermediates can decay by charge recombination, these reactions are expected to be relatively slow because the charges are farther apart than they are in the initial C-$P_{Zn}$-$P^{\bullet+}$-$NQ^{\bullet-}$-Q state, and electronic coupling is consequently weaker.

The two new intermediates can in turn undergo electron transfer reactions to yield yet another two intermediates, both of which decay to $C^{\bullet+}$-$P_{Zn}$-P-NQ-$Q^{\bullet-}$. This ultimate state may be readily detected spectroscopically by observation of the carotenoid radical cation absorption. In chloroform at ambient temperatures, 650-nm excitation produces the final state with an overall quantum yield of 0.83. As this yield is essentially the same as that of the initially formed C-$P_{Zn}$-$P^{\bullet+}$-$NQ^{\bullet-}$-Q species, the parallel electron transfer pathways compete very efficiently with charge recombination of all of the intermediates. The lifetime of $C^{\bullet+}$-$P_{Zn}$-P-NQ-$Q^{\bullet-}$ is $5.5 \times 10^{-5}$ s. In dichloromethane solution, the lifetime of $C^{\bullet+}$-$P_{Zn}$-P-NQ-$Q^{\bullet-}$ is increased to about $2.0 \times 10^{-4}$ s, and the quantum yield drops to 0.6.

Pentad **5** demonstrates a method for enhancing the yield of long-lived charge separation in an artificial reaction center that is not realizable in dyad- or triad-type systems. The most rapid charge recombination step in **5**, and therefore the step most likely to lead to a reduction in the overall yield of the final charge-separated state, is step 5, which is recombination of the initial charge-separated state formed by photoinduced electron transfer. In the pentad, two secondary electron transfers, steps 6 and 7, operate in parallel and compete with recombination step 5. Thus, as long as the yield of step 6 and/or step 7 is less than unity, introducing the parallel step necessarily increases the quantum yield of $C^{\bullet+}$-$P_{Zn}$-P-NQ-$Q^{\bullet-}$.

## 10.3.3
**Antenna–Reaction Center Complexes**

The field of artificial photosynthetic reaction centers is approaching its "adolescence," and a reasonable number of synthetic antenna systems have been prepared. Thus, it is becoming possible to prepare artificial antenna–reaction center complexes that combine the operations of the two units. In addition to the synthetic complexities introduced by the necessarily larger constructs, several other considerations come into play when one designs such complexes. Energy transfer quantum yields must be maximized not only within the antenna unit but also for the excitation transfer from the complex to the reaction center chromophore. Following charge separation in the artificial reaction center, migration of electrons or radical cations into the antenna can occur. This can be useful, as it can extend the lifetime of charge separation. However, electron transfer reactions involving the antenna system that compete with the desirable electron transfer events in the reaction center must be avoided by control of the redox potentials of the antenna chromophores and the kinetics of the electron transfer processes. In some cases, the kinetic factors may be controlled using the fact that singlet-singlet energy transfer usually occurs by the dipole-dipole mechanism, whose rate depends on the minus sixth power of the separation, whereas electron transfer usually has an exponential dependence on donor-acceptor separation.

Hexad **6** (Fig. 10.8) is an example of a covalently linked antenna–reaction center complex. The molecule consists of the antenna system **1** discussed above linked covalently to a fullerene, which plays the role of an electron acceptor, as in **3** and **4**. The photochemistry of **6** was elucidated using a combination of time-resolved spectroscopic methods [18]. As is the case with antenna **1**, excitation of a peripheral zinc porphyrin is followed by singlet-singlet energy transfer to the central zinc porphyrin with a time constant of $5.0 \times 10^{-11}$ s. Energy then migrates to the free-base porphyrin energy trap with a time constant of $3.0 \times 10^{-11}$ s. Because the free-base porphyrin bears the fullerene electron acceptor, it decays by photoinduced electron transfer to yield a $(P_{Zn})_4\text{-}P^{\bullet+}\text{-}C_{60}^{\bullet-}$ charge-separated state. This

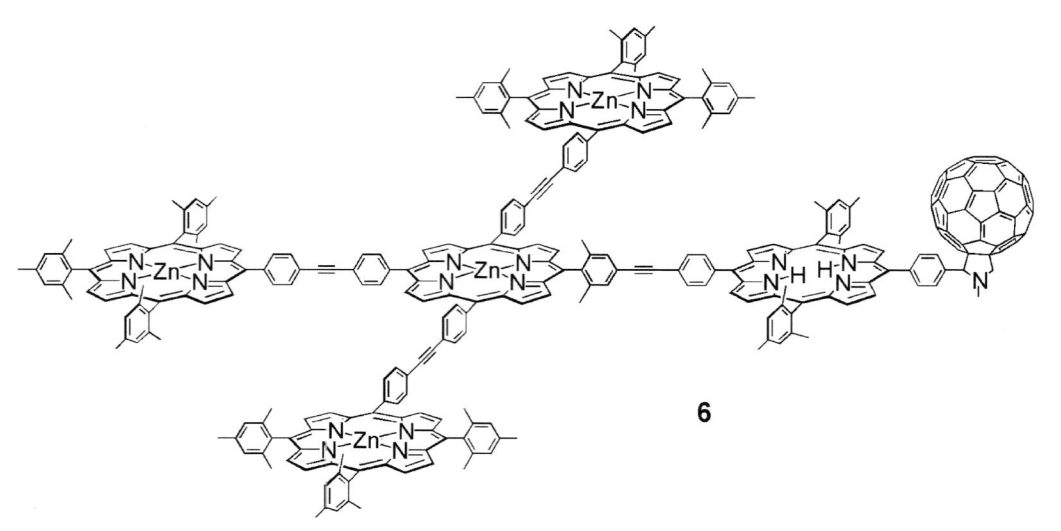

**Fig. 10.8** An artificial antenna–reaction center complex, **6**, is illustrated. The four zinc porphyrins comprise the antenna. The free-base porphyrin is linked to a fullerene electron acceptor to form the reaction center analogue.

occurs with a time constant of $2.5 \times 10^{-11}$ s, and therefore the quantum yield of the electron transfer is essentially unity. From model porphyrin-fullerene **3**, we know that charge recombination of this state will occur with a time constant of $\sim 3 \times 10^{-9}$ s. However, in **6**, electron transfer from the central zinc porphyrin, with a time constant of $3.8 \times 10^{-10}$ s, competes with charge recombination. This process moves the positive charge into the zinc porphyrin antenna array and hence retards charge recombination of the final state. Ultimately, charge recombination does occur, with a time constant of $2.4 \times 10^{-7}$ s. The overall quantum yield of long-lived charge separation in **6**, based on light absorbed in the antenna system, is $\sim 0.90$.

The preparation of artificial reaction centers bearing larger antenna systems is practical. However, these systems must be designed so that energy transfer among the antenna chromophores and between the antenna and reaction center is extremely rapid (a few picoseconds, or even sub-picoseconds). If this is not the case, migration of excitation energy from an antenna chromophore throughout the antenna and ultimately to an attached reaction center will not be able to compete successfully with loss of excitation energy by the usual photophysical processes involving the many antenna chromophores.

## 10.3.4
**Transmembrane Proton Pumping**

As mentioned earlier, photosynthetic bacteria use the electrochemical potential energy generated by the reaction center to power translocation of hydrogen ions across the membrane, generating a proton-motive force, which in turn supplies the energy needs of the organism. One of the goals of artificial photosynthesis research is mimicry of this process for producing energy in a biologically useful form. In contrast to the plethora of artificial antenna and reaction center systems that have been investigated, relatively few attempts to mimic photosynthetic proton pumping have been reported. A successful artificial light-driven pump [39] is based on the idea of a redox-sensitive shuttle molecule that can transport hydrogen ions across lipid bilayer membranes. Fig. 10.9 is a schematic representation of the system. The membrane is that of a liposome vesicle with a diameter of $\sim 150$ nm. The proton pump comprises carotenoid-porphyrin-quinone (C-P-Q) triad **7** and diphenylquinone **8** (Fig. 10.10). Triad **7** is an artificial photosynthetic reaction center that, upon absorption of light, undergoes a two-step electron transfer process to generate a $C^{\bullet +}$-P-$Q^{\bullet -}$ charge-separated state. The mechanism is similar to that described above for triad **4**. In order to function as a pump, the triad must be inserted vectorially into the membrane so that the majority of the molecules have the quinone near the external surface. This asymmetry is achieved by pre-forming the liposomes and then introducing the triad dissolved in a small amount of organic solvent into the vesicle solution. As the organic solvent dissipates, the water-insoluble triads dissolve into the membrane. The hydrophobic carotenoid portion can enter the hydrophobic core of the membrane, but the quinone moiety, which bears a carboxylate anion, remains near the hydrophilic exterior. Quinone **8** is soluble only in the

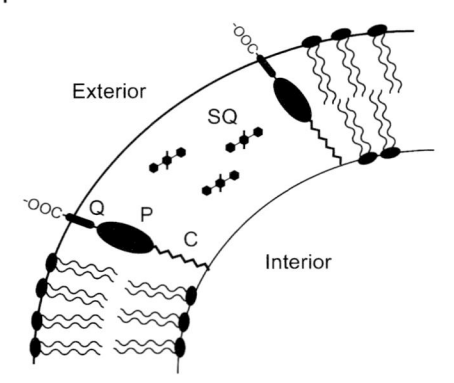

**Fig. 10.9** Schematic representation of a section of a liposome containing the elements of a light-powered, transmembrane proton pump. Interspersed among the phospholipids of the bilayer are a carotenoid (C)-porphyrin (P)-naphthoquinone (Q) triad artificial reaction center (**7**) and molecules of shuttle quinone (SQ) compound **8**. Note that the carboxylate end of the triad resides at the external interface of the bilayer with the aqueous environment. The ionic nature of this group at the pH values employed retards entry of this end of the triad into the hydrophobic lipid interior, thus helping to maintain the vectorial arrangement shown.

**Fig. 10.10** Chemical structures of triad **7** and shuttle quinone **8** used in transmembrane proton pumping and ATP synthesis.

hydrophobic membrane interior and has a higher reduction potential than the naphthoquinone moiety of triad **7**.

The pump is based on a redox loop [40] powered by light via triad **7**. Although all details of the function of the pump have not been elucidated, the following description illustrates the major aspects of the process. Excitation of the porphyrin with light generates $C^{\bullet+}$-P-$Q^{\bullet-}$ within the membrane. The quinone radical anion, located near the outer membrane surface, reduces a molecule of shuttle quinone **8** to yield the semiquinone anion, which is basic enough to accept a proton from the exterior aqueous environment. The resulting neutral semiquinone radical is able to diffuse within the bilayer, and when it

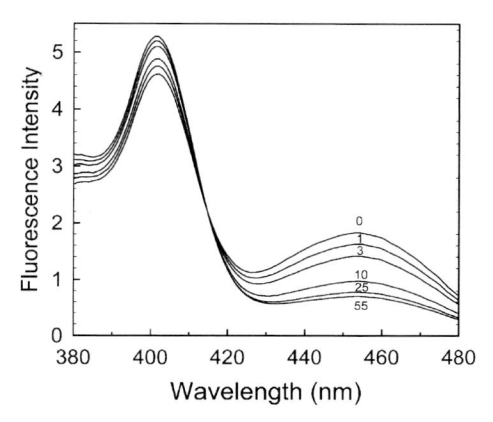

**Fig. 10.11** Fluorescence excitation spectrum of the pH-sensitive dye pyraninetrisulfonate contained in the inner aqueous volume of liposomes containing the components of the proton pump. Irradiation of the triad in the membrane leads to acidification of the inside of the liposome, and this is reflected in the change in the ratios of the absorption maxima at 406 nm and 456 nm. The curves are labeled with the total number of minutes of irradiation of the sample.

encounters a carotenoid radical cation near the inner membrane surface, it is oxidized back to the quinone. The protonated shuttle quinone is a strong acid ($pK_a$ ca. −6) that releases the hydrogen ion into the inner volume of the vesicle. The net result is proton translocation into the liposome interior and regeneration of the triad photocatalyst. The redox loop-based proton pump can in principle be driven from various combinations of redox levels of quinone 8.

The action of this pump acidifies the solution inside the liposome but has little effect on the pH of the much larger exterior volume. Therefore, characterization of the pump requires a method for detecting pH changes inside the liposomes. This can be done using the pH-sensitive fluorescent dye pyraninetrisulfonate. When this material is dissolved in the aqueous phase inside the liposomes, it reports the pH via the ratio of amplitudes of its fluorescence excitation spectrum at 406 nm and 456 nm. Fig. 10.11 shows the results of a typical experiment. Irradiation with light absorbed by the porphyrin of 7 leads to a change in the excitation spectrum that signals acidification of the liposome interior.

Measurement of the hydrogen ion gradient across the membrane using pH-sensitive dyes shows that under these conditions maximum $\Delta pH$ values up to $\sim 2$ units can be achieved. The transmembrane pH gradient is remarkably stable. When the liposomes are prepared from a suitable mix of phospholipids [41], the $\Delta pH$ decays over a period of several hours in the dark.

The proton-pumping photocycle transports protons into the liposome interior without translocating any compensating charges. This suggests that the pump should develop a transmembrane electrical potential, $\Delta \psi$. Membrane potentials may be detected using fluorescent dyes such as 8-anilino-1-naphthalenesulfonic acid. When this dye is added to the liposomal proton-pumping system, the fluorescence changes in concert with the developing $\Delta$pH, signaling a concurrent buildup of $\Delta$pH and $\Delta\psi$. Both gradients contribute to the proton-motive force, which is the total potential energy stored by the system. This suggests that if $\Delta\psi$ were relaxed, a larger maximum $\Delta$pH should be achievable. Addition of potassium and valinomycin (a potassium ionophore) to the system reduces $\Delta\psi$ by allowing potassium ions to flow out of the liposome as hydrogen ions are pumped in. Experimentally, relaxing $\Delta\psi$ in this way results in a 1.6-fold increase in the limiting $\Delta$pH. The maximum total free energy conserved by the proton-pumping system is on the order of 4 kcal mol$^{-1}$ of hydrogen ions. At this value of the proton-motive force, net hydrogen ion translocation across the membrane ceases.

### 10.3.5
### Synthesis of ATP

Proton-motive force, generated either by the photosynthetic apparatus or through other electron transport processes in energy-coupling membranes, is the major conduit for biological energy. Although proton-motive force may be used to directly power a variety of processes, an important application is the endergonic synthesis of ATP from ADP and inorganic phosphate (Pi). This is the case because ATP powers a large number of energy-requiring life processes. Thus, a next step in the mimicry of bacterial photosynthetic energy conversion is to use the artificial transmembrane light-driven proton pump to power ATP synthesis.

In photosynthetic membranes, ATP is prepared by ATP synthase enzymes, which consist of a membrane-spanning component ($F_0$) and a catalytic component projecting into the aqueous phase ($F_1$). The endergonic synthesis is powered by proton-motive force. Hydrogen ion flow through the enzyme is coupled to ATP production by a fascinating "mechanical" mechanism involving rotation of one of the subunits. Thus, the ATP synthase is not only a nanometer-scale chemical factory but also a molecular-scale rotary motor.

Biochemists have developed procedures for isolating intact ATP synthase molecules and reinserting them into the membranes of liposomes [42]. Using these methodologies, the $CF_0F_1$-ATP synthase from spinach chloroplasts has been inserted into the membranes of liposomes, along with the components of the light-driven proton pump [41]. The enzyme is incorporated by adding an aqueous detergent solution of ATP synthase to the pre-formed liposomes containing quinone **8** and slowly removing the detergent. The enzyme inserts into the membrane as illustrated schematically in Fig. 10.12, with the ATP-producing $F_1$ subunits in the aqueous phase. Once the enzyme is in place, the triad artificial reaction center **7** is incorporated as described earlier. The liposomal construct is thus set up for light-driven ATP production. The proton-pumping photocycle will translocate hydrogen ions into the liposome, establishing a proton-motive force. The enzyme can then transport the protons back out of the liposome, using the energy released to synthesize ATP.

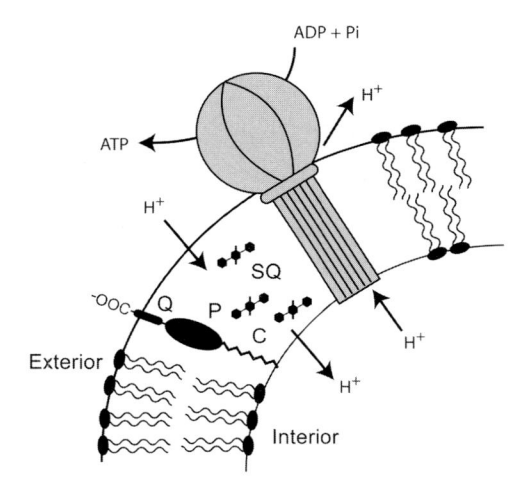

**Fig. 10.12** Schematic representation of biomimetic light-driven ATP synthesis. The proton-pumping construct diagrammed in Fig. 10.9 is augmented by the enzyme ATP synthase, which has been isolated from spinach chloroplasts and reconstituted into the bilayer membranes of liposomes. Irradiation of the system with red light leads to ATP synthesis, as described in the text.

In order to evaluate the function of such a construct, a method for assaying the ATP produced is required. A convenient bioluminescence-based assay uses the luciferin-luciferase system, in which consumption of one molecule of ATP is linked to emission of one photon in the 570-nm region. Alternatively, the reaction may be followed using $^{32}$P-labeled ADP and detecting the appearance of the label in the newly synthesized ATP.

Typical results are shown in Fig. 10.13. An aqueous suspension of liposomes containing the light-driven proton pump and the ATP synthase was prepared, and ATP (0.2 mM), ADP (0.2 mM), and Pi (5 mM) were added. Thioredoxin, which is necessary to activate the enzyme, was also present, and the solution was deoxygenated. The initial pH was 8.0 on both sides of the membrane. The sample was illuminated with a 5-mW laser at 633 nm, where the porphyrin of **7** absorbs. As irradiation continued, aliquots were removed and added to an assay solution containing luciferin, luciferase, and an appropriate buffer. The luminescence spectrum was measured immediately and the amount of ATP present was determined from a standardized curve. Fig. 10.13 shows the amount of ATP produced, relative to a control experiment with an identical but non-illuminated sample.

Also shown in Fig. 10.13 are the results of several additional control experiments. Addition of FCCP prevents ATP production by abolishing the proton-motive force. Omission of ADP or **8** also eliminates ATP production, as does the addition of tentoxin, an inhibitor of chloroplast ATP synthase.

At low light intensity, the quantum yield of ATP synthesis was estimated as one ATP molecule for every 14 incident photons. Because of light scattering by the liposomes, there is some uncertainty in determination of the amount of light absorbed by the triad, but spectroscopic measurements suggest that roughly 50% of the incident light was absorbed by the porphyrin, giving a quantum yield for ATP production of $\sim 0.15$. The number of protons transported across the mem-

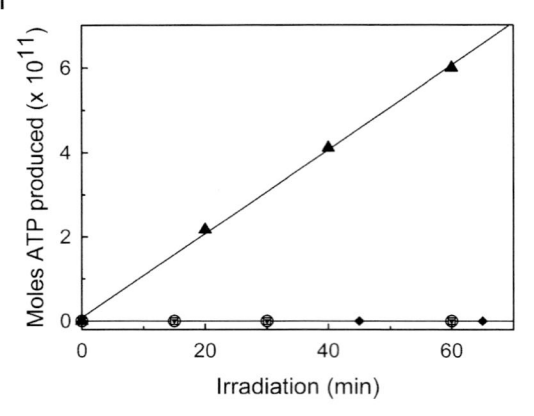

**Fig. 10.13** ATP production by the construct illustrated in Fig. 10.12, as monitored by the luciferin-luciferase luminescence assay. ATP is produced when the membranes and exterior aqueous environment of the liposomes contain all necessary components (▲). In the experiment shown, [ATP]=[ADP]= 0.2 mM and [Pi]=5 mM at the beginning of the experiment. Also shown are some control experiments. ATP was not synthesized when the proton ionophore FCCP was present (○), when no shuttle quinone **8** was present (◆), when ADP was absent (□), or when an inhibitor for the enzyme, 2 μM tentoxin, was added (▼).

brane by ATP synthase per ATP molecule produced is the subject of debate and may vary from species to species. Common estimates are three or four protons per ATP. Thus, the quantum yield of proton translocation by the light-driven pump in the ATP-synthesis system must be in the range of 0.5 to 0.6.

When the light intensity is increased, the rate of ATP synthesis saturates with respect to light intensity, and the turnover rate of the enzyme can be estimated. Values of about 100 ATP per second per molecule of ATP synthase can be achieved by the biomimetic system.

### 10.3.6
**Transmembrane Calcium Transport**

Calcium ion transport across biological membranes and against a thermodynamic gradient is essential to biological processes such as muscle contraction, glycogen metabolism, the citric acid cycle, vision, neurotransmitter release, signal transduction, and immune response. Light-powered transmembrane calcium pumping in artificial liposomal systems has been realized [43] using the ideas inherent in the proton-pumping redox loop discussed in Section 10.3.4. The system is constructed as shown in Fig. 10.9, with the exception that the proton shuttle quinone **8** is replaced by hydroquinone **9** (Fig. 10.14). Molecule **9** can lose a proton from the hydroxide group near the keto group and chelate $Ca^{++}$. Chelation of $Ca^{++}$ by two hydroquinone moieties yields a neutral, membrane-soluble complex. Oxidation of **9** gives a quinone, which does not chelate calcium. Thus, **9** can serve as a redox-sensitive shuttle for calcium ions.

**Fig. 10.14** Chemical structure of redox-sensitive calcium ion chelator and shuttle **9**.

The shuttle molecule **9** was designed to localize in the hydrophobic part of the lipid bilayer of a liposome, rather than the aqueous regions, due to the presence of the hydrocarbon "tail" on the molecule. Studies of this system using a calcium-sensitive dye located inside the liposome suggest that the following events occur upon excitation of the system with light. Chelation of $Ca^{++}$ by **9** allows transport of calcium ions from the external aqueous phase into the liposomal bilayer as a neutral complex. Excitation of the C-P-Q triad **7** with light generates the $C^{\bullet+}$-P-$Q^{\bullet-}$ charge-separated state, as discussed above, and the carotenoid radical cation has enough oxidizing potential to convert hydroquinone **9** into the quinone. This could occur either by sequential one-electron oxidations or by disproportion of the semiquinone intermediate. As a result of this conversion, $Ca^{++}$ is released, and some of the release is into the internal aqueous phase of the liposome. The quinone form of **9** can later be re-reduced by the quinone radical anion of triad **7**. The intimate details of the transport process are not fully understood, but the broad outlines of the mechanism seem to be as described above.

Fig. 10.15 shows the operation of the transmembrane calcium pump, as detected by the calibrated fluorescence emission of the calcium-sensitive dye fluo-3 located inside the liposome. With continuing irradiation, $Ca^{++}$ is transported

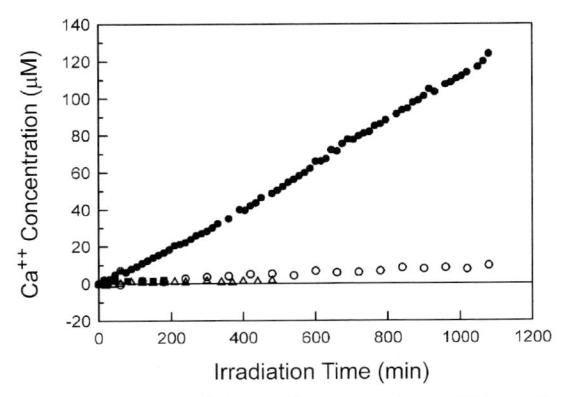

**Fig. 10.15** Transmembrane $Ca^{++}$ transport by the light-driven system described in the text (●). The calcium ion concentration inside the liposomes was monitored by the fluorescent dye fluo-3, and actinic radiation was at 633 nm. Control experiments shown are for similar experiments lacking the shuttle **9** (■), in the absence of actinic light (○), and when **9** is replaced by a related molecule incapable of binding $Ca^{++}$ (△).

into the liposome. The figure also shows the results of control experiments. No calcium is transported if the system is left in the dark, or if triad **7** or shuttle **9** is omitted. The quantum yield of $Ca^{++}$ pumping is only $\sim 1\%$.

Other experiments show that there is no net transport of hydrogen ions across the membrane, suggesting that calcium ion transport should be accompanied by a buildup of a transmembrane charge gradient ($\delta\psi$). Experiments with a suitable dye, and no fluo-3, show that a membrane potential, positive on the inside of the liposomes, does indeed develop during calcium ion transport. The light-driven pump not only develops a $\delta\psi$ but also functions when the initial calcium ion concentrations on both sides of the membrane are equal. Thus, the transport process can operate against a thermodynamic gradient. Some of the light energy absorbed by the porphyrin of the triad is ultimately converted to transmembrane electrochemical potential, and the system is a kind of photon energy transduction device, as are the other artificial photosynthetic constructs discussed in this chapter.

## 10.4
## Conclusions

In this chapter we have illustrated research progress in one area of artificial photosynthesis. Although we have used mainly examples from our laboratory, a large number of research groups around the world are contributing to understanding in this field. It has become clear that it is possible to mimic at least the gross features of photosynthetic solar energy conversion in totally artificial systems. The construction of "bionic" systems that are hybrids of artificial photosynthesis mimics and natural components is also feasible, as is demonstrated by the light-powered ATP synthesis system described above. In some cases, the performance of the artificial systems rivals that of the natural systems in the vital areas of quantum yield and energy conversion efficiency. The work from our laboratories and those of our colleagues in the field demonstrates that photosynthesis is not a "magic" biological process beyond the ken of humans, but rather a complex and exquisite natural implementation of basic chemical and physical principles, many of which are reasonably well understood.

However, a lot remains to be discovered. Although the broad outlines of natural photosynthesis are beginning to be understood, many subtle but important aspects of the process are still a mystery. For the development of artificial systems, there is a general understanding of what is needed, and theoretical guidelines for designing systems exist. However, "the devil is in the details." Molecular engineering to favorably balance yields and rate constants for all of the physical and chemical processes intervening between light absorption and the production of energy in useful forms is far from a trivial task. On the other hand, it is by no means an impossible task, and the next few years will see a continuation of the exciting progress being made in artificial photosynthesis.

Will artificial photosynthesis of the kind discussed here ever be technologically practical and useful? A great deal of research and development will be needed in order to find out. While it is clear that biomimetic solar conversion devices can now be made, questions of mass production, stability under real-world operating conditions, efficiencies, and, above all, economics remain. Practical solar energy harvesting in the future will likely consist of not one technology, but many, each optimized for the particular application. Artificial photosynthesis may well play a role.

## References

1  Ciamician, G. *Science* **1912**, *36*, 385–394.
2  Bard, A. J.; Fox, M. A. *Acc. Chem. Res.* **1995**, *28*, 141–145.
3  Bixon, M.; Fajer, J.; Feher, G.; Freed, J. H.; Gamliel, D.; Hoff, A. J.; Levanon, H.; Möbius, K.; Nechushtai, R.; Norris, J. R.; Scherz, A.; Sessler, J. L.; Stehlik, D. *Isr. J. Chem.* **1992**, *32*, 449–455.
4  Burrell, A. K.; Officer, D. L.; Plieger, P. G.; Reid, D. C. W. *Chem. Rev.* **2001**, *101*, 2751–2796.
5  Fox, M. A. *Acc. Chem. Res.* **1992**, *25*, 569–574.
6  Guldi, D. M. *Chemical Society Reviews* **2002**, *31*, 22–36.
7  Gust, D.; Moore, A. L.; Moore, T. A. Covalently linked systems containing porphyrin units; In *Electron Transfer in Chemistry Vol. 3, Biological and Artificial Supramolecular Systems*; Balzani, V., ed. Wiley-VCH, Weinheim, 2001; pp. 272–336.
8  Gust, D.; Moore, T. A. Intramolecular photoinduced electron transfer reactions of porphyrins; In *The Porphyrin Handbook*; Kadish, K. M., Smith, K. M., Guilard, R., eds. Academic Press: New York, 2000; pp. 153–190.
9  Gust, D.; Moore, T. A.; Moore, A. L. *Acc. Chem. Res.* **2001**, *34*, 40–48.
10  Imahori, H.; Sakata, Y. *European Journal Of Organic Chemistry* **1999**, 2445–2457.
11  Kurreck, H.; Huber, M. *Angew. Chem. Int. Ed. Engl.* **1995**, *34*, 849–866.
12  Lindsey, J. S. *NATO ASI Series, Series C: Mathematical and Physical Sciences* **1997**, *499*, 517–528.
13  Maruyama, K.; Osuka, A.; Mataga, N. *Pure Appl. Chem.* **1994**, *66*, 867–872.
14  Meyer, T. J. *Acc. Chem. Res.* **1989**, *22*, 163–170.
15  Sakata, Y.; Imahori, H.; Sugiura, K.-I. *Journal of Inclusion Phenomena and Macrocyclic Chemistry* **2001**, *41*, 31–36.
16  Wasielewski, M. R. *Chem. Rev.* **1992**, *92*, 435–461.
17  Förster, T. *Disc. Faraday Soc.* **1959**, *27*, 7–17.
18  Kodis, G.; Liddell, P. A.; de la Garza, L.; Clausen, P. C.; Lindsey, J. S.; Moore, A. L.; Moore, T. A.; Gust, D. *J. Phys. Chem. A* **2002**, *106*, 2036–2048.
19  Bensasson, R. V.; Land, E. J.; Moore, A. L.; Crouch, R. L.; Dirks, G.; Moore, T. A.; Gust, D. *Nature* **1981**, *290*, 329–332.
20  Dirks, G.; Moore, A. L.; Moore, T. A.; Gust, D. *Photochem. Photobiol.* **1980**, *32*, 277–280.
21  Gust, D.; Moore, T. A.; Moore, A. L.; Devadoss, C.; Liddell, P. A.; Hermant, R. M.; Nieman, R. A.; Demanche, L. J.; DeGraziano, J. M.; Gouni, I. *J. Am. Chem. Soc.* **1992**, *114*, 3590–3603.
22  Mariño-Ochoa, E.; Palacios, R.; Kodis, G.; Macpherson, A. N.; Gillbro, T.; Gust, D.; Moore, T. A.; Moore, A. L. *Photochem. Photobiol.* **2002**, *76*, 116–121.
23  Marcus, R. A. *J. Chem. Phys.* **1956**, *24*, 966–978.
24  Hush, N. S. *J. Chem. Phys.* **1958**, *28*, 962–972.
25  Hush, N. S. *Trans. Faraday Soc.* **1961**, *57*, 557–580.
26  Levich, V. *Adv. Electrochem. Electrochem. Eng.* **1966**, *4*, 249–371.
27  Jortner, J. *J. Chem. Phys.* **1976**, *64*, 4860–4867.
28  Jortner, J. *J. Am. Chem. Soc.* **1980**, *102*, 6676–6686.

29 Kong, J. L.; Loach, P. A. In *Frontiers of Biological Energetics: From Electrons to Tissues*; Dutton, P. L., Scarpa, H., eds. Academic Press, New York, 1978; p. 73.
30 Tabushi, I.; Koga, N.; Yanagita, M. *Tetrahedron Lett.* **1979**, 257.
31 Connolly, J. S.; Bolton, J. R. Intramolecular electron transfer. History and some implications for artificial photosynthesis; In *Photoinduced Electron Transfer, Part D*; Fox, M. A., Chanon, M., eds. Elsevier, Amsterdam, 1988; pp. 303–393.
32 Liddell, P. A.; Kodis, G.; de la Garza, L.; Bahr, J. L.; Moore, A. L.; Moore, T. A.; Gust, D. *Helv. Chim. Acta* **2001**, *84*, 2765–2783.
33 Gust, D.; Mathis, P.; Moore, A. L.; Liddell, P. A.; Nemeth, G. A.; Lehman, W. R.; Moore, T. A.; Bensasson, R. V.; Land, E. J.; Chachaty, C. *Photochem. Photobiol.* **1983**, *37S*, S46.
34 Moore, T. A.; Gust, D.; Mathis, P.; Mialocq, J.-C.; Chachaty, C.; Bensasson, R. V.; Land, E. J.; Doizi, D.; Liddell, P. A. *Nature (London)* **1984**, *307*, 630–632.
35 Nishitani, S.; Kurata, N.; Sakata, Y.; Misumi, S.; Karen, A.; Okada, T.; Mataga, N. *J. Am. Chem. Soc.* **1983**, *105*, 7771–7772.
36 Wasielewski, M. R.; Niemczyk, M. P.; Svec, W. A.; Pewitt, E. B. *J. Am. Chem. Soc.* **1985**, *107*, 5562–5563.
37 Gust, D.; Moore, T. A.; Moore, A. L.; Lee, S.-J.; Bittersmann, E.; Luttrull, D. K.; Rehms, A. A.; DeGraziano, J. M.; Ma, X. C.; Gao, F.; Belford, R. E.; Trier, T. T. *Science* **1990**, *248*, 199–201.
38 Gust, D.; Moore, T. A.; Moore, A. L.; Macpherson, A. N.; Lopez, A.; DeGraziano, J. M.; Gouni, I.; Bittersmann, E.; Seely, G. R.; Gao, F.; Nieman, R. A.; Ma, X. C.; Demanche, L. J.; Luttrull, D. K.; Lee, S.-J.; Kerrigan, P. K. *J. Am. Chem. Soc.* **1993**, *115*, 11141–11152.
39 Steinberg-Yfrach, G.; Liddell, P. A.; Hung, S.-C.; Moore, A. L.; Gust, D.; Moore, T. A. *Nature* **1997**, *385*, 239–241.
40 Mitchell, P. *Biol. Rev. Camb. Philos. Soc.* **1966**, *41*, 445–502.
41 Steinberg-Yfrach, G.; Rigaud, J.-L.; Durantini, E. N.; Moore, A. L.; Gust, D.; Moore, T. A. *Nature* **1998**, *392*, 479–482.
42 Richard, P.; Rigaud, J.-L.; Graber, P. *Eur. J. Biochem.* **1990**, *193*, 921–925.
43 Bennett, I. M.; Vanegas-Farfano, H. M.; Primak, A.; Liddell, P. A.; Otero, L.; Sereno, L.; Silber, J. J.; Moore, A. L.; Moore, T. A.; Gust, D. *Nature* **2002**, *420*, 398–401.

# Part IV
# Photohydrogen

# 11
Development of Algal Systems for Hydrogen Photoproduction: Addressing the Hydrogenase Oxygen-sensitivity Problem

*Maria L. Ghirardi, Paul King, Sergey Kosourov, Marc Forestier, Liping Zhang, and Michael Seibert*

## 11.1
## Introduction

Higher plants, green algae, and cyanobacteria are able to utilize light energy, water, and $CO_2$ to synthesize starch during the process of photosynthesis. Starch is oxidized in the dark to release chemical energy that can be used to fuel cellular metabolism, growth, and other cellular functions. The specific role of light in photosynthesis is to provide the energy to oxidize water, resulting in the evolution of $O_2$, release of protons, the generation of highly reduced NADPH molecules, and the establishment of a transmembrane proton gradient that drives the synthesis of ATP. Together with NADPH, ATP is required to fix $CO_2$ and produce starch. In green algae, it is possible to divert reductants from the Benson-Calvin $CO_2$ fixation pathway at the level of ferredoxin and use them instead to reduce protons. Hydrogen gas is released in a reaction catalyzed by the [Fe]-hydrogenase enzyme. In a real sense, this is an "artificial" light-dependent process, since it results in the stoichiometric release of one $O_2$ and two $H_2$ molecules from two $H_2O$ molecules only under conditions of sustained anaerobiosis. Anaerobiosis is required for the transcription of the hydrogenase gene [1, 2] and for the maintenance of hydrogenase functional activity [3–5]. Thus, if algal $H_2$ photoproduction is to provide enough clean, renewable $H_2$ to constitute a significant source of energy in a future hydrogen economy [6], the $O_2$ sensitivity of the enzyme is an important issue that must be addressed.

We are currently investigating two approaches to solve this problem. In the first approach, we use a physiological switch, sulfur deprivation, to reversibly cycle algal cultures from an aerobic, photosynthetic growth mode to an anaerobic, $H_2$-producing mode. In this case, anaerobic conditions are achieved by operation of the switch, and the cultures themselves maintain the requisite anaerobiosis during $H_2$ production. The second approach, in contrast, involves molecular engineering of the algal hydrogenase enzyme so that it can function in the presence of $O_2$, thus potentially eliminating the need for anaerobiosis during production. The current status of each area will be discussed in this chapter.

*Artificial Photosynthesis: From Basic Biology to Industrial Application*
Edited by Anthony F. Collings and Christa Critchley
Copyright © 2005 WILEY-VCH Verlag GmbH & Co. KGaA, Weinheim
ISBN: 3-527-31090-8

## 11.2
## Sulfur Deprivation and Hydrogen Photoproduction

### 11.2.1
### Background

Many techniques have been used in the past to remove $O_2$ from the algal culture medium in order to induce and maintain hydrogenase in an active state [7–11], but none of them has proven to be economically feasible. In 1998, Wykoff et al. [12] observed that sulfur-deprived cultures of the green alga *Chlamydomonas reinhardtii* underwent gradual and selective inactivation of photosynthetic water-oxidation activity over a short period of time. That observation led to the design and development of an $H_2$-production system [13], based on the temporal separation of $O_2$ and $H_2$ production. The mechanism of inhibition of water oxidation and subsequent induction of $H_2$ production has been described in several recent reviews [14–18]. Here, we present results on the metabolism of algal cells subjected to sulfur-deprivation conditions and discuss factors that may limit the $H_2$-production yields.

Under sulfur-deprived conditions, the rate of turnover of D1, one of the proteins that constitute the photosystem II (PSII) reaction center, decreases due to the lack of availability of sulfurylated amino acid residues [12, 15, 19]. As a consequence, the activity of PSII, measured as $O_2$ evolution [13], maximum PAM fluorescence yield [12, 20, 21], maximum flash-probe fluorescence yield [22], or change in absorbance at 320 nm [15], decays significantly within 24–48 h. Fig. 11.1A shows the effects of sulfur deprivation on the rates of photosynthetic $O_2$ evolution (open circles) and respiratory $O_2$ consumption (closed circles) by *C. reinhardtii* cultures. Extensive starch and protein buildup occurs during the initial 24 h [13, 15, 22] of sulfur deprivation. However, other cellular activities,

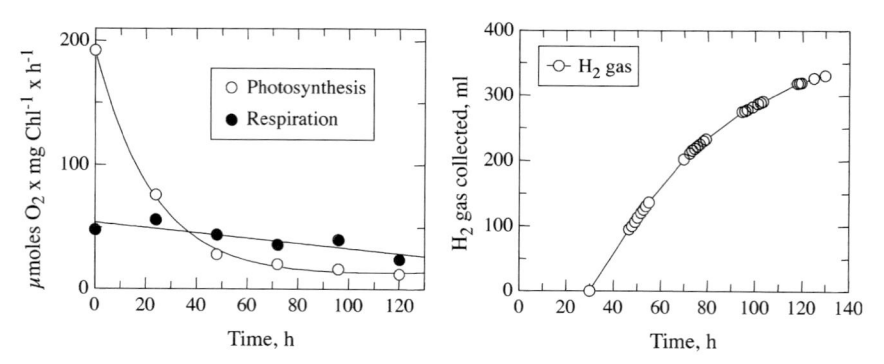

**Fig. 11.1** Effect of sulfur deprivation on the rates of photosynthetic $O_2$ evolution (left panel, open circles) and respiratory $O_2$ consumption (left panel, closed circles). Hydrogen gas production by sulfur-deprived *C. reinhardtii* cultures can be seen in the right panel.

such as respiratory $O_2$ consumption and electron transport through the cytochrome $b_6/f$ complex and PSI, are not affected significantly [13]. When the rate of photosynthetic $O_2$ evolution decreases below that of respiratory $O_2$ consumption (at about 24–40 h after the start of sulfur deprivation; see Fig. 11.1A), the cultures become anaerobic and the redox potential of the medium starts to decrease [23]. Following the establishment of anaerobiosis, starch and protein degradation commences [13, 15, 22, 24], and fermentation products such as formate and acetate are excreted into the growth medium [24, 25]. Simultaneously, the [Fe]-hydrogenase gene is expressed [1, 18, 19], and $H_2$ gas can be collected [13, 22, 24, 25] as shown in Fig. 11.1B. During the $H_2$-production phase, alterations in the composition and amount of LHCII also occur [15], suggesting that besides D1, sulfur deprivation eventually affects the synthesis of other algal proteins. After 120 h, all cellular activities decline and the cultures must be regenerated. This is accomplished by transferring the spent cells to fresh, sulfate-containing medium for normal photosynthetic growth. Multiple cycles of photosynthesis and $H_2$ production can be repeated in this manner [14, 16].

It is clear that, in order to support $H_2$ photoproduction, sulfur-deprived cultures must have both an active photosynthetic electron transport chain (PETC) from water to hydrogenase and an efficient oxidative respiratory system to remove any residual $O_2$ evolved by PSII. Moreover, a substantial amount of fermentation products are co-generated during sulfur deprivation [22, 25], demonstrating the simultaneous occurrence of both anaerobic and oxidative pathways for substrate degradation.

## 11.2.2
### Model of the Interactions Between Different Metabolic Pathways in Sulfur-deprived Cells

All of the observations described above suggest the model for $H_2$ metabolism in sulfur-deprived cells shown in Fig. 11.2. It is known that $H_2$ production in sulfur-deprived cultures occurs only in the light [14]. Accordingly, the model shows two pathways for $H_2$ photoproduction. The first pathway involves light-induced water oxidation associated with PSII, followed by electron transfer through the PETC to ferredoxin. The second pathway depends on the release of reductant (NADH) during the initial steps of glucose and amino acid degradation in the chloroplast. This reductant can be transferred to the PETC at the level of the plastoquinone (PQ) pool through the action of the NAD(P)H-PQ oxidoreductase [26]. From ferredoxin, reductants are normally used to reduce NADP to NADPH, which is then utilized along with ATP to fix $CO_2$. However, the level of Rubisco is extremely low in sulfur-deprived cells [19], and an alternative pathway for reductant utilization at the level of ferredoxin is induced (i.e., the [Fe]-hydrogenase enzyme system). As Rubisco levels decrease, the hydrogenase protein accumulates [19].

The main factor responsible for hydrogenase expression is the lack of $O_2$ [1–11]. However, it is necessary to assume that aerobic metabolism still occurs

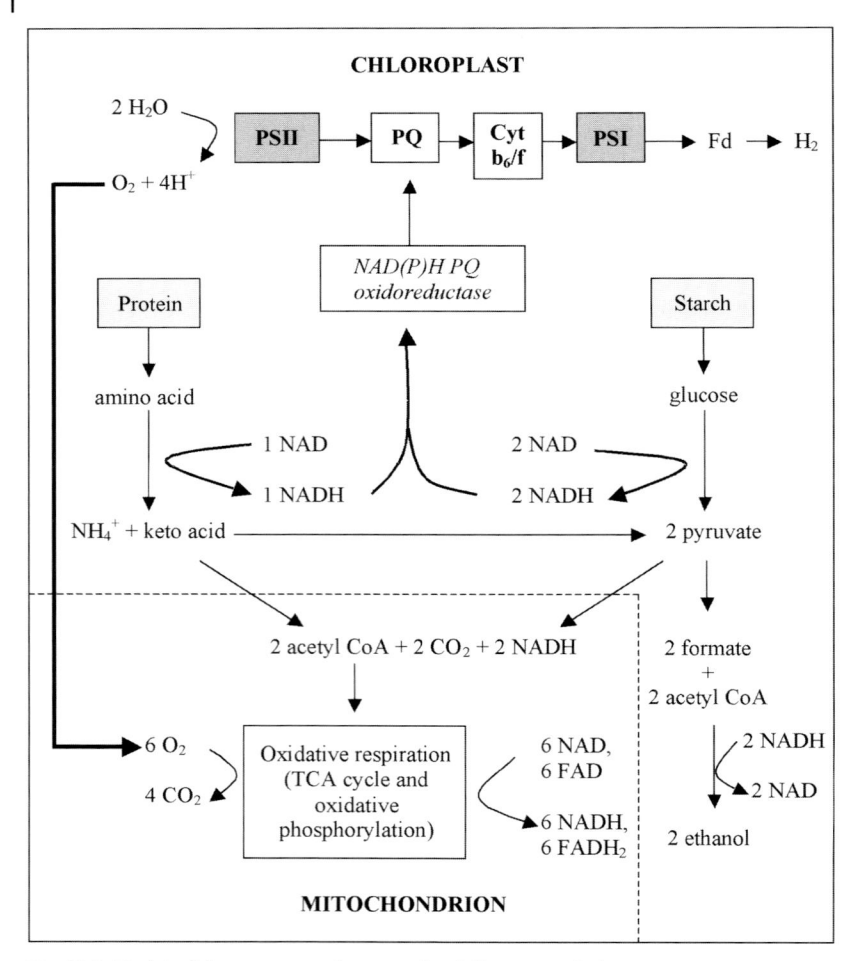

**Fig. 11.2** Model of the interaction between the different metabolic pathways in sulfur-deprived C. reinhardtii cultures.

during sulfur deprivation, and, in fact, it must remove the $O_2$ produced by residual photosynthesis. This role is performed by degradation of starch and protein, which results in the production of pyruvate and, following its transport to the mitochondria, acetyl-CoA. Acetyl-CoA can then be used to generate reductants and ATP (not shown) in the mitochondria. Fig. 11.2 shows the respiratory metabolism of acetyl-CoA in the bottom left corner and indicates that the pathway depends on the concomitant operation of the tricarboxylic acid (TCA) cycle and oxidative phosphorylation (with the simultaneous release of $CO_2$).

Fig. 11.2 also shows that, in C. reinhardtii, the anaerobic degradation of pyruvate (outside of the mitochondria) is performed mainly by the pyruvate-formate lyase pathway, which splits pyruvate into formate and acetyl-CoA [10]. Finally, acetyl-CoA can be reduced to ethanol, using NADH as the reductant, in a reaction catalyzed by alcohol dehydrogenase. It must be pointed out that ethanol production depends on the amount of NADH generated during the initial steps of glucose and amino acid degradation (see Fig. 11.2) and occurs preferentially at low pH [22].

### 11.2.3
### Confirmation of the Model

The two pathways for $H_2$ production, water oxidation, and substrate degradation can be inhibited by DCMU and DBMIB, respectively. The first inhibitor blocks electron transfer between PSII and the PQ pool, and the second interrupts electron flow from the PQ pool to the cyt $b_6/f$ complex. We tested the effect of both inhibitors on the $H_2$ production rates of sulfur-deprived cultures by adding them to the cultures a few hours after $H_2$ production started. DCMU inhibited 80% of the $H_2$ photoproduction rate, and DBMIB completely inhibited $H_2$ photoproduction [22]. These results demonstrate that most of the electrons for $H_2$ (80%) production come from water, whereas substrate degradation contributes the remaining 20%.

As shown in Fig. 11.2, starch and protein degradation should play a dual physiological role in sulfur-deprived cultures. First, the storage products contribute reductants to $H_2$ photoproduction, as shown above, but more importantly, they are ultimately responsible for removing photosynthetically evolved $O_2$. This allows the cultures to be maintained in an anaerobic state so that the hydroge-

**Table 11.1** Effects of the initial pH of the extracellular medium at the time of sulfur deprivation on different metabolic functions in sulfur-deprived cultures [22].

| Parameters measured | Initial pH | | | | |
|---|---|---|---|---|---|
| | 6.5 | 6.9 | 7.3 | 7.7 | 8.2 |
| $H_2$ gas produced, mmoles per photobioreactor | 1.7 | 3.6 | 5.9 | 7.7 | 0.6 |
| PSII capacity, $F_{max}$, relative units | 0.08 | 0.14 | 0.16 | 0.16 | 0.08 |
| PSI activity, μmoles $O_2 \cdot$ mg $Chl^{-1} \cdot h^{-1}$ | 238 | – | 524 | 648 | 800 |
| Protein degraded, mmoles of amino acid per photobioreactor | 2.8 | 2.4 | 1.4 | 1.0 | 0.9 |
| Starch degraded, mmoles of glucose per photobioreactor | 1.6 | 2.1 | 2.2 | 1.1 | 0.8 |
| Formate produced, mmoles per photobioreactor | 1.7 | 2.1 | 1.1 | 0.4 | 1.1 |
| Ethanol produced, mmoles per photobioreactor | 1.5 | 1.1 | 0 | 0 | 0.4 |

nase is induced and not immediately inactivated thereafter. However, not all substrate degradation results in aerobic respiration and generation of $CO_2$. Indeed, a significant accumulation of fermentation products by sulfur-deprived cultures has been reported previously [22]. We have shown that the initial pH of the medium at the start of sulfur-deprivation influences the amount of PSII and PSI activities, the rates of starch and protein degradation, and the rate of accumulation of fermentation products [22]. Those results are summarized in Table 11.1. The peak rate of $H_2$ production occurs at an initial pH of 7.7, which corresponds very closely to the peak rate of PSII activity. Photosystem I activity, measured as $O_2$ consumption by the Mehler reaction, increases as a function of pH. Starch and protein degradation, as well as formate accumulation, peaks at more acidic pHs, and formate production occurs at significantly decreased rates at pH 7.7. Finally, ethanol production does not occur at all in the pH 7.3–7.7 range, where maximum $H_2$ production is observed. These results show that algal metabolism can shift from anaerobic to aerobic as a function of the pH of the medium. The shifts in $H_2$ production are clearly correlated with the amount of residual PSII activity observed at each pH.

### 11.2.4
### Limiting Factors for $H_2$ Photoproduction under Sulfur Deprivation

The rates of $H_2$ photoproduction by sulfur-deprived cultures are maximal during the early stage of sulfur deprivation (about 5 µmoles $H_2 \cdot$ mg $Chl^{-1} \cdot h^{-1}$) and decrease over time, as illustrated in Fig. 11.3A. Preliminary economic analyses have indicated that, at the current rates, an algal $H_2$-production system based on sulfur deprivation would be too expensive to commercialize [27]. In order to lower the cost of $H_2$-gas production, it will be necessary among other things to increase the specific rate of $H_2$ production by a factor of at least 10, and thus we are attempting to identify the biochemical factors that limit the rate of $H_2$ production.

Since $H_2$ photoproduction is related to the capacity of PSII, we first determined the effect of sulfur deprivation on PSII function. This was done by measuring the rates of PSII-catalyzed $O_2$ evolution in cell aliquots taken directly from the photobioreactor at different times after sulfur deprivation. The measurements were performed under atmospheric $O_2$ and $CO_2$ conditions and represent the capacity of PSII, not the actual activity of PSII under the anaerobic conditions in the reactor. Fig. 11.3B shows that the capacity of PSII to generate $O_2$ and transfer electrons to PSI at $t=48$ h (50 µmoles electron pairs $\cdot$ mg $Chl^{-1} \cdot h^{-1}$) is 10 times higher than the actual rates of $H_2$ produced by the cultures at the same time. This demonstrates that PSII capacity per se does not limit $H_2$ production.

Next, we measured PSI capacity by performing Mehler reaction assays, where DCIP served as the electron donor to PSI and methyl viologen was the electron acceptor. Reduced methyl viologen reduces $O_2$, and the rate of PSI turnover was determined by the rate of $O_2$ consumption [28]. The maximum PSI capacity measured at $t=48$ h was 600 µmoles $O_2 \cdot$ mg $Chl^{-1} \cdot h^{-1}$ (not shown), which is

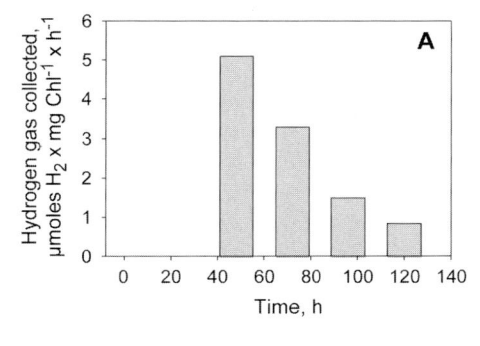

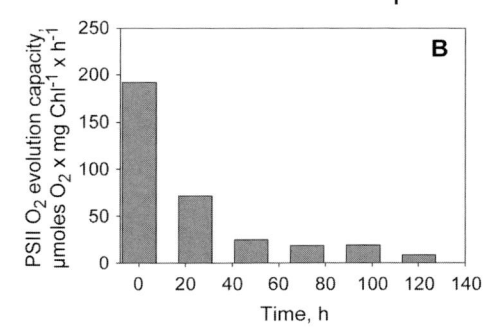

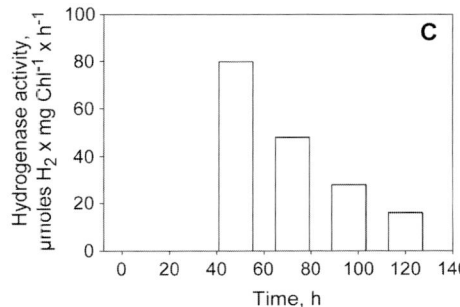

**Fig. 11.3** (A) Hydrogen-gas production, (B) PSII capacity for electron transport, and (C) light-induced hydrogenase activity of *C. reinhardtii* cultures as a function of time under sulfur-deprived conditions. The measurements were done with cultures sulfur-deprived at an initial pH of 7.7.

about two orders of magnitude higher than the measured rate of $H_2$ production. This demonstrates that, as was the case with PSII, PSI activity does not limit $H_2$ photoproduction.

Finally, the hydrogenase enzyme activity was measured with cells removed from the photo-bioreactor at different times after sulfur deprivation and dark-adapted for at least 2 min. The initial rates of light-induced $H_2$ production were measured with a Clark electrode. Fig. 11.3C shows that hydrogenase activity is high at 48 h but decreases gradually over the next three days. The hydrogenase activity at $t=48$ h is about 80 µmoles $H_2 \cdot$ mg $Chl^{-1} \cdot h^{-1}$, which is also 10 times higher than the measured rates of $H_2$ production by the cultures. The $H_2$ production rates measured as above must reflect not only the hydrogenase activity per se but also the number of electrons present in the PQ pool during the dark induction period. In contrast, when the rates of $H_2$ production were measured in vitro using dithionite-reduced methyl viologen as the electron donor [1], the hydrogenase activity actually increased from 24 h to 72 h. This substantiates our claim that $H_2$ production rates (which decrease over time) are not limited by hydrogenase activity.

Taken together, all of these results suggest that the rates of $H_2$ photoproduction by sulfur-deprived cultures are closely related to the residual PSII activity but that they must be further down-modulated by other factors. We have shown that the PQ pool becomes reduced following the establishment of anaerobiosis [20, 21] and that this reduction is immediately followed by an abrupt decrease in the photochemical activity of PSII. These results suggest that the reduction state of the PQ pool regulates the electron transport activity in anaerobic, sulfur-deprived algal cells and thus limits the amount of reductants that are utilized by hydrogenase to generate $H_2$.

## 11.2.5
### Mechanism of Regulation

To explain all of the above results, we propose a process under which the cultures initially switch from a photoheterotrophic growth mode at $t=0$ (at the start of sulfur deprivation) to the mixed metabolic state described in the model shown in Fig. 11.2. The switch in metabolic pathways entails shutting off acetate consumption and inducing the degradation and utilization of starch and/or protein. The shift from exogenous acetate to endogenous substrate utilization was shown by Kosourov et al. [22]. In that work, the amount of acetate in the growth medium decreased until about $t=48$ h. At that point, acetate excretion into the medium began, concomitant with endogenous substrate degradation. The occurrence of this metabolic shift occurs soon after anaerobiosis is established in the culture medium and may be triggered by it. Furthermore, the PSII photochemical activity is rapidly downregulated at the exact moment that anaerobiosis (defined as our inability to measure $O_2$ in the culture medium) is established [21]. The downregulation of electron transport from PSII was attributed to the over-reduction of the PQ pool [20, 21]. This over-reduction is due to two factors: (1) the lack of effective electron acceptors: Rubisco levels are extremely low at $t=24$ h [19], there is very little dissolved $O_2$ to accept electrons from the PQ oxidase and/or the reducing side of PSI, and the hydrogenase pathway is not yet induced [20, 21]; and (2) starch and protein degradation shuttles reductants to the PQ pool through the NAD(P)H-PQ oxidoreductase. Shortly after the establishment of anaerobiosis, hydrogenase activity is induced, and this relieves some of the PQ-pool over-reduction. This relief is partial, though, and may be explained by the inhibitory effect caused by the persistence of the proton gradient that occurs in the absence of $CO_2$ fixation [29]. Notably, sulfur-replete cultures flushed continuously with helium gas to maintain anaerobicity also produce $H_2$ at about the same rates as sulfur-deprived cultures [9], suggesting that they are also subject to similar limitations. Furthermore, we suggest that the over-reduction of the PQ pool modulates the distribution of carbon between the aerobic and anaerobic degradation pathways. As a consequence, efforts to increase the rate of $H_2$ production by sulfur-deprived cultures will require a much more detailed understanding of the regulatory mechanisms exerted by the over-reduction of the PQ pool and by the proton gradient on both photosynthetic electron transport and modulation of other metabolic pathways.

## 11.3
## Molecular Engineering of the Algal Hydrogenase

### 11.3.1
### Algal Hydrogenases and $H_2$ Production

Hydrogenases are enzymes that catalyze either $H_2$ uptake (providing electrons for the reduction of endogenous substrates) or $H_2$ production (eliminating excess reducing equivalents), and they are broadly classified as [FeFe]-only or [NiFe], according to the nature of their active site [17, 30]. The prosthetic group, or H-cluster, present in the [FeFe]-hydrogenase active site, is composed of an electron relay [4Fe4S]-center bridged by a cysteine to a catalytic [2Fe]-center. The unique [2Fe]-center is further coordinated to the active site by cyanide and carbon monoxide ligands [31, 32]. Hydrogen metabolism in the green alga *Chlamydomonas reinhardtii* is thought to be catalyzed, at least in part by a monomeric, 49-kDa reversible [FeFe]-hydrogenase enzyme, HydA1, which has been isolated to purity by Happe and Naber [3] and was shown to be rapidly inactivated by micromolar concentrations of $O_2$ [1–11]. This irreversible inactivation occurs by binding of $O_2$ to the H-cluster [2Fe]-center, as shown by EPR spectroscopy. Thus, one mechanism by which $H_2$ photoproduction in the alga ceases as the cultures become aerobic is inactivation of the [FeFe]-hydrogenase.

A new approach to achieving sustained $H_2$ production is to engineer an $O_2$-tolerant [FeFe]-hydrogenase and to express it in the alga. In order to perform in vitro engineering on the *C. reinhardtii* hydrogenase, we first had to clone, sequence, and characterize the gene(s) encoding for the enzyme, in order to gain insight into the enzyme structure and function.

### 11.3.2
### Cloning and Sequencing of the Two *C. reinhardtii* [FeFe]-Hydrogenases

Hydrogenase genes from *Scenedesmus obliquus* [34, 35], *C. reinhardtii* [1, 2, 36], and *Chlorella fusca* [18] have been cloned and sequenced. To date, all the cloned algal hydrogenases show homology to the other members of the [FeFe]-hydrogenase group. The first [FeFe]-hydrogenase gene from *C. reinhardtii*, *HydA1*, was cloned independently by three research groups (Genbank accession numbers CRE012098, AY055755, and AF289201). Purification and analysis of HydA1 have shown that it contains Fe and acid-labile S, and it catalyzes $H_2$ photoproduction using reduced ferredoxin as the electron donor [3].

To screen for additional *C. reinhardtii* hydrogenase genes, we used the amino acid sequence of the HydA1 catalytic site in a BLAST search of the EST database. The search identified a single positive EST clone. The cloned cDNA corresponding to the EST was designated *HydA2* [1] (Genbank accession numbers AY055756 and AY090770), and its sequence revealed a high degree of homology to *HydA1*. Comparison of the predicted polypeptide sequence of HydA2 to *C. reinhardtii* HydA1 [1, 2, 36], *S. obliquus* HydA1 [34], *C. fusca* HydA [18], and

**Table 11.2** Characteristics of the sequenced *hydA1* and *hydA2* cDNAs and their respective genomic DNAs.

| Type of DNA | HydA1 | HydA2 |
|---|---|---|
| cDNA | | |
| Total length (base pairs) | 2396 | 2527 |
| 5′ UTR (base pairs) | 158 | 139 |
| Coding region (base pairs) | 1491 | 1515 |
| Putative protein (amino acid residues) | 497 | 505 |
| 3′ UTR (base pairs) | 747 | 873 |
| Location of the polyadenylation site | 727 bp downstream from stop codon | 854 bp downstream from stop codon |
| Genomic DNA | | |
| TATA box | 187 bp upstream from 5′ UTR | 24 bp upstream from 5′ UTR |
| Introns | 7 | 9 |
| Exons | 8 | 10 |

*Clostridium pasteurianum* CpI [37] shows that HydA2 contains all three conserved motifs that define the catalytic H-cluster of all [FeFe]-hydrogenases; these are motifs 1 (PMFTSCCPxW), 2 (MPCxxKxxExxR), and 3 (FxExMACxGxCV). Each of these motifs contains a conserved cysteine residue that ligates the [4Fe4S]-center to the protein. An additional conserved cysteine residue in motif 3 bridges the [4Fe4S]-center to the [2Fe]-center.

Table 11.2 is a comparison of the structural characteristics of the cDNAs from the two *C. reinhardtii* hydrogenases. The *HydA1* cDNA has a 158-nucleotide 5′-UTR and a 747-nucleotide 3′-UTR (excluding the polyadenylated tail) [2], while the *HydA2* cDNA has a 139-nucleotide 5′-UTR and an 873 nucleotide 3′-UTR [1, 36]. A polyadenylation signal (TGTAA) that is characteristic of nuclear-encoded genes in *C. reinhardtii* [38] is located, respectively, 727 bp and 854 bp downstream from the stop codon in each cDNA. The ORFs for HydA1 and HydA2 encode proteins of, respectively, 497 and 505 amino acid residues. The genomic sequences of HydA1 and HydA2 reveal further differences between the two genes. A TATA-like sequence is not found in the promoter region of *HydA1* until 187 bp upstream from the 5′-UTR, while a characteristic TATA box is present 24 bp upstream from the 5′-UTR in *HydA2* [1]. Moreover, Fig. 11.4 indicates that *HydA1* contains only eight exons [2], while *HydA2* has 10 exons [4]. Clearly, genes for two different hydrogenase enzymes are encoded by the *C. reinhardtii* genome. Furthermore, their sequences are contained in different scaffolds (10 and 12, respectively) in the recently sequenced *C. reinhardtii* genome found in Genbank.

```
Cr HydA1:    MSALVLKPCAAVSIRGSSCRARQVAPRAPLAASTVRVALATLEAPARRLGNVACAA           56
Cr HydA2:    MALGLLAELRAGQAVACARRTNAPAHPAAVVPCLPSRAGKFFNLSQKVPSSQSARGSTIR       60

Cr HydA1:    AAPAAEAPLSHVQQALAELAKPKDDPTRKHVCVQVAPAVRVAIAETLGLAPGATTPKQLA      116
Cr HydA2:    VAATATDAVPHWKLALEELDKPKDG-GRKVLIAQVAPAVRVAIAESFGLAPGAVSPGKLA      119

Cr HydA1:    EGLRRLGFDEVFDTLFGADLTIMEEGSELLHRLTEHLEAHPHSDEPLPMFTSCCPGWIAM      176
Cr HydA2:    TGLRALGFDQVFDTLFAADLTIMEEGTELLHRLKEHLEAHPHSDEPLPMFTSCCPGWVAM      179

Cr HydA1:    LEKSYPDLIPYVSSCKSPQMMLAAMVKSYLAEKKGIAPKDMVMVSIMPCTRKQSEADRDW      236
Cr HydA2:    MEKSYPELIPFVSSCKSPQMMMGAMVKTYLSEKQGIPAKDIVMVSVMPCVRKQGEADREW      239

Cr HydA1:    FCVDADPTLRQLDHVITTVELGNIFKERGINLAELPEGEWDNPMGVGSGAGVLFGTTGGV      296
Cr HydA2:    FCVSE-PGVRDVDHVITTAELGNIFKERGINLPELPDSDWDQPLGLGSGAGVLFGTTGGV      298

Cr HydA1:    MEAALRTAYELFTGTPLPRLSLSEVRGMDGIKETNITMVPAPGSKFEELLKHRAAARAEA      356
Cr HydA2:    MEAALRTAYEIVTKEPLPRLNLSEVRGLDGIKEASVTLVPAPGSKFAELVAERLAHKVEE      358

Cr HydA1:    AAHGTPG---------PLAWDGGAGFTSEDGRGGITLRVAVANGLGNAKKLITKMQAGEA     407
Cr HydA2:    AAAAEAAAAVEGAVKPPIAYDGGQGFSTDDGKGGLKLRVAVANGLGNAKKLIGKMVSGEA     418

Cr HydA1:    KYDFVEIMACPAGCVGGGGQPRSTDKAITQKRQAALYNLDEKSTLRRSHENPSIRELYDT     467
Cr HydA2:    KYDFVEIMACPAGCVGGGGQPRSTDKQITQKRQAALYDLDERNTLRRSHENEAVNQLYKE     478

Cr HydA1:    YLGEPLGHKAHELLHTHYVAGGVEEKDEKK                                  497
Cr HvdA2:    FLGEPLSHRAHELLHTHYVPGGAEADA---                                  505
```

**Fig. 11.4** Alignment of the HydA1 and HydA2 amino acid sequences. A ClustalW alignment of the HydA1 and HydA2 amino acid sequences is shown with the known (HydA1) and predicted (HydA2) transit peptide sequences in gray text. The gray highlighted regions show where introns are located within the corresponding gene sequences of both hydrogenases. HydA1 introns 2, 3, 5, and 6 match the location of HydA2 introns 2, 3, 7, and 9, respectively.

### 11.3.3
### Anaerobic Expression of the Two *C. reinhardtii* Hydrogenases

The importance of $O_2$ in regulating algal $H_2$ metabolism is shown at the level of both hydrogenase gene expression [1, 2, 36] and enzyme activity [5, 39]. The detection of HydA1 and HydA2 expression by Western blots shows that the enzymes accumulate exclusively under anaerobic conditions [1, 39]. As shown in Fig. 11.5, hydrogenase activity (bars) peaks 90 min after anaerobic induction, and this coincides with peak *HydA1* and *HydA2* transcript levels (blots). The strong correlation between hydrogenase activity and transcript levels during anaerobic induction in photoheterotrophic-grown cells, however, is not observed in photoautotrophic-grown cells [1, 36]. This suggests that growth factors other than $O_2$ contribute to the regulation of hydrogenase activity.

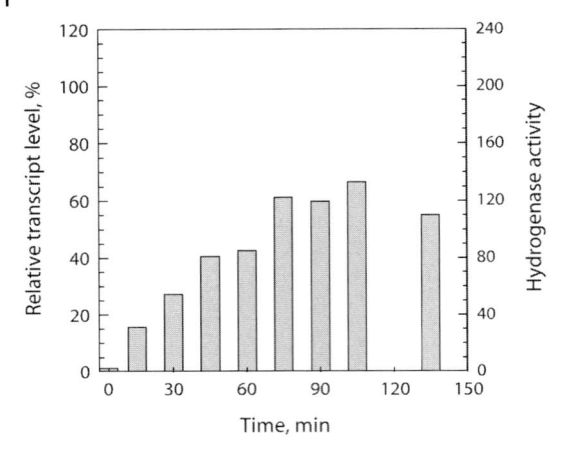

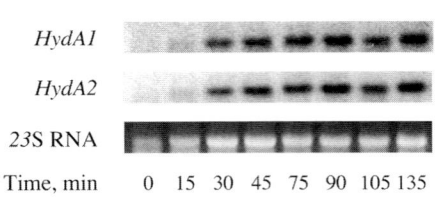

**Fig. 11.5** Anaerobic induction of $H_2$ photoproduction activity (bars) and representative blots showing the accumulation of the *HydA1* and *HydA2* transcripts in phototropically grown cells.

## 11.3.4
### Oxygen Inhibition of Hydrogenase Activity and Molecular Engineering for Increased $O_2$ Tolerance

Besides regulating hydrogenase expression, $O_2$ can also directly regulate enzyme activity by deactivating induced [FeFe]-hydrogenases in *C. reinhardtii* whole cells [3, 5, 40]. By comparison, purified bacterial [FeFe]-hydrogenases also show high $O_2$ lability, resulting from irreversible inactivation of the active site H-cluster [33]. An initial assessment of the bacterial enzyme structure has led to the observation that $O_2$ is able to gain access to the catalytic site through the proposed $H_2$ channel, which functions in the diffusion of $H_2$ from the active site to the surface [33, 41]. Indeed, the narrower region between the $H_2$ channel and the active site of $H_2$-sensing [NiFe]-hydrogenases has been proposed to explain their greater $O_2$ resistance compared to catalytic [NiFe]-hydrogenases [42, 43]. Our structural modeling of HydA2 suggested that a similar gas channel was present in algal [FeFe]-hydrogenases as well [1]. Based on this observation, we now suggest that resistance to $O_2$ inactivation in algal [FeFe]-hydrogenases might be gained by engineering a steric restriction to $O_2$ diffusion through the $H_2$ channel. Physical restriction of $O_2$ access (either by reducing the diffusional rate or by shielding the active site) should protect the active site and prevent hydrogenase inactivation.

To this end, we generated a structural model of HydA1 by homology modeling the HydA1 protein sequence to the solved X-ray structure of [FeFe]-hydro-

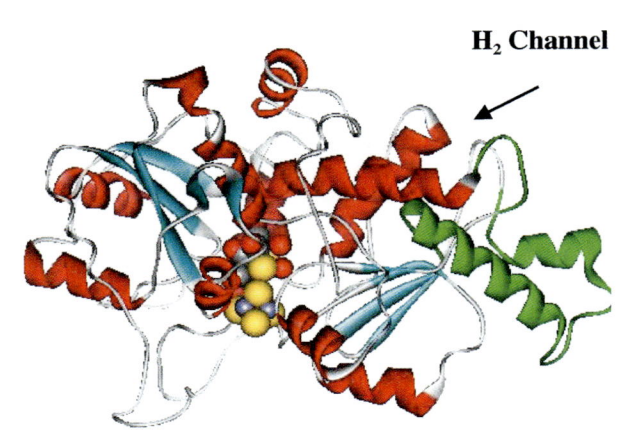

**Fig. 11.6** Homology structural model of the *C. reinhardtii* HydA1 hydrogenase. The H-cluster is identified in CPK colors as space-filled atoms. The backbone colors correspond to secondary structure type (red = $\alpha$-helix, blue = $\beta$-sheet, and gray = random coil). The Fe-S centers are represented as ball-and-stick diagrams. Domains unique to HydA1 and not found in CpI are represented in green. Images of the structures were made with ViewerLite software (Accelrys).

genase, CpI (Fig. 11.6), as done previously for HydA2 [1]. The high degree of amino acid sequence homology between HydA1 and CpI was particularly evident within the $H_2$-channel (62% identity and 92% similarity) and active-site (80% identity and 89% similarity) domains. The channel environment is primarily composed of small hydrophobic residues (i.e., glycine, alanine, valine) organized in two $\alpha$-helices and two $\beta$-sheets, resulting in an estimated diameter larger than the effective diameters of both $H_2$ (2.8 Å) and $O_2$ (3.5 Å) [44]. As a result, the predicted size of the HydA1 channel is sufficient to promote the diffusion of $H_2$ from the active site to the surface, but it is also large enough to allow for the diffusion of the inhibitor, $O_2$.

The above results suggest that engineering $O_2$ tolerance into HydA1 might be accomplished by altering the residues that line the interior of the $H_2$ channel to reduce its diameter and, in turn, limit $O_2$ diffusion to the active site [45]. Without detailed X-ray structural analysis of HydA1, the possibility also remains that alternative channels might exist for gas access to and from the surface to the active site at the point where ferredoxin binds to the membrane. Should alternative channels exist that are sufficient in size to allow for inhibitor access to the active site, the methodology suggested in this study could also be used to target these channels to improve tolerance to inhibitors. This work thus constitutes the foundation for further research aimed at generating site-directed mutants that can photoproduce $H_2$ gas under atmospheric $O_2$ conditions.

## Acknowledgments

This work was supported by the U.S. DOE Hydrogen, Fuel Cells, and Infrastructure Technologies Program.

## References

1 Forestier M., King P., Zhang, L., Posewitz, M., Schwarzer, S., Happe, T., Ghirardi, M. L., Seibert M. *Eur. J. Biochem.* **2003**, *270*, 2750–2758.
2 Happe T., Kaminski A. *Eur. J. Biochem.* **2002**, *269*, 1–11.
3 Happe T., Naber J. D. *Eur. J. Biochem.* **1993**, *214*, 475–481.
4 Urbig T., Schulz R., Senger H. Z. *Naturforsch.* **1993**, *48*, 41–45.
5 Ghirardi M. L., Togasaki R. K., Seibert M. *Appl. Biochem. Biotechnol.* **1997**, *63–65*, 141–151.
6 Melis A. *Int. J. Hydrogen Energy* **2002**, *27*, 1217–1228.
7 Healey F. P. *Plant Physiol.* **1970**, *45*, 153–159.
8 Greenbaum E. *Science* **1982**, *196*, 879–880.
9 Greenbaum E., Guillard R. R. L., Sunda W. G. *Photochem. Photobiol.* **1983**, *37*, 649–655.
10 Gfeller R. P., Gibbs M. *Plant Physiol.* **1984**, *75*, 212–218.
11 Randt C., Senger H. *Photochem. Photobiol.* **1985**, *42*, 553–557.
12 Wykoff, D. D., Davies J. P., Melis A., Grossman A. R. *Plant Physiol.* **1988**, *117*, 129–139.
13 Melis A., Zhang L., Forestier M., Ghirardi M. L., Seibert M. *Plant Physiol.* **2000**, *122*, 127–135.
14 Ghirardi M. L., Zhang L., Lee J. W., Flynn T., Seibert M., Greenbaum E., Melis A. *Trends Biotech.* **2000**, *18*, 506–511.
15 Zhang L., Happe T., Melis A. *Planta* **2002**, *214*, 552–561.
16 Melis A. *Int. J. Hydrogen Energy* **2002**, *27*, 1217–1228.
17 Boichenko V. A., Greenbaum E., Seibert M. **2004**, in *Photoconversion of Solar Energy: Molecular to Global Photosynthesis*, Vol. 2, M. D. Archer and J. Barber (Eds), Imperial College Press, London, pp. 397–452.
18 Winkler M., Heil B., Heil B., Happe T. *Biochim. Biophys. Acta*, **2002**, *1576*, 330–334.
19 Zhang L., Happe T., Melis A. *Planta* **2002**, *214*, 552–561.
20 Antal T. K., Krendeleva T. E., Laurinavichene T. V., Makarova V. V., Tsygankov A. A., Seibert M., Rubin A. *Proc. Russian Acad. Sci.* **2001**, *381*, 371–375.
21 Antal T. K., Krendeleva, T. E., Laurinavichene T. V., Makarova V. V., Ghirardi M. L., Rubin A. B., Tsygankov A. A., Seibert M. *Biochim. Biophys. Acta* **2003**, *1607*, 153–160.
22 Kosourov S., Seibert M., Ghirardi M. L. *Plant Cell Physiol.* **2003**, *44*, 146–155.
23 Kosourov S., Tsygankov A., Seibert M., Ghirardi M. L. *Biotechnol. Bioeng.* **2002**, *78*, 731–740.
24 Kosourov S., Tsygankov A., Ghirardi, M. L., Seibert M. **2001**, in *Proceedings of the 12$^{th}$ International Congress on Photosynthesis*, 18–23 August 2001, Brisbane, Australia, http://www.publish.csiro.au/ps2001, CSIRO Publishing, Melbourne, Australia, S37–009.
25 Tsygankov A., Kosourov S., Seibert M., Ghirardi M. L. *Int. J. Hydrogen Energy* **2002**, *27*, 1239–1244.
26 Gfeller R. P., Gibbs M. *Plant Physiol.* **1985**, *77*, 509–511.
27 Ghirardi M. L., Amos W. *Biocycle* **2004**, in press.
28 Izawa S. *Meth. Enzymol.* **1980**, *69*, 413–434.
29 Lee, J., Greenbaum, E. *Appl. Biochem. Biotechnol.* **2003**, *105–108*, 303–313.
30 Vignais P. M., Billoud B., Meyer, J. *FEMS Micro. Rev.* **2001**, *25*, 455–501.

31 Peters J. W., Lanzilotta W. N., Lemon B. J., Seefeldt, L.C. *Science* **1998**, *282*, 1853–1858.
32 Nicolet Y., Piras C., Legrand P., Hatchikian C. E., Fontecilla-Camps, J. C. *Structure* **1998**, *7*, 13–23.
33 Adams M. W. W. *Biochim. Biophys. Acta* **1990**, *1020*, 115–145.
34 Florin L., Tsokoglou A., Happe T. *J. Biol. Chem.* **2001**, *276*, 6125–6132.
35 Wünschiers R., Stangier K., Senger H., Schulz R. *Current Microbiol.* **2001**, *42*, 353–360.
36 Forestier M., Zhang L., King P., Plummer S., Ahmann D., Seibert M., Ghirardi M. **2001**, in *Proceedings of the 12th International Congress on Photosynthesis*, 18–23 August 2001, Brisbane, Australia, http://www.publish.csiro.au/ps2001, CSIRO Publishing, Melbourne, Australia, S37–003.
37 Meyer J., Gagnon J. *Biochemistry* **1991**, *30*, 9697–9704.
38 Silflow C. D., Chisholm R. L., Conner T. W., Ranum L. P. *Mol. Cell Biol.* **1985**, *5*, 2389–2398.
39 Happe T., Mosler B., Naber J. D. *Eur. J. Biochem.* **1994**, *222*, 769–774.
40 Bamberger E. S., King D., Erbes D. L., Gibbs M. *Plant Physiol.* **1982**, *69*, 1268–1273.
41 Montet Y., Amara P., Volbeda A., Vernede X., Hatchikian E. C., Field M. J., Frey M., Fontecilla-Camps, J. C. *Nat. Struc. Biol.* **1997**, *4*, 523–526.
42 Volbeda A., Montet Y., Vernede X., Hatchikian E.C., Fontecilla-Camps, J. C. *Int. J. Hydrogen Energy* **2002**, *27*, 1449–1461.
43 Bernhard M., Buhrker, T., Bleijlevens B., De Lacey A. L., Fernandes V. M., Albracht S. P. J., Friedrich, B. *J. Biol. Chem.* **2001**, *76*, 15592–15597.
44 Nenoff T. M. **2000**, in *Proceedings of the 2000 Hydrogen Program Review*, NREL/CP-570-28890, http://www.eere.energy.gov/hydrogenandfuelcells/pdfs/28890q.pdf.
45 Cohen, J., Kim, K., Posewitz, M., Ghirardi, M. L., Schulten, K., Seibert, M., King, P. *Biochem. Soc. Transact.* **2005**, *33*, 102–110.

# 12
# Bioengineering of Green Algae to Enhance Photosynthesis and Hydrogen Production

*Anostasios Melis*

## 12.1
## Introduction

Photosynthesis and hydrogen production in unicellular green algae can operate with a nearly 100% photon utilization efficiency (Ley and Mauzerall 1982; Greenbaum 1988), making these microorganisms an efficient biocatalyst for the generation of hydrogen from sunlight and $H_2O$. However, green microalgal cultures under direct sunlight show rather poor light utilization efficiency in photosynthesis. The reason for this shortcoming is that, under bright sunlight, the rate of photon absorption by the chlorophyll (Chl) antenna arrays in photosystem II (PSII) and photosystem I (PSI) far exceeds the rate at which photosynthesis can utilize them. Excess photons cannot be stored in the photosynthetic apparatus but are dissipated (lost) as fluorescence or heat. Up to 80% of absorbed photons could thus be wasted (Melis et al. 1999), decreasing light utilization efficiency and compromising cellular productivity to unacceptably low levels. Thus, in a high-density mass culture, cells at the surface over-absorb and waste sunlight, whereas cells deeper in the culture are deprived of much-needed irradiance, as this is strongly attenuated due to filtering.

Biotechnological applications of green algae in the areas of biomass accumulation, carbon sequestration, and energy production require utilization of green algae that are not subject to this optical pitfall and suboptimal utilization of sunlight in mass culture, and that operate with maximal light utilization efficiency and photosynthetic productivity under bright sunlight. To attain these performance characteristics, it is necessary to minimize the absorption of sunlight by individual cells so as to permit greater transmittance of irradiance through the high-density green alga mass culture. This requirement was recognized long ago (Kok 1953; Myers 1957; Radmer and Kok 1977) but could not be satisfied because algae with a "truncated light-harvesting chlorophyll antenna size" are not encountered in nature. Until recently, this problem could not be addressed and solved in the laboratory either, due to the lack of the necessary technologies by which to approach it. The advent of molecular genetics in com-

*Artificial Photosynthesis: From Basic Biology to Industrial Application.*
Edited by Anthony F. Collings and Christa Critchley
Copyright © 2005 WILEY-VCH Verlag GmbH & Co. KGaA, Weinheim
ISBN: 3-527-31090-8

bination with sensitive absorbance-difference kinetic spectrophotometry for the precise measurement of the Chl antenna size in green algae now offers a valid approach by which to pursue a reduction in the number of photosynthetic Chl antenna molecules. Accordingly, this chapter summarizes recent progress and the state of the art in this field.

## 12.2
## Rationale and Approach

The rationale for attempting a bioengineering approach to truncate the Chl antenna size in green algae is that such modification will prevent individual cells at the surface of a high-density culture from over-absorbing sunlight and wastefully dissipating most of it (Fig. 12.1). A truncated Chl antenna size will permit sunlight to penetrate deeper into the culture, thus enabling many more cells to contribute to useful photosynthesis, culture productivity, and $H_2$ production (Fig. 12.2). It has been shown that a truncated Chl antenna size will enable a three- to fourfold greater solar energy conversion efficiency and photosynthetic productivity than could be achieved with fully pigmented cells (Melis et al. 1999). Such bioengineering, therefore, would enhance photosynthetic productivity of microalgae in mass culture.

A systematic approach to this problem is to identify genes that regulate the Chl antenna size of photosynthesis and, further, to manipulate such genes so as to confer a permanently truncated Chl antenna size to the model green alga *Chlamydomonas reinhardtii*. Identification of such genes in *Chlamydomonas* will permit a subsequent transfer of this trait to other microalgae that are of interest to the alga biotechnology industry. This objective has been approached in the laboratory of the author upon application of DNA insertional mutagenesis tech-

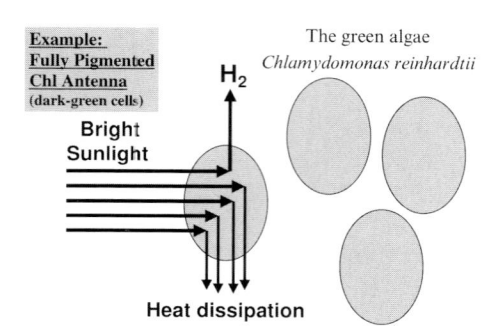

**Fig. 12.1** Schematic presentation of the fate of absorbed sunlight in fully pigmented (dark green) algae. Individual cells at the surface of the culture over-absorb incoming sunlight (i.e., they absorb more than can be utilized by photosynthesis) and "heat dissipate" most of it. Note that a high probability of absorption by the first layer of cells would cause shading, i.e., would prevent cells deeper in the culture from being exposed to sunlight.

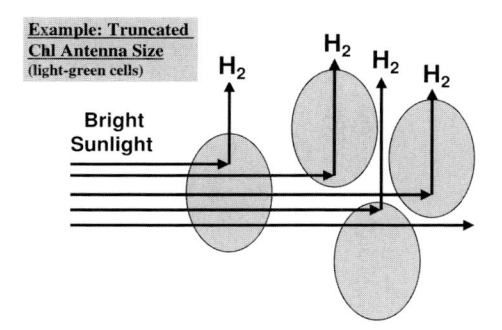

**Fig. 12.2** Schematic presentation of sunlight penetration through cells with a truncated chlorophyll antenna size. Individual cells have a diminished probability of absorbing sunlight, thereby permitting deeper penetration of irradiance and $H_2$ production by cells deeper in the culture.

niques (Kindle 1990; Gumpel and Purton 1994), screening, biochemical/molecular/genetic, and absorbance-difference kinetic spectrophotometry analyses of *Chlamydomonas reinhardtii* cells. Bioengineered strains of green algae, acquired in the course of such effort, may find direct application in biomass accumulation (Vazquezduhalt 1991; Brown and Zeiler 1993; Westermeier and Gomez 1996), $H_2$ production (Zaborsky 1998; Melis et al. 2000; Zhang et al. 2002), and carbon sequestration (Mulloney 1993; Nakisenovic 1993).

## 12.3
## Physiological State of the Chl Antenna Size in Green Algae

Green algal photosynthesis depends on the absorption of sunlight by chlorophyll (Chl) molecules in photosystem I and photosystem II. In each photosystem (PS), chlorophylls and other accessory pigments act cooperatively in the absorption of incoming electromagnetic radiation (Emerson and Arnold 1932a, b). The concept of the photosynthetic unit, first proposed by Gaffron and Wohl (1936), stipulates that distinct assemblies, or arrays, of photosynthetic pigments serve as antennae for the collection of light energy and as a conducting medium for excitation migration toward a photochemical reaction center. Distinct pigment-protein complexes are contained within PSI and PSII and perform the functions of light absorption and excitation energy transfer to a photochemical reaction center (Simpson and Knoetzel 1996; Pichersky and Jansson 1996). Up to 350 chlorophyll *a* (Chl *a*) and Chl *b* molecules can be found in association with PSII, whereas the Chl antenna size of PSI may contain up to 300 mainly Chl *a* molecules (Melis 1991, 1996). Most of these Chl molecules are organized within 10 peripheral subunits of the so-called auxiliary chlorophyll *a-b* light-harvesting complex (Lhc). There are six such subunits for PSII (Lhc b1–b6) and four for PSI (Lhc a1–a4) (Jansson et al. 1992). These peripheral Lhc subunits are not essential for the process of photosynthesis (Glick and Melis 1988), and, in principle, the corresponding 10 genes (*Lhc b1–b6* and *Lhc a1–a4*) could be deleted from the genome of the organism in order to limit the size of the Chl antenna. In practice, this gene-deletion approach may be difficult because of the possible existence of multiple copies for each of these genes, all of which would have to be deleted. An additional difficulty is that in the absence of one of the Lhc subunits, the algae can recruit another protein subunit for the assembly of the fully pigmented Chl antenna (Sukenik et al. 1988; Ruban et al. 2003).

The wasteful dissipation of excitation energy by the photosynthetic apparatus is not the only problem associated with the large arrays of chlorophyll antenna molecules in green microalgae. Due to the high rate of photon dissipation under bright sunlight, cells at the surface of the mass culture are subject to photoinhibition of photosynthesis (Powles 1984; Melis 1999), an adverse reaction that further compounds losses in productivity. Meanwhile, cells deeper in the culture are deprived of much-needed sunlight, as this is strongly attenuated due to filtering (Naus and Melis 1991; Neidhardt et al. 1998; Melis et al. 1999). A genetic tendency of the algae to assemble large arrays of light-absorbing chlorophyll antenna molecules in their photosystems is a survival strategy and a competitive advantage in the wild, where light is often limited (Kirk 1994). Obviously, this property of the algae is detrimental to the yield and productivity in a mass culture. A truncated Chl antenna would clearly compromise the ability of the strain to compete and survive in the wild. However, in a controlled mass culture in photo-bioreactors, this would help to diminish the over-absorption and wasteful dissipation of excitation energy by individual cells, and it would also diminish photoinhibition of photosynthesis at the surface while permitting greater transmittance of light deeper into the culture. Such altered optical properties of the cells would result in greater photosynthetic productivity and enhanced solar energy conversion efficiency by the mass culture. In support of this contention, preliminary experiments (Neidhardt et al. 1998; Melis et al. 1999) confirmed that a smaller Chl antenna size would result in a relatively higher light intensity for the saturation of photosynthesis in individual cells, while allowing for an overall greater productivity by the mass culture (Nakajima and Ueda 1997, 1999; Polle et al. 2003). Thus, genetic approaches by which to permanently truncate the Chl antenna size of photosynthesis in green algae merit serious consideration.

## 12.4
### The Genetic Control Mechanism of the Chl Antenna Size in Green Algae

The Chl antenna size of the photosystems is not fixed but could vary substantially depending on developmental, genetic, physiological, and even environmental conditions (Fig. 12.3). It is recognized in the field that a genetic regulatory mechanism dynamically modulates the Chl antenna size of photosynthesis (Anderson 1986; Escoubas et al. 1995; Melis 1991, 1996, 2002). Physiological consequences of the function of this mechanism are well understood (Smith et al. 1990; LaRoche et al. 1991; Maxwell et al. 1995; Falbel et al. 1996; Webb and Melis 1995; Ohtsuka et al. 1997; Tanaka and Melis 1997). However, little is known about the genes and proteins and their mode of action in the genetic regulation of the Chl antenna size. This Chl antenna size regulatory mecha-

## 12.5 Effect of Pigment Mutations on the Chl Antenna Size of Photosynthesis

| **Large Chl Antenna Size** |
| PSII: 350 Chl *a* and *b* |
| PSI: 300 Chl *a* and *b* |

↑ (Limiting light)

**Molecular mechanism**

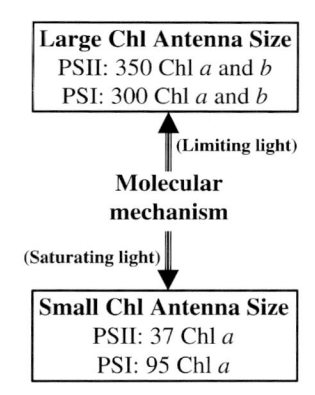

(Saturating light) ↓

| **Small Chl Antenna Size** |
| PSII: 37 Chl *a* |
| PSI: 95 Chl *a* |

**Fig. 12.3** Schematic of a molecular mechanism for the regulation of the Chl antenna size of photosynthesis. This sensory and signal transduction pathway is highly conserved in all photosynthetic organisms and regulates the Chl antenna size of the photosystems. Within limits for the PSII and PSI Chl antenna size, defined by genetic and structural considerations, the Chl antenna size could vary in response to environmental, genetic, developmental, or physiological conditions. Depicted is the effect of irradiance on the Chl antenna size, which is essentially a compensation response to the level of irradiance.

nism is highly conserved and functions in all organisms of oxygenic and anoxygenic photosynthesis (Anderson 1986; Nakada et al. 1995; Escoubas et al. 1995; Huner et al. 1998; Yakovlev et al. 2002; Masuda et al. 2003). Thus, identification of the relevant genes and elucidation of the genetic mechanism for the regulation of the Chl antenna size in *Chlamydomonas reinhardtii* could apply to all photosynthetic organisms. As such, this topic is of fundamental importance to the field and of practical importance to the alga biotechnology industry. Thus, in order to alleviate over-absorption and loss of sunlight energy by green algae, a genetic interference with the regulation of this highly conserved mechanism is required. The objective is to genetically develop microalgae with a permanently truncated light-harvesting chlorophyll antenna size.

## 12.5
### Effect of Pigment Mutations on the Chl Antenna Size of Photosynthesis

Recent efforts have contributed to the identification of pigment mutants and the corresponding genes, conferring a partially truncated Chl antenna size in *C. reinhardtii*. A Chl *b*-less mutant was examined that lacked a functional chlorophyll *a* oxygenase (*CAO*) gene (Tanaka et al. 1998) and that failed to synthesize this auxiliary pigment. This strain was derived via DNA insertional mutagenesis in which the chlorophyll *a* oxygenase gene was interrupted by the transforming plasmid (Tanaka et al. 1998). The advantage of such tagged genetic transformation for the generation of mutants is that genes responsible for a given property can then be isolated.

The Chl *b*-less mutant of *C. reinhardtii* had a significantly truncated Chl antenna size for PSII and a slight increase in the Chl antenna size for PSI (Table 12.1; see also Polle et al. 2000). The PSII Chl antenna size in the Chl *b*-less mutant (Chl-PSII = ~96 Chl molecules) was substantially smaller than that in the wild type (Chl-PSII = 230 *a* and *b* molecules). Nevertheless, it remained significantly larger than that of the PSII-core antenna (Chl-PSII = ≈37 Chl molecules)

**Table 12.1** Chl antenna size of PSII and PSI in *C. reinhardtii* wild type, Chl *b*-less, and lutein-less mutants. Numbers show the functional Chl antenna size, i.e., the Chl (*a* and *b*) molecules specifically associated with each photosystem, as determined by sensitive absorbance-difference kinetic spectrophotometry analysis.

|  | Wild type | Chl *b*-less | *npq2 lor1* (lutein-less) | Minimum Chl antenna size |
|---|---|---|---|---|
| Chl-PSII | 230 | 90 | 125 | 37 |
| Chl-PSI | 240 | 289 | 294 | 95 |

(Glick and Melis 1988). Analysis of the LHCII polypeptide composition confirmed the presence of lower amounts of CP26 and CP29 and a smaller portion of the major LHCII proteins in the thylakoid membrane of the Chl *b*-less mutant (Polle et al. 2000). Obviously, there is some stable integration of LHCII polypeptides in the thylakoid membrane of *C. reinhardtii* in the absence of Chl *b*, consistent with previous reports on this matter (Michel et al. 1983; Allen and Staehelin 1994; Plumley and Schmidt 1995). In contrast to PSII, the Chl antenna size of PSI in the Chl *b*-less mutant (Chl-PSI = 290 *a* and *b* molecules) was similar to or even slightly larger than that of the wild type (Chl-PSI = 240 *a* molecules). In agreement, LHCI proteins that were present in the wild type were also found at comparable levels in the Chl *b*-less mutant. It was concluded that the PSI auxiliary light-harvesting Chl antenna can fully assemble and be functionally connected with the reaction center P700, even in the absence of Chl *b*. Additional studies were conducted to determine the role of the *CAO* gene in the function of the Chl antenna size regulatory mechanism (Fig. 12.3). It was determined that *CAO* gene expression is highly regulated according to the Chl antenna-size needs of the organism (Masuda et al. 2002, 2003). Modulation of *CAO* gene expression exerts a significant effect on the Chl antenna size of PSII, but not on that of PSI. The *CAO* gene may thus be a target for a truncated PSII Chl antenna size.

A *Chlamydomonas reinhardtii* double mutant *npq2 lor1* (Niyogi et al. 1997) lacked the $\beta,\varepsilon$-carotenoids lutein and loroxanthin as well as all $\beta,\beta$-epoxycarotenoids derived from zeaxanthin (e.g., violaxanthin and neoxanthin). The only carotenoids present in the thylakoid membranes of the *npq2 lor1* cells were $\beta$-carotene and zeaxanthin. The effect of these pigment mutations on the Chl antenna size of photosynthesis was investigated (Polle et al. 2001). Table 12.1 shows that the *npq2 lor1* mutant had a significantly smaller PSII light-harvesting Chl antenna size and a slightly larger PSI Chl antenna size than the wild type. SDS-PAGE and Western blot analyses revealed that some of the LHCII and most of the LHCI were assembled and functionally connected with PSII and PSI reaction centers, respectively. Light-saturation curves of photosynthesis revealed that a significantly greater light-intensity was required for the saturation of photosynthesis in the Chl antenna mutant than in the wild type. It was concluded that a lesion in the *lycopene ε-cyclase* gene prevents synthesis of lutein in green algae and that this specific mutation affects the functional Chl antenna size of PSII but not that of PSI (Polle et al. 2001). Xanthophyll-biosynthesis genes in general, and the *lycopene ε-cyclase* gene in particular, may be suitable targets for a truncated PSII Chl antenna size.

## 12.6
## Genes for the Regulation of the Chl Antenna Size of Photosynthesis

The ability of the photosynthetic apparatus to regulate the size of the functional Chl antenna was first recognized more than 30 years ago in pioneering work by Bjorkman (Bjorkman et al. 1972). In spite of the substantial number of physiological and biochemical studies on this phenomenon since then (reviewed in Anderson 1986; Melis 1996), genes for the regulation of the Chl antenna size of photosynthesis remained unknown. Recent bioengineering efforts by which to truncate the Chl antenna size of photosynthesis contributed to the first-time cloning of a Chl antenna size regulatory gene in photosynthesis. This was achieved through the application of DNA insertional mutagenesis with the green alga *C. reinhardtii*. Briefly, 6500 DNA insertional mutagenesis transformants were generated and screened. Based on the screening protocol applied (Polle et al. 2003), only one mutant having a truncated light-harvesting chlorophyll antenna size (*tla1*) was identified out of this batch of transformants. Genetic crosses and mapping of the DNA around the insertion site confirmed that the exogenous plasmid interfered with a novel gene, termed by us *Tla1*. DNA, mRNA, and protein sequences of *Tla1* were elucidated and deposited in the GenBank (Accession No. AF534570 and AF534571). Evidence was presented that the *Tla1* gene is responsible for the regulation of the Chl antenna size in green algae (Polle et al. 2003). Sensitive absorbance-difference kinetic spectrophotometry revealed that interference with *Tla1* gene expression resulted in a truncated PSII Chl antenna size, down to 50% of that in the wild type, and a truncated PSI Chl antenna size, down to 67% of that in the wild type (Table 12.2). Thus, in the *tla1* strain, both PSII and PSI had a smaller Chl antenna

**Table 12.2** *Chlamydomonas reinhardtii* Chl content and photosystem Chl antenna size in wild type and *tla1* mutant as determined by sensitive absorbance-difference kinetic spectrophotometry analysis. The long-term goal (limiting values of the Chl antenna size) is 37 Chl for PSII and 95 Chl for PSI.

|  | Wild type | *tla1* | % change | Long-term goal |
|---|---|---|---|---|
| Chl-PSII | 230 | 115 | 50% | 37 |
| Chl-PSI | 240 | 160 | 67% | 95 |

Sources: Glick and Melis 1988; Zouni et al. 2001; Jordan et al. 2001.

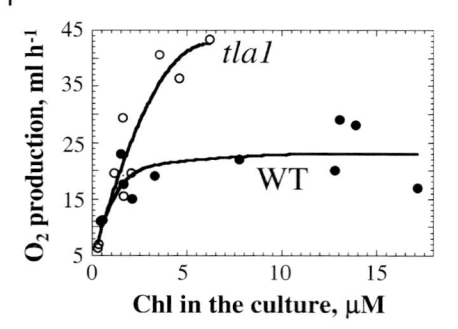

**Fig. 12.4** Photosynthetic productivity ($O_2$ evolution) measurements were conducted in the greenhouse under mass culture conditions with the wild type and the *tla1* mutant of *Chlamydomonas reinhardtii*. Productivity was measured as a function of chlorophyll concentration in the mass culture photo-bioreactor (Polle et al. 2003) at a solar incident intensity (photosynthetically active radiation) of about 1500 µmol photons $m^{-2}$ $s^{-1}$.

size relative to the wild type, which is an improvement over the result of pigment mutations on the Chl antenna size of photosynthesis (Table 12.1). SDS-PAGE analysis revealed a lack of specific LHCII and LHCI proteins from the antenna of the *tla1* (Polle et al. 2003). Measurements of the light-saturation curve of photosynthesis revealed a lower absorbance of sunlight, a greater light intensity for the saturation of photosynthesis, and diminished dissipation of excitation energy as heat in the *tla1* mutant relative to that in the wild type.

As anticipated, the *tla1* strain had enhanced photosynthetic productivity and improved light utilization efficiency under mass culture conditions in the greenhouse (Polle et al. 2003). Fig. 12.4 shows photosynthetic productivity results of *C. reinhardtii* cultures conducted in the greenhouse under ambient mass culture conditions (Polle et al. 2003). It is seen that, under bright sunlight, the rate of photosynthetic $O_2$ accumulation is a function of the Chl concentration in the respective culture at the time of measurement. Results from this detailed analysis showed that productivity of green algae in a mass culture increases linearly as a function of cell density and Chl concentration in both the wild type and the *tla1* mutant. The linear increase is observed under conditions when the amount of the biomass, but not irradiance, is the yield-limiting factor. This initially linear increase in the yield of the culture levels off as the green alga biomass reaches a certain density. The "saturation" occurs because, at a threshold Chl concentration, cells in the culture would absorb all incoming irradiance. From that point on, light utilization efficiency would define yield (Polle et al. 2003). In both the wild type and the *tla1* mutant, productivity appeared to become saturated at a Chl concentration of about 5 µM in the culture, with the rate of the *tla1* mutant being substantially (twofold) greater than that of the wild type. The lower yield of the wild type was attributed directly to the greater fraction of photons that are absorbed but not utilized (Polle et al. 2003), resulting in dissipation and loss of the excess photons as fluorescence or heat (Melis et al. 1999). This is apparently alleviated to some extent by the smaller Chl antenna size in the *tla1* mutant.

## 12.7
## Conclusions

It is the objective of this work to minimize, or truncate, the chlorophyll antenna size in green algae in order to maximize photosynthetic light utilization efficiency of green algae in mass culture. Further, the work seeks to demonstrate that a truncated Chl antenna size would minimize absorption and wasteful dissipation of sunlight by individual cells, resulting in better light utilization efficiency and greater photosynthetic productivity by the green alga culture. The approach has been to employ DNA insertional mutagenesis, screening, and biochemical and molecular genetic analyses for the isolation of "truncated Chl antenna size" strains in the green alga *Chlamydomonas reinhardtii*. Efforts are under way to clone and characterize the genes that affect the "Chl antenna size" in *Chlamydomonas reinhardtii*. Eventually, the work seeks to apply such genes to generate "truncated Chl antenna size" strains in *Chlamydomonas reinhardtii* and other green algae of interest to the alga biotechnology industry.

In the course of this effort, *Tla1*, the first "Chl antenna size regulatory gene," was cloned. It was shown that the partially truncated chlorophyll antenna size of the *tla1* mutant alleviates the over-absorption of incident sunlight by individual cells and the wasteful dissipation of over-absorbed irradiance. A truncated light-harvesting chlorophyll antenna size in the *tla1* mutant diminishes the severe cell shading that occurs in normally pigmented wild type, permits a more uniform illumination of the cells in a mass culture, and results in greater light utilization efficiency and photosynthetic productivity of the algae under bright sunlight conditions. Results from this work apply directly to green alga $H_2$ production, biomass accumulation, and carbon sequestration efforts.

## Acknowledgements

This work was supported by DOE-UCB Cooperative Agreement DE-FC36-00GO10536.

## References

Allen KD, Staehelin LA (1994) Polypeptide composition, assembly and phosphorylation patterns of the photosystem II antenna system of *Chlamydomonas reinhardtii*. Planta 194: 42–54.

Anderson JM (1986) Photoregulation of the composition, function and structure of thylakoid membranes. Annu Rev Plant Physiol 37: 93–136.

Bjorkman O, Boardman NK, Anderson JM, Thorne SW, Goodchild DJ, Puliotis NA (1972) Effect of light intensity during growth *of Atriplex patula* on the capacity of photosynthetic reactions, chloroplast components and structure. Carnegie Institution Yearbook 71: 115–135.

Brown LM, Zeiler KG (1993) Aquatic biomass and carbon dioxide trapping. Energy Conversion and Management 34: 1005–1013.

Emerson R, Arnold W (1932a) A separation of the reactions in photosynthesis by means of intermittent light. J Gen Physiol 15: 391–420.

Emerson R, Arnold W (1932b) The photochemical reactions in photosynthesis. J Gen Physiol 16: 191–205.

Escoubas JM, Lomas M, LaRoche J, Falkowski PG (1995) Light intensity regulation of cab gene transcription is signalled by the redox state of the plastoquinone pool. Proc Nat Acad Sci 92: 10237–10241.

Falbel T, Meehl JB, Staehelin LA (1996) Severity of mutant phenotype in a series of chlorophyll-deficient wheat mutants depends on light intensity and the severity of the block in chlorophyll synthesis. Plant Physiol 112: 821–832.

Gaffron H, Wohl K (1936) Zur theorie der assimilation. Naturwissenschaften 24: 81–90.

Glick RE, Melis A (1988) Minimum photosynthetic unit size in system-I and system-II of barley chloroplasts. Biochim Biophys Acta 934: 151–155.

Greenbaum E (1988) Energetic efficiency of $H_2$ photoevolution by algal water-splitting. Biophys J 54: 365–368.

Gumpel NJ, Purton S (1994) Playing tag with *Chlamydomonas*. Trends in Cell Biology 4: 299–301.

Huner NPA, Oquist G, Sarhan F (1998) Energy balance and acclimation to light and cold. Trends in Plant Science 3: 224–230.

Jansson S, Pichersky E, Bassi R, Green BR, Ikeuchi M, Melis A, Simpson DJ, Spangfort M, Staehelin LA, Thornber JP (1992) A nomenclature for the genes encoding the chlorophyll a/b-binding proteins of higher plants. Plant Mol Biol Rep 10: 242–253.

Jordan P, Fromme P, Witt HT, Klukas O, Saenger W, Krauss N (2001) Three-dimensional structure of cyanobacterial photosystem I at 2.5 angstrom resolution. Nature 411(6840): 909–917.

Kindle KL (1990) High-frequency nuclear transformation of *Chlamydomonas reinhardtii*. Proc Natl Acad Sci USA 87: 1228–1232.

Kirk JTO (1994) Light and photosynthesis in aquatic ecosystems, 2nd Edition. Cambridge University Press, Cambridge, England.

Kok B (1953) Experiments on photosynthesis by *Chlorella* in flashing light. In: Burlew JS (ed), Algal culture: from laboratory to pilot plant. Carnegie Inst. of Washington, Washington DC, pp. 63–75.

LaRoche J, Mortain-Bertrand A, Falkowski PG (1991) Light-intensity-induced changes in cab mRNA and light-harvesting complex II apoprotein levels in the unicellular chlorophyte *Dunaliella tertiolecta*. Plant Physiol 97: 147–153.

Ley AC and Mauzerall DC (1982) Absolute absorption cross sections for photosystem II and the minimum quantum requirement for photosynthesis in *Chlorella vulgaris*. Biochim Biophys Acta 680: 95–106.

Masuda T, Polle JEW, Melis A (2002) Biosynthesis and distribution of chlorophyll among the photosystems during recovery of the green alga *Dunaliella salina* from irradiance stress. Plant Physiol 128: 603–614.

Masuda T, Tanaka A, Melis A (2003) Chlorophyll antenna size adjustments by irradiance in *Dunaliella salina* involve coordinate regulation of chlorophyll *a* oxygenase (*CAO*) and *Lhcb* gene expression. Plant Mol Biol 51: 757–771.

Maxwell DP, Falk S, Huner NPA (1995) Photosystem II excitation pressure and development of resistance to photoinhibition. 1. Light harvesting complex II abundance and zeaxanthin content in *Chlorella vulgaris*. Plant Physiol 107: 687–694.

Melis A (1991) Dynamics of photosynthetic membrane composition and function. Biochim Biophys Acta (Reviews on Bioenergetics) 1058: 87–106.

Melis A (1996) Excitation energy transfer: functional and dynamic aspects of *Lhc* (*cab*) proteins. In: Oxygenic Photosynthesis: The Light Reactions (DR Ort and CF Yocum, eds), Kluwer Academic Publishers, Dordrecht, The Netherlands, pp. 523–538.

Melis A (1999) Photosystem-II damage and repair cycle in chloroplasts: what modulates the rate of photodamage in vivo? Trends in Plant Science 4: 130–135.

Melis A (2002) Green alga hydrogen production: progress, challenges and prospects. Intl. J. Hydrogen Energy 27: 1217–1228.

Melis A, Neidhardt J, Benemann JR (1999) *Dunaliella salina* (Chlorophyta) with small chlorophyll antenna sizes exhibit higher photosynthetic productivities and photon use efficiencies than normally pigmented cells. J Appl Phycol 10: 515–552.

Melis A, Zhang L, Forestier M, Ghirardi ML, Seibert M (2000) Sustained photobiological hydrogen gas production upon reversible inactivation of oxygen evolution in the green alga *Chlamydomonas reinhardtii*. Plant Physiol 122: 127–136.

Michel H-P, Tellenbach M, Boschetti A (1983) A chlorophyll *b*-less mutant of *Chlamydomonas reinhardtii* lacking in the light-harvesting chlorophyll a/b protein complex but not in its apoproteins. Biochim Biophys Acta 725: 417–424.

Mulloney JA (1993) Mitigation of carbon dioxide releases from power production via sustainable agri-power – the synergistic combination of controlled environmental agriculture (large commercial greenhouses) and disbursed fuel cell. Energy Conversion and Management 34: 913–920.

Myers J (1957) Algal culture. In: Kirk RE, Othmer DE (eds), Encyclopedia of chemical technology. Interscience, New York, NY, pp. 649–668.

Nakada E, Asada Y, Arai T, Miyake J (1995) Light penetration into cell suspensions of photosynthetic bacteria and relation to hydrogen production. J Ferment Bioengin 80: 53–57.

Nakajima Y, Ueda R (1997) Improvement of photosynthesis in dense microalgal suspension by reduction of light harvesting pigments. J Appl Phycol 9: 503–510.

Nakajima Y, Ueda R (1999) Improvement of microalgal photosynthetic productivity by reducing the content of light harvesting pigment. J Appl Phycol 11: 195–201.

Nakicenovic N (1993) Carbon dioxide mitigation measures and options. Environ Sci and Tech 27: 1986–1989.

Naus J, Melis A (1991) Changes of photosystem stoichiometry during cell growth in *Dunaliella salina* cultures. Plant Cell Physiol 32: 569–575.

Neidhardt J, Benemann JR, Zhang L, Melis A (1998) Photosystem-II repair and chloroplast recovery from irradiance stress: relationship between chronic photoinhibition, light-harvesting chlorophyll antenna size and photosynthetic productivity in *Dunaliella salina* (green algae). Photosynth Res 56: 175–184.

Niyogi KK, Björkman O, Grossman AR (1997) *Chlamydomonas* xanthophyll cycle mutants identified by video imaging of chlorophyll fluorescence quenching. Plant Cell 9: 1369–1380.

Ohtsuka T, Ito H, Tanaka A (1997) Conversion of chlorophyll *b* to chlorophyll *a* and the assembly of chlorophyll with apoproteins by isolated chloroplasts. Plant Physiol 113: 137–147.

Pichersky E, Jansson S (1996) The light-harvesting chlorophyll *a/b*-binding polypeptides and their genes in angiosperm and gymnosperm species. In: Ort DR and Yocum CF (eds), Oxygenic Photosynthesis: The Light Reactions, pp. 507–521, Kluwer Academic Publishers, Dordrecht, The Netherlands.

Plumley FG, Schmidt GW (1995) Light-harvesting chlorophyll a/b complexes: Interdependent pigment synthesis and protein assembly. The Plant Cell 7: 689–704.

Polle JEW, Benemann JR, Tanaka A, Melis A (2000) Photosynthetic apparatus organization and function in wild type and a Chl *b*-less mutant of *Chlamydomonas reinhardtii*. Dependence on carbon source. Planta 211: 335–344.

Polle JEW, Kanakagiri S, Melis A (2003) *tla1*, a DNA insertional transformant of the green alga *Chlamydomonas reinhardtii* with a truncated light-harvesting chlorophyll antenna size. Planta 217: 49–59, DOI: 10.1007/s00425-002-0968-1.

Polle JEW, Niyogi KK, Melis A (2001) Absence of lutein, violaxanthin and neoxanthin affects the functional chlorophyll antenna size of photosystem-II but not that of photosystem-I in the green alga *Chlamydomonas reinhardtii*. Plant Cell Physiol 42: 482–491.

Powles S (1984) Photoinhibition of photosynthesis induced by visible light. Annu Rev Plant Physiol 35: 15–44.

Radmer R, Kok B (1977) Photosynthesis: Limited yields, unlimited dreams. Bioscience 29: 599–605.

Ruban AV, Wentworth M, Yakushevska AE, Andersson J, Lee PJ, Keegstra W, Dekker

JP, Boekema EJ, Jansson S, Horton P (2003) Plants lacking the main light-harvesting complex retain photosystem II macro-organization. Nature 421: 648–652.

Simpson DJ, Knoetzel J (1996) Light-harvesting complexes of plants and algae: introduction, survey and nomenclature. In: Ort DR and Yocum CF (eds), Oxygenic Photosynthesis: The Light Reactions, pp. 493–506, Kluwer Academic Publishers, Dordrecht, The Netherlands.

Smith BM, Morrissey PJ, Guenther JE, Nemson JA, Harrison MA, Allen JF, Melis A (1990) Response of the photosynthetic apparatus in *Dunaliella salina* (green algae) to irradiance stress. Plant Physiol 93: 1433–1440.

Sukenik A, Bennett J, Falkowski PG (1988) Changes in the abundance of individual apoproteins of light-harvesting chlorophyll *a/b*-protein complexes of photosystem I and II with growth irradiance in the marine chlorophyte *Dunaliella tertiolecta*. Biochim Biophys Acta 932: 206–215.

Tanaka A, Ito H, Tanaka R, Tanaka N, Yoshida K, Okada K (1998) Chlorophyll *a* oxygenase (CAO) is involved in chlorophyll *b* formation from chlorophyll *a*. Proc Natl Acad Sci USA 95: 12719–12723.

Tanaka A, Melis A (1997) Irradiance-dependent changes in the size and composition of the chlorophyll *a-b* light-harvesting complex in the green alga *Dunaliella salina*. Plant Cell Physiol 38: 17–24.

Vazquezduhalt R (1991) Light-effect on neutral lipids accumulation and biomass composition of *Botryococcus sudeticus* (Chlorophyceae). Cryptogamie Algologie 12: 109–119.

Webb MR, Melis A (1995) Chloroplast response in *Dunaliella salina* to irradiance stress. Effect on thylakoid membrane assembly and function. Plant Physiol 107: 885–893.

Westermeier R, Gomez I (1996) Biomass, energy contents and major organic compounds in the brown alga *Lessonia nigrescens* (Laminariales, Phaeophyceae) from Mehuin, south Chile. Botanica Marina 39: 553–559.

Yakovlev AG, Taisova AS, Fetisova ZG (2002) Light control over the size of an antenna unit building block as an efficient strategy for light harvesting in photosynthesis. FEBS Lett 512: 129–132.

Zaborsky OR (1998) BioHydrogen. Plenum Publishing Corporation, New York, NY.

Zhang L, Happe T, Melis A (2002) Biochemical and morphological characterization of sulfur-deprived and $H_2$-producing *Chlamydomonas reinhardtii* (green alga). Planta 214: 552–561.

Zouni A, Witt HT, Kern J, Fromme P, Krauss N, Saenger W, Orth P (2001) Crystal structure of photosystem II from *Synechococcus elongatus* at 3.8 angstrom resolution. Nature 409: 739–743.

# Part V
# The Carbon Connection

# 13
# Manipulating Ribulose Bisphosphate Carboxylase/Oxygenase in the Chloroplasts of Higher Plants [1]

*T. John Andrews and Spencer M. Whitney*

## 13.1
## Introduction

Literature concerning ribulose bisphosphate carboxylase/oxygenase (Rubisco), the $CO_2$-fixing enzyme that enables most forms of photo- and chemoautotrophic life, has been reviewed extensively in recent years [1–4]. Its structure and catalytic mechanism [5], the interactions between the constituent large and small subunits of the predominant Form I type of the enzyme [6], and its regulation by formation of a specific lysyl carbamate within the active site under the influence of the ancillary protein Rubisco activase [7] are covered elsewhere. Previously published reviews have touched on the genetic manipulation of Rubisco in plants and microbes to varying extents, but none has focused primarily on this topic. This review addresses that gap in the context of higher plants.

## 13.2
## Why Manipulate Rubisco in Plants?

### 13.2.1
### Genetic Manipulation of Higher-plant Rubisco Is Now Feasible

Thirteen years ago, the advent of technology for genetically transforming the small circular genome of the plastids of tobacco [8] opened the door to molecular manipulation of higher-plant Rubisco. Previously, such manipulation was limited to varying the content of Rubisco in leaf cells by transforming the nucle-

---

[1] This chapter was originally published in *Archives of Biochemistry and Biophysics* 414 (2003) 159–169. Wiley-VCH thanks Elsevier Science for kind permission to reprint this article. Abbreviations: RuBP, D-ribulose-1,5-bisphosphate; Rubisco, RuBP carboxylase/oxygenase; 5'-UTR, 5'-untranslated region; $k_{cat}^c$, $CO_2$-saturated carboxylase activity; $k_{cat}^o$, $O_2$-saturated oxygenase activity; $K_c$, Michaelis constants for $CO_2$; $K_o$, Michaelis constant for $O_2$; $S_{c/o}$, specificity for $CO_2$ relative to $O_2$ ($=[k_{cat}^c/K_c]/[k_{cat}^o/K_o]$).

*Artificial Photosynthesis: From Basic Biology to Industrial Application*
Edited by Anthony F. Collings and Christa Critchley
Copyright © 2005 WILEY-VCH Verlag GmbH & Co. KGaA, Weinheim
ISBN: 3-527-31090-8

ar genome with antisense genes directed at the RbcS message for the nuclear-encoded small subunits [9, 10]. Even the less ambitious manipulation of higher-plant Rubisco in heterologous hosts, such as *Escherichia coli*, was (and still is) blocked by unknown requirements for folding and assembly that are not satisfied in the bacterial host and result in the production of aggregated, nonfunctional protein [11, 12]. Plastid transformation has circumvented this impediment, allowing deletion [13], mutation [14], and replacement [15, 16] of the rbcL gene for the large, catalytic subunit. This review covers recent progress enabled by this technology and canvasses the opportunities it presents for studying Rubisco's catalytic mechanism, activity regulation, subunit interactions, synthesis and assembly, and the impact of these on the physiology of the plant.

### 13.2.2
### The Advantages of "Ecological" Studies of Rubisco "at Home" in Its Physiological Context

Although plastid transformation is now a routine procedure, at least with tobacco, it is time-consuming. Usually, several months are required before the primary transformant plants are available. Successive backcrossing with wild-type pollen donors to remove any unwanted nuclear mutations that might have been introduced during tissue culture takes even longer. The result, however, is worth the wait. Provision of the mutated or substituted Rubisco in vivo enables detailed assessment of not only the effects of the mutation on the protein but also the physiological consequences for the plant as a whole. The surgical precision of the homologous-recombination mechanism exploited by plastid transformation leaves no unchanged *rbc*L genes to confuse interpretations. The physiology of the whole plant reports the integrated consequences of the genetic alteration quite unambiguously. The plant also can produce copious amounts of the altered Rubisco protein for structural and biochemical studies.

### 13.2.3
### A Compelling Example of Genome–Phenome Interactions

Rubisco's central, enabling role in all schemes of aerobic $CO_2$ assimilation – now standard textbook material – and the pervading influence of its properties on photosynthesis and growth make it an obvious target for manipulation. The connection between the information residing in the genes for Rubisco and its ancillary proteins and the phenome of plants could hardly be more direct and explicit. Therefore, the challenge to test the adequacy of knowledge about Rubisco – its catalytic processes, the way its activity is regulated, its synthesis and assembly, etc. – by making precise changes to that information and observing how well the phenomic consequences match predictions based on that knowledge becomes inescapable.

## 13.2.4
### An Improvement in the Resource-use Efficiency of Photosynthesis?

A second motivation is more utilitarian. It derives from the realization that Rubisco's catalytic performance dictates the maximum efficiency of photosynthesis in its use of light, water, and fertilizer N resources. Thus, the properties of Rubisco determine, among other things, the size of the biosphere. Despite this critical role, Rubisco's performance falls far short of the near-perfect speed and specificity attained by many other enzymes [17]. The possibility of achieving profound improvements in such fundamental efficiencies, through manipulating Rubisco in the chloroplasts of higher plants, focuses the attention of Rubisco researchers and their funding agencies.

## 13.3
### What Constitutes an Efficient Rubisco?

### 13.3.1
### Key Kinetic Parameters

An enzyme perfectly adapted for speed and specificity would convert its substrate as fast as it diffused into its active site and would never mistake non-substrate for substrate. There are many examples of enzymes that approach this standard [18]. A "perfect" Rubisco would have a $k^c_{cat}/K_c$ quotient (symbols defined in Abbreviations) of at least $10^8$ $M^{-1}$ $s^{-1}$. If both $k^c_{cat}/K_c$ and the carboxylation rate have been maximized during evolution [18], it would also have a $K_c$ higher than the $CO_2$ concentration in the chloroplast stroma (commonly approximately 8 µM, which is a little less than that of air-equilibrated water due to diffusional barriers). Finally, its specificity for $CO_2$ relative to $O_2$ ($S_{c/o}$) would be at infinity. Real Rubiscos fulfill the second criterion but fail the other two miserably. Table 13.1 lists just how far short of perfection natural Rubiscos fall, and Fig. 13.1 illustrates how these shortcomings affect photosynthetic $CO_2$ assimilation. Rubisco's tendency to mistake the product of photosynthesis, $O_2$, for the

**Table 13.1** Rubisco kinetic parameters used in Fig. 13.1 [a]

| Rubisco | $k^c_{cat}$ ($s^{-1}$) | $K_c$ (µM) | $k^c_{cat}/K_c$ ($M^{-1}$ $s^{-1}$) | $K_o$ (µM) | $S_{c/o}$ |
|---|---|---|---|---|---|
| Tobacco | 3.4 | 10.7 | $3.2 \times 10^5$ | 295 | 82 |
| *Griffithsia monilis* | 2.6 | 9.3 | $2.8 \times 10^5$ | 710 | 167 |
| *Rhodospirillum rubrum* | 7.3 | 89 | $8.2 \times 10^4$ | 406 | 12 |
| "Perfect" [b] | 1070 | 10.7 | $10^8$ | ∞ | ∞ |

a) Data taken from [21] and [74–76].
b) No oxygenation and $k^c_{cat}/K_c = 10^8$ $M^{-1}$ $s^{-1}$.

substrate, $CO_2$, which results in wasteful oxygenation of D-ribulose-1,5-bisphosphate (RuBP) and concomitant photorespiratory metabolism, has been reviewed extensively [1–4]. $S_{c/o}$ $(=[(k^c_{cat}/K_c)/(k^o_{cat}/K_o)]$ [19]) represents the ratio of carboxylation to oxygenation at equal $CO_2$ and $O_2$ concentrations.

## 13.3.2
### Physiological Consequences of Rubisco Efficiency

The impact of Rubisco's confused substrate specificity on photosynthetic $CO_2$ assimilation may be modeled (Fig. 13.1) as two intersecting curves describing the way the $CO_2$-assimilation rate is limited at low $CO_2$ by the activity of Rubisco and at higher $CO_2$ by electron transport-dependent regeneration of RuBP [20]. Both branches of the curve are affected by RuBP oxygenation; the slope of the lower branch is reduced by the competitive inhibition by $O_2$, and the curvature of the upper branch is induced by the increased energy demands of photorespiratory metabolism as the $CO_2:O_2$ ratio decreases. Thus, if transferred into plants in active form, the more $CO_2$-specific Rubisco from the red alga *Griffithsia monilis* [21] would support higher rates of photosynthesis per unit Rubisco than the plant enzyme over the whole $CO_2$ concentration range shown (Fig. 13.1). Moreover, it could support a given assimilation rate at a lower stromal $CO_2$ concentration and with a reduced requirement for photosynthetic electron transport. In the example shown by the dotted and dashed lines in Fig. 13.1, the lower $CO_2$ requirement of the red-algal enzyme would allow a 49% reduction in the leaf's conductance to $CO_2$ (and water-vapor) diffusion, with a proportionate saving in water transpiring. Furthermore, this could be achieved in dimmer illumination because of the 15% decrease in the required electron transport rate. Such large improvements in resource-use efficiencies, which have such direct impacts on growth and, potentially, yield, could hardly be achieved by any other form of genetic manipulation. Substitution of a more efficient Rubisco into plant chloroplasts will be a challenging task, however, as we will discuss later.

Large though the potential improvements conferred by the red-algal Rubisco are, they are insignificant in comparison with those conferred by a hypothetical Rubisco with a $k^c_{cat}/K_c$ of $10^8$ $M^{-1}$ $s^{-1}$ and no tendency to mistake $O_2$ for $CO_2$. Relieved of the energy demands of oxygenation and photorespiration, the upper branch of the curve becomes $CO_2$-independent. Furthermore, the ability to fix $CO_2$ nearly as fast as it arrives in the active site makes the lower branch so steep that it cannot be resolved from the ordinate unless the amount of Rubisco is reduced by 99% (Fig. 13.1). This "perfect" Rubisco could support the given photosynthesis rate with 86% less water loss, 35% less light, and 99% less protein investment in Rubisco.

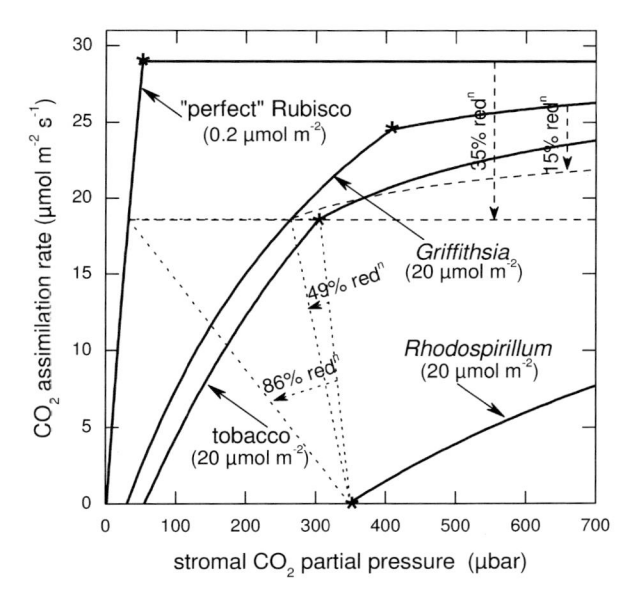

**Fig. 13.1** The effects of the kinetic efficiency of Rubisco on the dependency of the photosynthetic $CO_2$ assimilation rate on the chloroplast stromal $CO_2$ partial pressure modeled according to Farquhar et al. [20]. The asterisks indicate the points of transition between limitation by Rubisco activity and by RuBP regeneration. The kinetic parameters listed in Table 13.1 were used, together with a photosynthetic electron transport rate ($J$) of 120 µmol m$^{-2}$ s$^{-1}$, a rate of non-photorespiratory respiration ($R_d$) of 1 µmol m$^{-2}$ s$^{-1}$, and the contents of Rubisco active sites ($B$) shown. The Rubisco-activity-limited assimilation rate ($A$) below the transition point is modeled according to the equation

$$A = \frac{B(p_c \cdot s_c - 0.5 \cdot o/S_{c/o})k_{cat}^c}{p_c \cdot s_c + K_c(1 + o/K_o)} - R_d$$

where $p_c$ is the $CO_2$ partial pressure in the chloroplast, $o$ is the $O_2$ concentration in the chloroplast, and $s_c$ is the solubility of $CO_2$ in water (0.0334 M bar$^{-1}$). The RuBP-regeneration-limited value of $A$ at higher $CO_2$ concentrations is given by

$$A = \frac{J(p_c \cdot s_c - 0.5 \cdot o/S_{c/o})}{4(p_c \cdot s_c + o/S_{c/o})} - R_d$$

The slopes of the dotted lines indicate the leaf conductance to $CO_2$ diffusion required to support the same $A$ as achieved by tobacco Rubisco at its transition point in an atmosphere containing 350 µbar of $CO_2$. The dashed lines indicate the RuBP-regeneration-limited $A$ after reducing $J$ as indicated to adjust the transition points to occur at the same $A$ as for tobacco Rubisco.

### 13.3.3
### Regulatory Properties

This is not meant to imply that catalytic power and specificity are the only requirements for an efficient Rubisco. Effective regulation of its activity to match the demands of photosynthetic metabolism, for example, may be just as important. This regulation and its mediation by the helper protein, Rubisco activase, is reviewed in [7].

### 13.3.4
### Evolution of Rubisco Efficiency

The modeled Rubiscos (Fig. 13.1 and Table 13.1) span the full range of diversity in Rubisco's kinetic properties and include both the Form-I ($L_8S_8$) and the small-subunit-lacking Form-II ($L_2$) enzymes. More than a 10-fold range of $S_{c/o}$ values is encompassed. The Form-II enzyme comes from an anaerobic organism (Rhodospirillum rubrum); therefore, it is no surprise that its properties are not suited to the conditions within a leaf. Indeed, it would barely be able to support positive carbon gain in normal air (Fig. 13.1). However, natural selection in the leaf environment obviously has not resulted in an enzyme "perfectly" suited to that environment (Fig. 13.1). From the point of view of catalytic effectiveness, the red-algal enzyme appears to be better suited than the plant enzyme to the leaf environment. How the algal enzyme has been able to evolve further or faster than the plant enzyme to function efficiently in an aerobic, $CO_2$-limited environment is mysterious. It may be relevant that the plant ("green") and red-algal ("red") enzymes are very divergent phylogenetically [4, 22]. Although both large and small subunits have the same basic folded structures in the red and green subtypes, there are important structural differences, particularly in the small subunits [5, 6, 23]. Perhaps the structure of the "red" Rubisco type is different in ways that allow it to be more responsive to natural selection.

## 13.4
## How to Find a Better Rubisco?

### 13.4.1
### In Nature?

The discovery that Form-I Rubiscos from some non-green algae have higher $CO_2/O_2$ specificities ($S_{c/o}$) [24, 25] and, at least in one case [21], globally better combinations of kinetic properties than do higher-plant Rubiscos illustrates the natural diversity in these critically important properties. There is also a report of a Form-II Rubisco from a thermophilic archaebacterium with an even greater $S_{c/o}$, albeit at high temperature [26]. It seems quite probable that the most efficient natural Rubiscos are yet to be discovered. A systematic search for efficient Rubiscos is warranted. It might be profitable to sample all classes of organisms that use the RuBP-based carbon cycle and all branches of the expanding Rubisco phylogeny [4].

### 13.4.2
### By Rational Design?

An ever-increasing number of high-resolution crystal structures of Rubiscos of all major types from several species, complexed with substrates, products, and intermediate analogues [5] coupled with decades of biochemical and directed mutagenic studies has led to detailed formulations of the sequence of reaction intermediates involved in both carboxylation and oxygenation reactions. This information suggests the roles of critical active-site residues that are conserved in all Rubiscos [1, 27, 28]. These insights have been supplemented by computational simulation of the catalytic chemistry in vacuo [29] and within the active site [30–33]. Nevertheless, we are still a long way from being able to design ways of improving the active site rationally. The essential features in the active site that govern critical parameters of catalytic performance, such as $S_{c/o}$, $k^c_{cat}$, $K_c$, and $K_o$, are too subtle to be resolved crystallographically. Obviously, the small subunits are important, particularly the loop that projects into the lumen of the barrel created by the octamer of large subunits of the high-specificity green [34–36] and red [23] Rubiscos from eukaryotes. The way these interactions are transmitted over considerable distances to the active sites and how they influence the active site's structure are presently mysterious. Subtle, but critical, aspects such as these are obvious areas of interest in post-crystallographic Rubisco research.

### 13.4.3
**By in Vitro Evolution?**

Directed molecular evolution in vitro – coupling random mutagenesis (particularly the highly recombinogenic strategies embodied by DNA shuffling [37] and "family" shuffling [38]) with organismal or high-throughput selection for desired properties [39] – is a highly effective means of conferring desired non-natural properties upon enzymes, such as altered substrate specificity, changed stereospecificity, thermostability, etc. Requiring no prior knowledge about the mechanistic basis of the desired property, it is free from preconceptions. Its application to Rubisco has been hampered by difficulties with devising suitable selection systems, although the substitution of *R. rubrum* Rubisco into the cyanobacterium *Synechocystis* PCC 6803 [40, 41] was engineered with this prospect in view. To have a realistic chance of improving the speed and specificity of Rubisco, the selection system would need to be capable of screening more Rubisco variants than nature has already screened, because these are the same properties that nature has selected throughout 3.5 Gy of Rubisco evolution. This is probably a very tall order, and there are no clear examples where directed evolution has improved any enzyme property that has been optimized already by natural selection. There are, however, secondary possibilities where directed evolution is likely to be an essential tool. Screening for the non-natural ability of naturally efficient (or rationally designed) Rubiscos to express and assemble in foreign hosts is one example.

A combination of all three approaches – prospecting for more kinetically efficient Rubiscos throughout nature, engineering them rationally to the limits of advancing knowledge, and recombining and refining the products by artificial evolution to acclimate them to the new host – may be most likely to achieve the goal of better Rubisco efficiency in plants.

## 13.5
## How to Manipulate Rubisco in Plants?

### 13.5.1
### Nuclear Transformation

In plants, Rubisco's small subunits are encoded by a multigene family of *RbcS* genes in the nuclear genome. This restricts genetic manipulation to suppression of the expression of the small subunit by antisense RNA or supplementation with additional sense *RbcS* genes.

*RbcS* expression has been suppressed with antisense genes in tobacco [9, 10], rice [42], and *Flaveria bidentis* [43] (Fig. 13.2 A). The amounts of sense *RbcS* transcript, small subunits, and holoenzyme were reduced, sometimes to very low levels. For plants grown in dim illumination, whole-leaf photosynthesis measured in the same illumination was not affected until more than half of the control's Rubisco content had been removed [44]. However, photosynthesis measured at higher illumination was more linearly correlated with Rubisco content [45]. A similar near-linear correlation was observed when the plants were grown in strong natural illumination and measured at high light [10, 42, 46]. Reduced Rubisco content was compensated only partly by an increase in the degree of activation of the remaining enzyme [47], and the reduced Rubisco activity resulted in lower steady-state pools of its product, 3-phosphoglycerate [47, 48]. Reduction of the Rubisco content to 35% of that of the wild type reduced photosynthesis sufficiently that, at 1000 µmol quanta $m^{-2} s^{-1}$ illumination, it was never limited by RuBP regeneration, even at high $CO_2$. This allowed measurement of Rubisco's kinetic parameters in vivo by leaf gas exchange, and the resulting data agreed quite well with in vitro measurements with isolated Rubisco [49].

**Fig. 13.2** Types of manipulation of Rubisco in tobacco conducted so far. (A) Suppression of *RbcS* by nuclear antisense [9, 10, 42, 43]. (B) Deletion of plastomic *rbcL* [13]. (C) Complementation of "c" with nuclear *RbcL* [13]. (D) Supplementation of *RbcS* with a plastomic hepta-His-tagged *rbcS*H$_7$ copy [56]. (E) Replacement of tobacco *rbcL* with *rbcM* of *R. rubrum* [16]. (F) Directed mutation of *rbcL* [14]. (G) Supplementation with plastomic copies of red-type *rbcLS* operons. The example of the red-type Rubisco from a diatom is shown [21]. (H) Replacement of tobacco *rbcL* with sunflower *rbcL* [15]. (I) Replacement of tobacco *rbcL* with *Synechococcus* PCC 6301 *rbcL* [15]. (J) Wild-type tobacco (*Nicotiana tabacum* cv. Petit Havana). (K) Autotrophic plants obtained from experiment "e" growing in a 1.3% $CO_2$ v/v atmosphere. (L) Autotrophic plants obtained from experiment "h" growing in 1.3% $CO_2$. (M) Non-autotrophic plants obtained from experiment "i" growing on sucrose-containing medium. nm, not measured; IR$_A$, IR$_B$, inverted repeat regions A and B; LSC, SSC, large and small single-copy regions. The yellow background indicates the genome(s) manipulated. Amounts of Rubisco and mRNA are indicated as percentages of the amounts in the wild type. *n* denotes the multiple members of the *RbcS* nuclear gene family. The plastomic *psbA* gene, which encodes the D1 protein of photosystem II, produces one of the most abundant mRNA transcripts in chloroplasts.

## 13.5 How to Manipulate Rubisco in Plants?

| Experiment | Nucleus | Plastome | Autotrophic? | Leaf mRNA content (% of wild type) | | Leaf Rubisco content (% of wild type) | |
|---|---|---|---|---|---|---|---|
| (a) Antisense RbcS | RbcS (xn), antisense-RbcS (x1) | rbcL (N. tabacum GenBank: Z00044) | Yes | RbcS<br>rbcL | 5-100<br>100 | 10-100 | |
| (b) rbcL deletion | RbcS (xn) | ΔrbcL | No | RbcS<br>rbcL | nm<br>0 | 0 | |
| (c) Nuclear RbcL complement | RbcS (xn), RbcL (with pea SSu TP) (x1) | ΔrbcL | Yes | RbcS<br>RbcL | nm<br>nm | <10 | |
| (d) Plastome rbcSH₇ supplement | RbcS (xn) | rbcL, rbcSH₇ | Yes | RbcS<br>rbcSH₇<br>rbcL | 100<br>200-1000<br>100 | ~92 | ~8 |
| (e) Bacterial rbcM replacement | RbcS (xn) | rbcM (R. rubrum) | No (air)<br>Yes (↑CO₂) | RbcS<br>rbcM | >100<br>~65 | ~30 | |
| (f) rbcL mutagenesis | RbcS (xn) | MrbcL (L335V mutation) | No (air)<br>Yes (↑CO₂) | RbcS<br>MrbcL | 100<br>100 | 100 | |
| (g) Diatom rbcLS supplement | RbcS (xn) | rbcL, (Phaeodactylum tricornutum) rbcLS | Yes | psbA<br>rbcLS | 100<br>40% of psbA | 100 | (unassembled) 25-100 |
| (h) Sunflower rbcL replacement | RbcS (xn) | rbcL (Helianthus annuus) | No (air)<br>Yes (↑CO₂) | RbcS<br>rbcL | nm<br>~30 | ~30 | |
| (i) Cyano rbcL replacement | RbcS (xn) | rbcL (Synech. PCC6301) | No | RbcS<br>rbcL | nm<br><10 | 0 | |

(j) Wild-type

(k) Bacterial rbcM

(l) Sunflower rbcL

(m) Cyano rbcL

Introduction of a pea *Rbc*S gene into the nuclear genome of *Arabidopsis thaliana* resulted in production of the heterologous small subunit, which was transported to the plastid and assembled into the Rubisco holoenzyme. However, the catalytic rate of the hybrid enzyme and its ability to bind ligand appeared to be impaired equally, and the extent of the impairment was approximately proportional to the amount of foreign small subunits present [50]. This illustrates the incompatibility of mismatched Rubisco subunits.

These data serve to highlight the impediments to manipulating Rubisco in plants by nuclear transformation only. Not only is the manipulation restricted to the small subunit, but the presence of multiple *Rbc*S genes also makes precise mutagenesis or targeted replacement cumbersome and arduous.

### 13.5.2
### Plastid Transformation

The surgical precision of the homologous-recombination mechanism that enables plastome manipulation [51] stands in marked contrast to the limitations of nuclear transformation. Both site-specific mutagenesis [14] and total replacement [15, 16] of the plastid-encoded large subunits are now routine. Although tobacco is the only higher plant to which plastid-transformation technology is applicable routinely, efforts to develop the procedure for other species are showing promise [52–54]. In the meantime, tobacco plastid transformation provides an excellent vehicle for studying a wide range of aspects of Rubisco biology.

## 13.6
## What Have We Learned So Far?

### 13.6.1
### Both Nuclear and Plastidic Genomes Are Able to Express Both *rbc*L and *Rbc*S Genes

Higher-plant Rubisco provides an excellent, and comparatively simple, example of the migration of genes from the original prokaryotic ancestor of the plastid to the nucleus. Migration of *Rbc*S, but not *rbc*L, to the nucleus must have required solutions to problems of coordinated expression of the two genes, targeting of cystoplasmically synthesized small subunits to the plastid, and coordinated folding and assembly of the subunits synthesized in different cellular compartments. Separate transplantation of *rbc*L and *Rbc*S to the opposite genome has provided unique insight into some aspects of these processes.

Kanevski and Maliga [13] excised the *rbc*L gene from the tobacco plastome, producing plants that contained neither Rubisco subunit and required supplementation with sucrose to grow (Fig. 13.2 B). They then transformed the nucleus of these plants with the same *rbc*L gene fused to the transit pre-sequence of pea *Rbc*S under the transcriptional control of the cauliflower mosaic virus 35S promoter. The introduced nuclear gene complemented the plastomic *rbc*L deletion, allowing Rubisco to accumulate to approximately 3% of the wild type's content, which was enough to permit slow autotrophic growth (Fig. 13.2 C). This shows that there is no impediment to nuclear expression of *rbcL* and that large-subunit precursors can be imported to the plastid and processed, folded, and assembled by the existing machinery. Expression of a bacterial *rbcL* similarly incorporated into the nuclear genome was unsuccessful, however [55].

Whitney and Andrews [56] performed the reciprocal transplantation, transferring one member of the tobacco nuclear *RbcS* family, both with and without its transit pre-sequence and equipped with tobacco plastid *psbA* promoter and terminator elements, back to its endosymbiotic origin in the plastome (Fig. 13.2 D). In this case, the limitations to engineering the nuclear genome mentioned earlier prevented removal of the nuclear copies, but attachment of a C-terminal polyhistidine tag allowed the small subunits synthesized in the plastid to be distinguished. Once again, the transplanted gene was active in its new compartment, producing abundant transcript, and its product was correctly processed and assembled into hexadecameric Rubisco. However, again the amount of tagged small subunits found assembled into the Rubisco complex was not large (approximately 1% of the total small subunits). Competition from nucleus-encoded small subunits does not appear to be the cause of the scarcity of plastid-synthesized small subunits in this system because suppression of the abundance of nucleus-encoded small subunits with a nuclear antisense small-subunit gene did not improve the result [57].

Both sets of transplantation experiments agree that there is no qualitative deficiency preventing expression of either subunit gene from either genome. However, the quantitative shortcomings are glaring in both cases. For the transplanted small subunits, at least, the limitation is not transcriptional [56, 57]. More study of both of these systems will be required to pinpoint the causes of the quantitative restrictions. Translational difficulties such as codon-usage bias, plastid targeting and import, processing, folding, and assembly are all potential hurdles that could impede the expression of the transplanted gene. One possible post-translational bottleneck might apply, in reciprocal senses, to both transplanted subunits: the newly translated or newly transported and processed subunits might need to be presented to the plastid chaperone machinery in particular contexts. Thus, the chaperone complexes that draw the small-subunit precursor through the translocon complex in the plastid envelope may be different from those that receive the nascent large subunits from the plastid ribosomes – despite, perhaps, having some subunits, such as Hsp70, in common. Although both types of chaperone complexes may pass their unfolded clients on to the plastid chaperonin 60/21 system, they may not carry out their roles with equal efficiency when their clients are exchanged. Any unfolded subunits lost from, or denied access to, the folding pathway would be degraded.

### 13.6.2
### Photosynthesis and Growth Can Be Supported by a Foreign Rubisco

It is easy to imagine potential reasons why foreign Rubiscos, even if folded and assembled correctly in plastids, might not be able to support the photosynthesis of the plant. For example, a foreign Rubisco's regulatory mechanism might not receive appropriate signals from the rest of chloroplast metabolism, or its physical properties might lead to detrimental interactions with other components of the concentrated milieu of the stroma. These types of complex interactions are not easily predictable. The demonstration by Whitney and Andrews [16] that tobacco photosynthesis and growth can be supported by the structurally simple Rubisco from *R. rubrum* (Fig. 13.2 E, K) provides proof that replacement of the plant Rubisco is feasible in principle, encouraging further experimentation.

*R. rubrum* is a photosynthetic anaerobe and the kinetic properties of its Rubisco are adapted to a high-$CO_2$/low-$O_2$ environment and are not at all suited to leaf chloroplasts (Table 13.1 and Fig. 13.1). Little is known about the regulation of this Rubisco. It is a Form-II type, lacking small subunits, almost at the opposite pole of the Rubisco phylogeny [58] to the green-type, Form-I enzyme of higher plants. Nevertheless, substitution of the *R. rubrum rbc*M gene in place of the tobacco plastomic *rbc*L resulted in fully autotrophic plants that grew and reproduced quite successfully, provided that the bacterial enzyme's $CO_2$ requirements were satisfied [16]. As expected, no trace of either the large or small subunit of the plant enzyme could be detected in the tobacco-*rubrum* transformants. The apparently simple folding and assembly requirements of this Rubisco, known previously to be readily accommodated in *E. coli* [59] and the cyanobacterium *Synechocystis* PCC 6803 [40], are obviously also well satisfied in plastids. Little sign of unassembled, aggregated bacterial subunits was detected [16].

### 13.6.3
### The Properties of a Mutated or Foreign Rubisco Are Reflected in the Leaf's Gas-exchange Properties

A real advantage of manipulation of Rubisco in vivo is that the phenomic consequences of the manipulation are so readily measurable by leaf gas exchange. This was tested first by directed mutagenesis of tobacco *rbc*L (Fig. 13.2 F). Substitution of any of the canonical active-site residues that participate directly in catalysis and are conserved throughout all Rubisco large subunits [28] is catastrophic, resulting in otherwise uninformative inactive enzymes. Guided by previous mutagenesis of prokaryotic Rubiscos [60, 61], Whitney et al. [14] chose residue Leu-335, which is not involved in catalysis directly but is located near the apex of an important loop of the large subunit (loop 6). This loop folds against the substrate after it binds and positions the catalytically critical adjacent residue, Lys-334. Shortening the side chain by one methylene unit to Val impaired, but did not totally inactivate, carboxylation and enhanced oxygenation so that $S_{c/o}$ decreased fourfold. These changes were directly reflected in the phenotype of the plants, which were unable to grow in normal air but were fully autotrophic when the atmosphere was supplemented with 0.3% (v/v) $CO_2$. The responses of the $CO_2$ assimilation rate of the leaves to varying $CO_2$ and $O_2$ concentrations were very close to those predicted by the kinetic properties of the isolated mutant Rubisco [14]. As expected, in partial compensation for the catalytic impairment, the degree of activation (carbamylation) of the mutant enzyme in vivo was nearly doubled compared to that in the wild-type plants under the same growth conditions.

Total substitution of tobacco Rubisco with its *R. rubrum* counterpart was a much more drastic change. Again, the consequences were reported faithfully by the growth and leaf gas-exchange phenotype of the transformed plants [16]. The sevenfold reduction in $S_{c/o}$ and eightfold increase in $K_c$ (compared to the tobacco enzyme, Table 13.1), coupled with a threefold reduction in Rubisco content, caused the tobacco-*rubrum* plants to need much higher atmospheric $CO_2$ concentrations (2.5% v/v) for satisfactory growth (Fig. 13.2 E, K). The feeble catalytic performance of the *R. rubrum* enzyme at low $CO_2$ concentrations, and its strong inhibition by $O_2$, caused leaf $CO_2$ assimilation to be barely detectable with an infrared gas analyzer. However, mass-spectrometric measurements at the higher growth $CO_2$ concentration revealed rates of 50–70% of those seen in the wild type, consistent with the content and properties of the bacterial enzyme. Although it is unlikely that tobacco Rubisco activase can interact with *R. rubrum* Rubisco, the *R. rubrum* enzyme was nearly fully carbamylated under the growth conditions. The little that is known about the carbamylation characteristics of the *R. rubrum* enzyme is consistent with this observation [62, 63]. Obviously, lack of responsiveness of Rubisco activity to the regulatory signals from chloroplast metabolism is not catastrophic.

### 13.6.4
### The Requirements for Folding and Assembly of the Subunits of Red-type, Form-I Rubisco Are Not Accommodated in Chloroplasts

As a first step towards exploiting the desirable kinetic properties of some red-type, Form-I Rubiscos from non-green algae (Table 13.1 and Fig. 13.1) in higher plants, the genes for both large and small subunits of Rubisco from a rhodophyte (*Galdieria sulfuraria*) and a diatom (*Phaeodactylum tricornutum*) were incorporated into the tobacco plastome (diatom experiment illustrated in Fig. 13.2G) [21]. In non-green algae, the *rbc*L and *rbc*S genes occur together as an operon in the plastome. These operons were inserted into the inverted repeat regions of the tobacco plastome without disturbing the tobacco *rbc*L gene in the large single-copy region. Both introduced operons were very active transcriptionally and translationally. The diatom *rbc*LS operon directed the synthesis of both large and small subunits of the algal Rubisco in tobacco leaves in amounts that rivaled the abundance of the endogenous tobacco Rubisco subunits. However, the foreign subunits were recovered exclusively in the insoluble fraction of leaf extracts. Apparently, they were not able to assemble into the $L_8S_8$ Rubisco complex to any extent. This indicates that one or more processes associated with the folding and/or assembly of the red-type Rubiscos is completely blocked in chloroplasts. The folding and assembly mechanisms of red-type Rubisco will need to be understood much better, and perhaps transplanted into chloroplasts as well, if this type of Rubisco transplantation is to be successful.

### 13.6.5
### A Better Strategy for Directed Mutagenesis of *rbc*L

As well as being physiologically informative, the tobacco-*rubrum* transformants provide a much-needed means for circumventing the problem of marker unlinking that previously has made directed mutagenesis of *rbc*L in plants very tedious. Homologous recombination into the plastome of a point-mutated *rbc*L gene linked to a selectable-marker gene results in a situation where both transformed and wild-type plastomes coexist for a period during the sorting-out process that precedes homoplasmy. If the mutation impairs the function of a critical protein like Rubisco, there is strong selection during this phase for a secondary recombination where one of the crossovers occurs in the region of exact homology that unavoidably must exist between the site of the mutation and the marker gene. The product of this recombination acquires the marker without the Rubisco mutation, defeating the purpose of the experiment (Fig. 13.3 A). This very probable result greatly diminishes the frequency with which the desired mutations become homoplasmic [14]. One way of avoiding this problem is to delete the *rbc*L gene completely first; however, null-Rubisco plants are bleached and grow very slowly, even with sucrose supplementation (e.g., Fig. 13.2 M). They are not ideal recipients for plastid transformation. By contrast, tobacco-*rubrum* plants grow vigorously in high-$CO_2$ atmospheres and there is less than 25% nucleotide homology between the *rbc*L and *rbc*M genes, virtually eliminating the likelihood of the unwanted crossovers that would unlink the selection marker (Fig. 13.3 B). The procedure requires the use of a second selection marker, different from the one used to construct tobacco-*rubrum*, such as *npt*II [64] or, alternatively, a marker-free version of tobacco-*rubrum* where the *aad*A gene is excised following transformation (Fig. 13.3 C) [65–67].

### 13.6.6
### Subunit Hybrids Can Be Formed *in vivo*

Plastid transformation was also used to construct, in vivo, a uniform hybrid of Rubisco with large and small subunits derived from different species. Replacement of tobacco *rbc*L with the homologous gene derived from sunflower resulted in non-autotrophic plants from which a hybrid Rubisco with sunflower large subunits and tobacco small subunits could be isolated (Fig. 13.2 H) [15]. The hybrid enzyme was crippled catalytically (fourfold reduced $k_{cat}^c$, sixfold increased $K_c$ and $K_m$ for RuBP) but was able to bind ligand and was not completely inactive, seemingly contradicting the conclusion drawn by Getzoff et al.

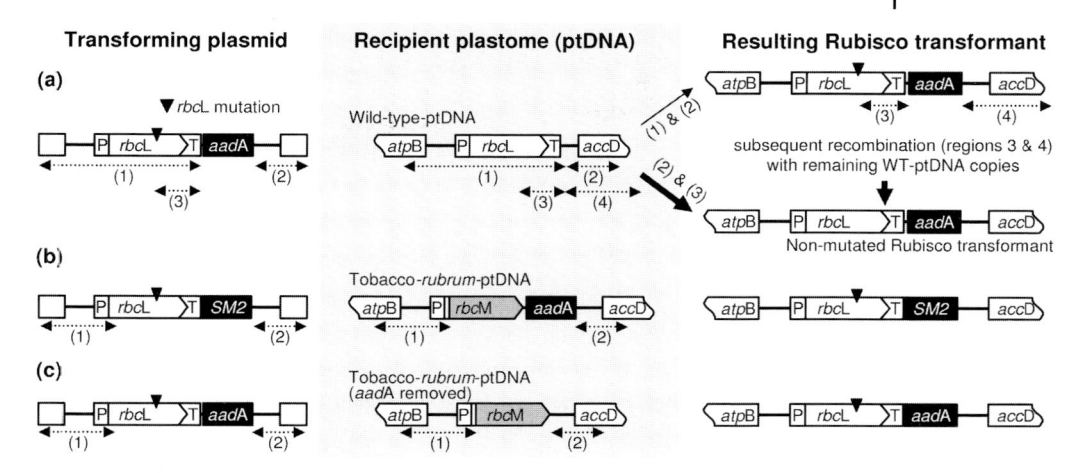

**Fig. 13.3** Strategies for mutating tobacco *rbc*L specifically in vivo. (A) Point mutations generated in tobacco *rbc*L (B) are very vulnerable to becoming unlinked from the selectable marker gene (*aad*A) by recombination with the wild-type plastome, resulting in a crossover in the intervening sequence between the mutation and the marker [14]. (B) This can be avoided by using tobacco-*rubrum* [16] as the recipient because the *rbc*M has <25% nucleotide homology with *rbc*L, but a second selectable marker gene (*SM2*) is required. (C) The requirement for the second selectable marker can be avoided if the *aad*A gene used in constructing tobacco-*rubrum* is removed [65–67]. The dotted, double-arrowed lines numbered (1)–(4) indicate the regions of the transforming DNA that are homologous to the recipient plastome.

[50] from observations with the partial subunit hybrid constructed in the reverse sense by nuclear transformation. Recently, we have discovered that these sunflower-*rbc*L/tobacco-*rbc*S plants are capable of slow growth with $CO_2$ enrichment (unpublished data, Fig. 13.2 L). This system provides an important tool for studying Rubisco's subunit interactions. A similar replacement of tobacco *rbc*L with a cyanobacterial *rbc*L (Fig. 13.2 I) resulted in non-autotrophic plants (Fig. 13.2 M) that did not produce any large subunits or hybrid enzyme [15].

## 13.7
## Priorities for Future Manipulation of Rubisco *in vivo*

### 13.7.1
### The Structural Foundations of Efficient Properties

The ability to construct directed mutants of higher-plant Rubisco will enable ideas about the structural foundations of catalytic properties to be checked with the higher-plant enzyme. Previously, such studies were restricted to microbial Rubiscos. Use of the $CO_2$-assimilation properties of the intact leaf to report the consequences, for photosynthesis as a whole, of subtle changes that do not completely inactivate the enzyme will be very informative. As demonstrated by the sunflower *rbc*L replacement [15], grosser changes that cripple the enzyme may still be studied at the level of the isolated protein, even if non-autotrophic plants are obtained.

### 13.7.2
### Regulation of Rubisco Gene Expression

For manipulation of Rubisco in plants to be conducted with any precision, a means of manipulating the amount of the foreign or altered Rubisco produced is required. Although much is known about chloroplast gene expression, we are not yet able control it with reliable predictability. Therefore, arriving at a suitable level of the product of a plastomic transgene is still largely a matter of trial and error [68]. Further detailed research in this area is, therefore, a priority.

It is generally considered that transcription does not have as dominating an influence on gene expression in plastids as it does in the nucleus [51, 69]. Nevertheless, there are some examples where expression of plastomic transgenes appears to correlate to some degree with the amounts of transcript present [21, 56]. This suggests that transcript abundance has some role in determining plastid gene expression and that the factors that control it – such as promoter strength and transcript stability – should not be ignored.

The rate at which a protein is synthesized in the plastid appears to be the product of transcript abundance and transcript translatability, moderated further by the influence of nucleus-encoded, translation-regulating factors, possibly feedback regulation by the product itself, and the availability of its assembly partners [51, 69]. The 5′-untranslated region (5′-UTR) of the transcript is important in determining translation rate, both by its interactions with the 3′ sequence of the 16S rRNA and because *trans*-activating proteins bind to it. In some cases, including the *rbc*L 5′-UTR, these interactions extend into the 5′ region of the coding sequence [51]. Transcript length and secondary structure may also influence these interactions, leading to variations in translation rate between different coding sequences, even when the 5′ and 3′ elements of the transcript are the same. The frequency with which various codons are used in a particular transcript appears to have less influence; optimization of codon usage to match that of plastid genes induces only modest improvements in expression level [70]. These several variables produce a sizeable matrix of possibilities that need to be investigated to optimize the expression level of a foreign protein in plastids. With further research, the increasing size of the database of tested combinations may eventually bring some welcome predictability to this process.

### 13.7.3
### Folding and Assembly of Rubisco Subunits

So far, no eukaryotic Rubisco has been expressed in functional form in any foreign host. This is the major obstacle currently facing manipulation of Rubisco in higher plants, and it frustrated the attempt to introduce red-type Rubiscos to chloroplasts [21]. Other than a general suspicion that incompatibilities between the foreign Rubisco and the plastid chaperone system are somehow involved, nothing is known about the causes of the problem. Beginning with the realization that the so-called Rubisco-binding protein in plastids was a molecular chaperone [71], the obligatory involvement of the chaperonin 60/21 system in folding and assembling Rubisco has been established by considerable research (reviewed in [1]). However, other chaperone complexes are also involved in Rubisco assembly. The likely necessity of the Hsp70-DnaJ complex was indicated by a maize nuclear mutant lacking a DnaJ homologue that produces no functional Rubisco [72]. Incompatibility between any of these complexes and the foreign Rubisco would preclude production of functional enzyme.

Further detailed study of the folding and assembly mechanisms of Form-I Rubiscos, both in vitro and using plastid extracts [1], is obviously warranted, as are further mutant-screening and genomic efforts to discover whether any other products of nuclear genes are involved in Rubisco assembly. The possibility that post-translational modifications of the Rubisco large subunit [73] might be a prerequisite for assembly can now be studied by directed mutagenesis in vivo. Furthermore, as mentioned earlier, it might also be possible to devise strategies for acclimating foreign Rubiscos to assemble in bacterial hosts, at least, by directed molecular evolution.

## 13.8 Conclusions

With the functional replacement of Rubisco in tobacco by plastid-transformation technology (Fig. 13.2 E, K) [16], a milestone in the engineering of photosynthesis has been passed that demonstrates the potential of Rubisco manipulation. Further progress will require better understanding of many aspects of Rubisco – its catalytic mechanism, its regulation, its natural diversity, and its synthesis and assembly in plants and other organisms. Better understanding of gene expression in chloroplasts and how to manipulate it predictably will also be beneficial.

## Acknowledgments

We thank H. Kane and D. Yellowlees for reading the manuscript critically.

## References

1 H. Roy, T.J. Andrews, in: R.C. Leegood, T.D. Sharkey, S. von Caemmerer (Eds.), Photosynthesis: Physiology and Metabolism. Kluwer Academic Publishers, Dordrecht, 2000, pp. 53–83.
2 R.J. Spreitzer, M.E. Salvucci, Annu. Rev. Plant Biol. **53** (2002) 449–475.
3 R.J. Spreitzer, Photosynth. Res. **60** (1999) 29–42.
4 F.R. Tabita, Photosynth. Res. **60** (1999) 1–28.
5 I. Andersson, T.C. Taylor, Arch. Biochem. Biophys. (2002) in press.
6 R.J. Spreitzer, Arch. Biochem. Biophys. (2002) in press.
7 A.R. Portis, Jr., R.L. Houtz, Arch. Biochem. Biophys. (2002) in press.
8 Z. Svab, P. Hajdukiewicz, P. Maliga, Proc. Natl Acad. Sci. USA **87** (1990) 8526–8530.
9 S.R. Rodermel, M.S. Abbott, L. Bogorad, Cell **55** (1988) 673–681.
10 G.S. Hudson, J.R. Evans, S. von Caemmerer, Y.B.C. Arvidsson, T.J. Andrews, Plant Physiol. **98** (1992) 294–302.
11 A.A. Gatenby, Eur. J. Biochem. **144** (1984) 361–366.
12 L.P. Cloney, D.R. Bekkaoui, S.M. Hemmingsen, Plant Mol. Biol. **23** (1993) 1285–1290.
13 I. Kanevski, P. Maliga, Proc. Natl Acad. Sci. USA **91** (1994) 1969–1973.
14 S.M. Whitney, S. von Caemmerer, G.S. Hudson, T.J. Andrews, Plant Physiol. **121** (1999) 579–588.
15 I. Kanevski, P. Maliga, D.F. Rhoades, S. Gutteridge, Plant Physiol. **119** (1999) 133–141.
16 S.M. Whitney, T.J. Andrews, Proc. Natl Acad. Sci. USA **98** (2001) 14738–14743.
17 M.K. Morell, K. Paul, H.J. Kane, T.J. Andrews, Aust. J. Bot. **40** (1992) 431–441.
18 A. Fersht, Enzyme Structure and Mechanism. W.H. Freeman and Company, New York, 1985.
19 W.A. Laing, W.L. Ogren, R.H. Hageman, Plant Physiol. **54** (1974) 678–685.
20 G.D. Farquhar, S. von Caemmerer, J.A. Berry, Planta **149** (1980) 78–90.
21 S.M. Whitney, P. Baldet, G.S. Hudson, T.J. Andrews, Plant J. **26** (2001) 535–547.
22 C.F. Delwiche, J.D. Palmer, Mol. Biol. Evol. **13** (1996) 873–882.
23 H. Sugawara, H. Yamamoto, N. Shihata, T. Inoue, S. Okada, C. Miyake, A. Yokota, Y. Kai, J. Biol. Chem. **274** (1999) 15655–15661.
24 B.A. Read, F.R. Tabita, Arch. Biochem. Biophys. **312** (1994) 210–218.
25 K. Uemura, Anwaruzzaman, S. Miyachi, A. Yokota, Biochem. Biophys. Res. Commun. **233** (1997) 568–571.
26 S. Ezaki, N. Maeda, T. Kishimoto, H. Atomi, T. Imanaka, J. Biol. Chem. **274** (1999) 5078–5082.
27 T.C. Taylor, I. Andersson, J. Mol. Biol. **265** (1997) 432–444.
28 W.W. Cleland, T.J. Andrews, S. Gutteridge, F.C. Hartman, G.H. Lorimer, Chem. Rev. **98** (1998) 549–561.
29 O. Tapia, J. Andres, V.S. Safont, J. Chem. Soc. – Faraday Trans. **90** (1994) 2365–2374.
30 W.A. King, J.E. Gready, T.J. Andrews, Biochemistry **37** (1998) 15414–15422.
31 M. Oliva, V.S. Safont, J. Andres, O. Tapia, J. Phys. Chem. **105** (2001) 4726–4736.
32 M. Oliva, V.S. Safont, J. Andres, O. Tapia, J. Phys. Chem. **105** (2001) 9243–9251.
33 H. Mauser, W.A. King, J.E. Gready, T.J. Andrews, J. Amer. Chem. Soc. **123** (2001) 10821–10829.
34 I. Andersson, J. Mol. Biol. **259** (1996) 160–174.
35 T.C. Taylor, A. Backlund, K. Bjorhall, R.J. Spreitzer, I. Andersson, J. Biol. Chem. **276** (2001) 48159–48164.
36 E. Mizohata, H. Matsumura, Y. Okano, M. Kumei, H. Takuma, J. Onodera, K. Kato, N. Shibata, T. Inoue, A. Yokota, Y. Kai, J. Mol. Biol. **316** (2002) 679–691.
37 W.P.C. Stemmer, Proc. Natl Acad. Sci. USA **91** (1994) 10747–10751.
38 A. Crameri, S.-A. Raillard, E. Bermudez, W.P.C. Stemmer, Nature **391** (1998) 288–291.

39 N. Cohen, S. Abramov, Y. Dror, A. Freeman, Trends Biotechnol. **19** (2001) 507–510.
40 J. Pierce, T. J. Carlson, J. G. K. Williams, Proc. Natl Acad. Sci. USA **86** (1989) 5753–5757.
41 D. Amichay, R. Levitz, M. Gurevitz, Plant Mol. Biol. **23** (1993) 465–476.
42 A. Makino, T. Shimada, S. Takumi, K. Kaneko, M. Matsuoka, K. Shimamoto, H. Nakano, M. Miyao-Tokutomi, T. Mae, N. Yamamoto, Plant Physiol. **114** (1997) 483–491.
43 R. T. Furbank, J. A. Chitty, S. von Caemmerer, C. L. D. Jenkins, Plant Physiol. **111** (1996) 725–734.
44 W. P. Quick, U. Schurr, K. Fichtner, E.-D. Schulze, S. R. Rodermel, L. Bogorad, M. Stitt, Plant J. **1** (1991) 51–58.
45 M. Stitt, W. P. Quick, U. Schurr, E.-D. Schulze, S. R. Rodermel, L. Bogorad, Planta **183** (1991) 555–566.
46 A. Krapp, M. M. Chaves, M. M. David, M. L. Rodriques, J. S. Pereira, M. Stitt, Plant Cell Environ. **17** (1994) 945–953.
47 W. P. Quick, U. Schurr, R. Scheibe, E.-D. Schulze, S. R. Rodermel, L. Bogorad, M. Stitt, Planta **183** (1991) 542–554.
48 C. J. Mate, S. von Caemmerer, J. R. Evans, G. S. Hudson, T. J. Andrews, Planta **198** (1996) 604–613.
49 S. von Caemmerer, J. R. Evans, G. S. Hudson, T. J. Andrews, Planta **195** (1994) 88–97.
50 T. P. Getzoff, G. H. Zhu, H. J. Bohnert, R. G. Jensen, Plant Physiol. **116** (1998) 695–702.
51 P. Maliga, Curr. Opin. Plant Biol. **5** (2002) 164–172.
52 S. R. Sikdar, G. Serino, S. Chaudhuri, P. Maliga, Plant Cell Rep. **18** (1998) 20–24.
53 S. Ruf, M. Hermann, I. J. Berger, H. Carrer, R. Bock, Nature Biotechnol. **19** (2001) 870–875.
54 V. A. Sidorov, D. Kasten, S. Z. Pang, P. T. J. Hajdukiewicz, J. M. Staub, N. S. Nehra, Plant J. **19** (1999) 209–216.
55 P. J. Madgwick, S. P. Colliver, F. M. Banks, D. Z. Habash, H. Dulieu, M. A. J. Parry, M. J. Paul, Ann. Appl. Biol. **140** (2002) 13–19.
56 S. M. Whitney, T. J. Andrews, Plant Cell **13** (2001) 193–205.
57 X. H. Zhang, R. G. Ewy, J. M. Widholm, A. R. Portis, Jr., Plant Cell Physiol. **43** (2002) 1302–1313.
58 C. F. Delwiche, Amer. Natural. **154** (1999) S164–S177.
59 C. R. Somerville, S. C. Somerville, Mol. Gen. Genet. **193** (1984) 214–219.
60 B. E. Terzaghi, W. A. Laing, J. T. Christeller, G. B. Petersen, D. F. Hill, Biochem. J. **235** (1986) 839–846.
61 G. J. Lee, K. A. McDonald, B. A. McFadden, Protein Sci. **2** (1993) 1147–1154.
62 J. T. Christeller, W. A. Laing, Biochem. J. **173** (1978) 467–473.
63 W. A. Laing, J. T. Christeller, Biochem. J. **159** (1976) 563–570.
64 H. Carrer, T. N. Hockenberry, Z. Svab, P. Maliga, Mol. Gen. Genet. **241** (1993) 49–56.
65 S. Iamtham, A. Day, Nature Biotechnol. **18** (2000) 1172–1176.
66 P. T. J. Hajdukiewicz, L. Gilbertson, J. M. Staub, Plant J. **27** (2001) 161–170.
67 S. Corneille, K. Lutz, Z. Svab, P. Maliga, Plant J. **27** (2001) 171–178.
68 P. Maliga, Trends Biotechnol. **21** (2003) 20–28.
69 Y. Choquet, F. A. Wollman, FEBS Lett. **529** (2002) 39–42.
70 G. N. Ye, P. T. J. Hajdukiewicz, D. Broyles, D. Rodriguez, C. W. Xu, N. Nehra, J. M. Staub, Plant J. **25** (2001) 261–270.
71 R. J. Ellis, Nature **328** (1987) 378–379.
72 T. P. Brutnell, R. J. H. Sawers, A. Mant, J. A. Langdale, Plant Cell **11** (1999) 849–864.
73 R. L. Houtz, L. Poneleit, S. B. Jones, M. Royer, J. T. Stults, Plant Physiol. **98** (1992) 1170–1174.
74 M. K. Morell, H. J. Kane, T. J. Andrews, FEBS Lett. **265** (1990) 41–45.
75 D. B. Jordan, W. L. Ogren, Nature **291** (1981) 513–515.
76 H. J. Kane, J. Viil, B. Entsch, K. Paul, M. K. Morell, T. J. Andrews, Aust. J. Plant Physiol. **21** (1994) 449–461.

## 14
## Defining the Inefficiencies in the Chemical Mechanism of the Photosynthetic Enzyme Rubisco by Computational Simulation

*Jill E. Gready*

### 14.1
### Introduction

Life on earth depends critically on $CO_2$ fixation by the photosynthetic enzyme D-ribulose-1,5-bisphosphate (RuBP) carboxylase/oxygenase (Rubisco): it is responsible for virtually all of the carbon in the biosphere [1–4]. Hence, it is a very great puzzle that this enzymic process is so woefully inefficient. Why has evolutionary adaptation, apparently, "failed"? We might expect Rubisco's catalytic properties to have been under the severest selective pressure possible. Deciphering the reasons underlying these deficiencies is, thus, a classic problem in understanding how enzymes evolve to fulfill their biological requirements, especially those such as Rubisco, which catalyze complex reactions with non-redundant functions. Such knowledge is also of immense potential benefit in re-engineering more efficient Rubiscos or implementing some of Rubisco's mechanistic features into simpler model catalytic systems [5, 6].

#### 14.1.1
#### Catalytic Inefficiencies

Rubisco exhibits multiple inefficiencies covering all aspects of its action: activation/inactivation by carbamoylation of the active-site lysine residue, a low turnover rate, poor selectivity for its substrate $CO_2$ over $O_2$, and inhibition by dead-end complexes, including unwanted reaction byproducts. The consequence of the slow rate of only about one complete cycle per second, even for the best Rubiscos, is a requirement for huge amounts of protein: Rubisco constitutes 20–60% leaf protein, a major drain on metabolism and plant nitrogen. Because of the poor selectivity for $CO_2$, Rubisco mistakenly binds and fixes $O_2$ in a competing oxygenation reaction, which wastes energy in recycling of the phosphoglycolate product in photorespiration. These compounded inefficiencies compromise water usage and produce high transpiration ratios (500–1000 $H_2O$ molecules per $CO_2$ fixed) because

*Artificial Photosynthesis: From Basic Biology to Industrial Application*
Edited by Anthony F. Collings and Christa Critchley
Copyright © 2005 WILEY-VCH Verlag GmbH & Co. KGaA, Weinheim
ISBN: 3-527-31090-8

of the increased need for leaf stomata to be open for $CO_2$ entry. A faster more discriminating Rubisco could fix $CO_2$ at lower $CO_2$ concentration.

### 14.1.2
### Evolutionary Constraints?

Reference to the main chemical steps for both the carboxylation and oxygenation reactions shown in Fig. 14.1 provides some initial insights. Rubisco catalyzes a multi-step reaction involving as many as four enzyme-bound intermediates, whose instabilities give rise to multiple side reactions that further compromise its efficiency. One could argue that current Rubiscos are very good solutions for effecting such difficult chemistry: Rubisco's complex structure and mechanism is a compromise solution, but at least it works. With respect to evolutionary improvement, one could hypothesize that no evolutionary pathway for gradual improvement exists via single mutations: mutations that might assist some steps could fatally compromise others. This naïve view of evolution ignores the question of why more complex genetic mechanisms for exploring "catalytic space," such as larger or multiple changes through gene duplication, recombination, or other mechanisms, have not been used (successfully) for Rubisco. However, together these views hint at a possibility that if the catalytic bottlenecks in the Rubisco mechanism could be mapped in molecular detail, then they might be "repaired" by engineering multiple changes into the Rubisco

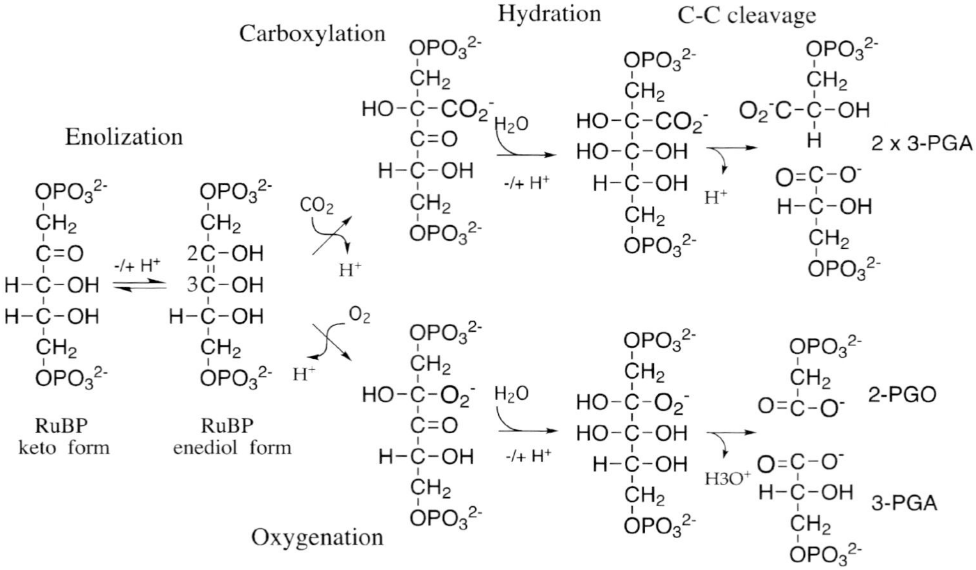

**Fig. 14.1** Schematic showing chemistry in the multi-step Rubisco reactions for both carboxylation and oxygenation, including production and use of protons and water molecules.

sequence and structure with techniques now available, including expression in higher plants, thus short-circuiting the need for a mutational evolutionary pathway. Encouragement for these ideas is provided by reported natural variability in Rubisco's catalytic abilities to suit particular environments [1–3, 7–9], suggesting that evolutionary improvement is still possible, and by the observation that atmospheric concentrations of both $CO_2$ and $O_2$ have changed drastically over biological time, suggesting that Rubisco's capacity for evolutionary change is too slow to accommodate such changing conditions.

## 14.1.3
### Experimental Limitations

Rubisco has been studied intensively experimentally. In combination with crystallographic structures, for which more than 20 have been solved for all Rubisco subtypes and in complexes with reaction-intermediate analogues (see [10]), the catalytic mechanism and characterization of reaction intermediates and byproducts has been studied by kinetic analysis, spectroscopic procedures, and directed mutagenesis [1, 2, 7–9]. The unique features of the $Mg^{2+}$-stabilized active-site carbamoylated lysine, both its role in activating the enzyme [11, 12] and its possible catalytic roles in the multi-step reaction [4, 9], have attracted much attention. However, although a detailed molecular picture of the active-site organization and the identities of key groups has emerged, the orchestration of the catalysis – i.e., the specific roles and protonation states of the active-site groups and water molecules in the different steps of the reaction, the sources and fates of the numerous protons involved, and the influence of the wider protein environment, including dynamical motions – remains undefined. Elucidation of these questions requires energetic and thermodynamic analysis using computational simulations: the effective energetic roles of all the players in the catalysis are not approachable by experiment [13, 14].

## 14.1.4
### Goals of Simulations

We first define the goals for the simulation studies in terms of the above unanswered questions, and then consider what simulation methods might be necessary to provide useful information, i.e., predictions as starting points for further experiments or protein engineering. Ideally, we wish to define a complete mechanistic profile for the carboxylation reaction cycle and an analogous profile also for, at least, the initial steps of the oxygenation cycle, as these are critical for understanding substrate selectivity. Key issues are:

1. The specific roles of the active-site groups and how they work in concert or sequentially in different reaction steps. As relevant, this includes their protonation states during the course of the reaction.
2. The origins and fates of protons that are added to and abstracted from reactive species throughout the reaction cycle and from different "directions" of

the active site. As shown in Fig. 14.1, there are net two protons produced in the carboxylation reaction and one water consumed, while in the oxygenation reaction there is additionally one water produced.
3. The definition of networks for channeling protons produced in the reaction away from the reaction center.
4. The elucidation of the unique roles of the carbamylated lysine, including evidence for suggested acid-base roles.
5. The definition of the reaction-cycle energetics, including the possibility for suggested concerted steps (e.g., gas addition and hydration [4]), the relative energies of intermediates and the likelihood of back reactions, and the possibility for mechanistic flexibility in some steps, i.e., energetically comparable mechanisms on a relatively flat free-energy surface. It is increasingly recognized that there is an ensemble of enzyme reactive-complex states that can lead to reaction [15–17]; consequently, the reaction pathway should not be defined too narrowly.

### 14.1.5
### Simulation Options

While, in principle, computer simulation provides a means to address these enzyme mechanistic problems [18], accurate representation of the chemical events and partitioning of energetic and dynamical components to understand the roles of groups and the whole enzyme are beyond the feasibility of any current computational method. Certainly no current method can provide this reliably at the level now routinely achievable in small-molecule computational chemistry. Fig. 14.2 summarizes the "windows" accessible by current methods. As shown, these windows can be divided into three main classes depending on the quality of the electronic description of the molecule, the quality of the representation of its chemical environment, and the degree of dynamic sampling. With the supercomputer resources and the highly developed software now available (see Section 14.2), typical size-dependent limitations of these methods are: 50–100 atoms for ab initio quantum mechanics (QM) calculations with full geometry optimization; several hundred atoms for conventional semi-empirical QM (SE-QM) calculations with full geometry optimization; 10,000 atoms or more for linear-scaling SE-QM calculations with partial geometry optimization [19]; 40,000 atoms or more (enzyme+solvent) for multiple trajectory MD simulations of nanoseconds (ns) each; and 10,000–20,000 atoms for multiple trajectory simulations of about 0.5 ns each with a QM/MM potential (up to 200 atoms in the SE-QM region) [20–22].

For our work on Rubisco and other enzyme problems, we have pioneered a multi-method approach that allows us to use all of these methods to explore different aspects of the mechanistic questions within their particular capability ranges. Here we present results using two of them: ab initio QM calculations on active-site fragment complexes and MD simulations with an SE-QM/MM potential. The fragment-complex calculations were used first to map the complete

```
                Accurate ab initio and DFT methods (gradient minimization)
Reaction center in enzyme active site  ⟹  Approximate representation of enzyme
(approximately 100 atoms in vacuum)   STATIC   -Dielectric continuum SCRF models
                                              -Point charge and QM/MM models
```

```
      ⇓ DYNAMIC          →  Pathway for enzyme-catalysed reaction  ⇐

Approximate semiempirical QM methods          Approximate semiempirical methods
(MD simulations, gradient minimization)       (gradient minimization)
Accurate representation of enzyme             Accurate representation of enzyme
- SE QM/MM + MD models                        -Linear scaling (MOZYME, LocalSCF)
```

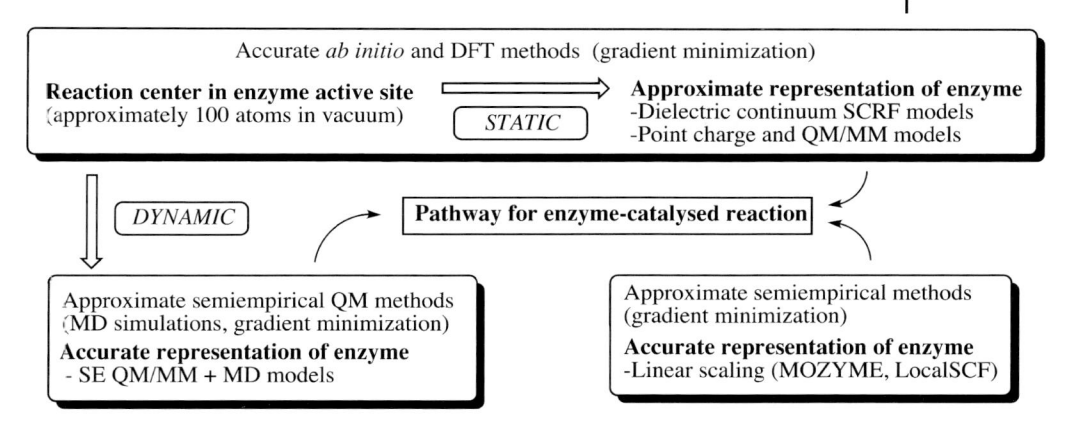

**Fig. 14.2** Relationship between uses and accuracy of QM, linear scaling QM, and mixed-method "divide-and-conquer" approaches for studying enzyme-catalyzed reactions.

carboxylation reaction cycle in order to define the basic energetic (enthalpic) profile and structures of transition states and intermediates. QM/MM+MD calculations were then undertaken to mirror the steps of the fragment calculations in order to study the effect of the rest of the enzyme environment on the reaction and to enable computation of free energies.

## 14.2 Computational Methods

### 14.2.1 Computational Programs

Ab initio QM calculations on fragment complexes were performed using the Gaussian 94 and 98 program packages [23, 24] at SCF, MP2, and DFT levels, using the B3LYP functional for DFT calculations. Geometry optimizations for stationary points (reactant, product, transition state, and intermediate complexes) were performed at SCF or DFT levels using the 6-31G(d) basis set. Single-point calculations at the optimized geometries were performed with the 6-311G+(2d,p) or 6-31+G(d,p) basis sets at DFT or MP2 levels, respectively, in order to calculate reliable reaction-path enthalpies. Full details have been reported [25, 26]. QM/MM MD simulations were performed with the MOPS program package [20]. The semi-empirical PM3 Hamiltonian [27] was used for the QM region, the AMBER protein force field [28] and TIP3P water model [29] were used for the MM region, and H atoms were used as link atoms to cap broken peptide bonds.

## 14.2.2
### Enzyme Models

Spinach Rubisco is a large complex of 4720 residues composed of eight large (467 residues) and eight small (123 residues) subunits, with eight active sites each formed from two large subunits. This complete system is too large, and unnecessary, for simulations. For preparation of starting structures for both fragment calculations and QM/MM MD simulations, initial coordinates were taken from the 1.6-Å crystal structure [30] of a complex (Rubisco · $Mg^{2+}$ · Lys-201-$CO_2$ · CABP complex) of spinach Rubisco co-crystallized with 2′-carboxyarabinitol 1,5-bisphosphate (CABP), an analogue of the $\beta$-keto acid intermediate (**3**, Fig. 14.6). For the fragment calculations, coordinates were extracted for two large and two small subunits (1180 residues total), which reproduced one complete active site and included all protein within 22 Å of the magnesium ion and 279 water molecules. After replacement of CABP by RuBP (or other ligand) and addition of a further 238 water molecules within an approximately 28-Å sphere of the Mg ion, the system was equilibrated by MD (simulation time 1 ns, initial temperature 10 K, simulation temperature 300 K, coupled to a heat bath with the time constant of 0.2 ps; see [26] for more details). For the QM/MM calculations, we extracted coordinates for a dimer of two large subunits (934 residues, 16,350 atoms, 3040 water molecules) and added an additional 262 water molecules within the 22-Å sphere. The CABP ligand was removed and replaced by coordinates for the DFT-optimized geometry of the $\beta$-keto acid **3** from the fragment-complex study and water to produce the initial structure. The second active site was constrained according to the crystal structure during all calculations. The DFT geometry was constrained, and the two phosphate groups of the RuBP were fixed at their crystal-structure positions. After relaxation of hydrogen positions only with a 2-ns MD simulation, the structure was minimized by 800 cycles of steepest descent and then used as input for the QM/MM simulations. Structures for other intermediates and TSs were modeled similarly by replacement of relevant atoms with their respective DFT-optimized fragments.

## 14.2.3
### Active-site Fragment Complexes

Starting coordinates for 29- or 30-atom fragment models of each active-site complex were constructed from six sets of enzyme-complex coordinates sampled at 5-ps simulation intervals, as described previously [25, 26]. For the first step of enolization (see Fig. 14.4), the coordination sphere around Mg was composed of two formate groups ($HCOO^-$) representing Asp203 and Glu204, $CH_3NHCOO^-$ representing carbamylated Lys201, $HC(2)OC(3)H_2OH$ representing the initial (i.e., RuBP) substrate C(2)C(3) fragment of the five-carbon sugar, and a water to fill the sixth valency. Some additional calculations were performed with an ammonium group representing Lys175 [25]. For the study of later steps of the reaction starting from the enediolate form of RuBP, Asp203 and Glu204 were repre-

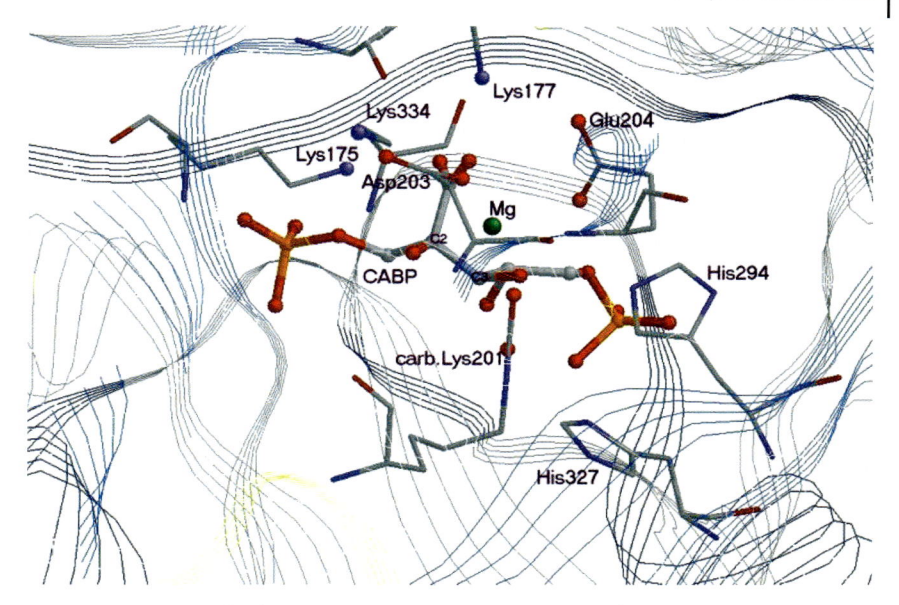

**Fig. 14.3** Active site of Rubisco according to the X-ray structure of activated spinach Rubisco [30]. The residues close to the Mg ion are shown explicitly; for other residues only the backbone is drawn. (After [26] with permission).

sented by water molecules rather than carboxylate groups to provide a more suitable (neutral) representation of the electrostatics within the active site [26]. This was done for reasons of computational stability [31] and recognition of the contributions of positive groups (Lys175, Lys177, and His294) in the second co-ordination sphere (see Fig. 14.3) [30].

## 14.2.4
### QM/MM Simulations

As indicated above, DFT-optimized geometries for the active-site fragments representing the first coordination region around Mg were re-embedded in the active site. A QM region of 165 atoms that included these Mg-bound residues (carbamylated Lys201, Asp203, Glu204), second coordination sphere residues (Lys175, Lys177, His294, Lys334), the Mg ion, and the complete RuBP analogue was defined and it formed an ellipsoid of 7–9 Å. A 16-Å sphere centered on the Mg ion was allowed for the mobile region during geometry minimizations and MD simulations. A residue-based cutoff of 20 Å for non-bonded electrostatic interactions and of 10 Å for van der Waals interactions was applied to the MM region. For each stationary point in the carboxylation reaction, 240-ps equilibration simulations were first run (temperature = 300 K, temperature relaxation time = 0.1 ps, time step = 1 fs), followed by calculation of free energies for each complex by averaging over 800 runs (corresponding to 160 ps) using the linear response approximation [32].

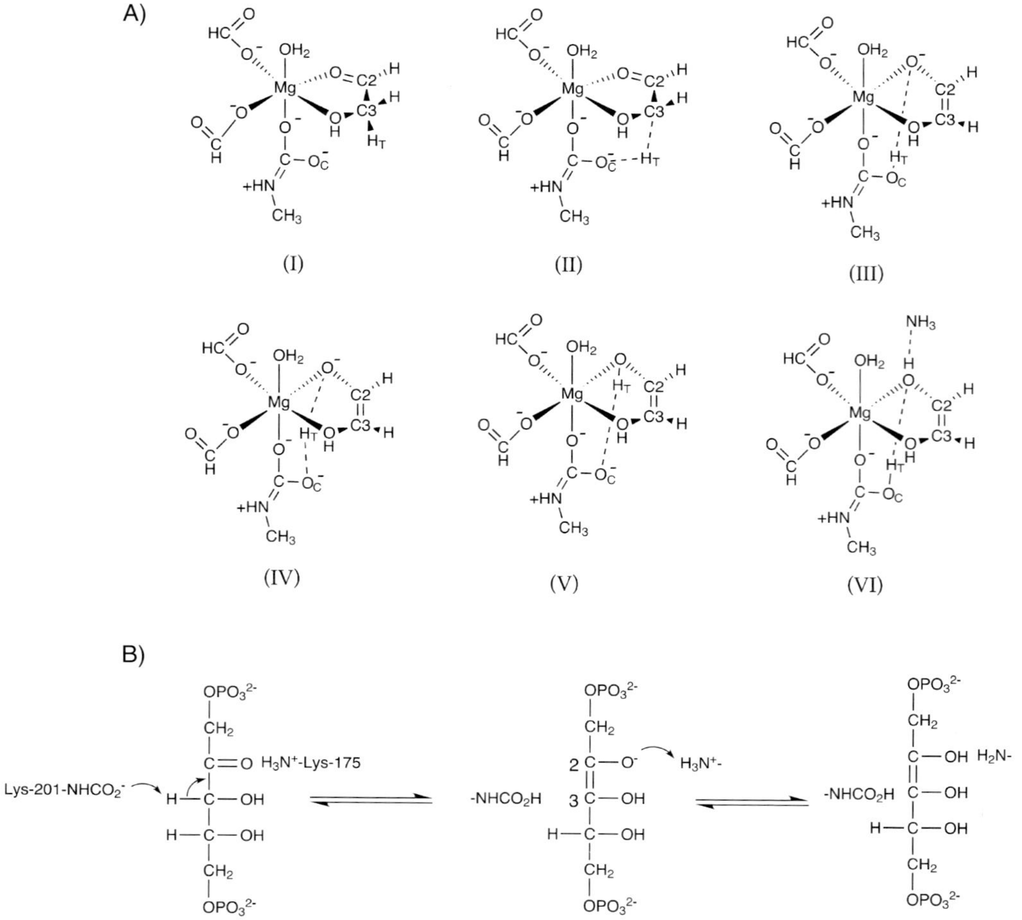

**Fig. 14.4** States and mechanism for the initial enolization step. (A) States for the 29-atom fragment used to model the enolization of RuBP. Atoms representing the C2 and C3 atoms of RuBP are labeled. (I) Substrate in ketone form. (II) Transition-state structure for the removal of the C3 proton. (III) Protonated carbamate H-bonded to the O2 atom. (Note that no energy minimum was found for the state with protonated carbamate associated with the C3 atom, i.e., the closest structural analogue to state (II)). (IV) Transition-state structure for the reprotonation of the substrate on O2. (V) Substrate in enol form. (VI) Enol form (i.e., analogous to (V)) for fragment in the presence of an ammonium ion representing Lys175, with protonated carbamate and ammonium proton transferred to O2. (B) The proposed overall mechanism. The carbamate group of Lys201 is the initial base that deprotonates the C3 atom, while the amino group of Lys175 initially stabilizes the resultant enolate and then acts as the acid that protonates the O2 atom.

## 14.3
## Results and Discussion

### 14.3.1
### Fragment-complex Calculations

#### 14.3.1.1 Enolization Step

The first step in the Rubisco reaction is conversion of the ketone form of RuBP to the less stable enediol form. This tautomerization requires removal of the proton from carbon C3 $\alpha$ to the carbonyl group on carbon C2, and protonation of the carbonyl group. The protonation is essential for correct direction of the $CO_2$ addition in the subsequent step to C2 rather than C3. The identities of the acidic and basic groups have not been established experimentally. X-ray structures show that the only group suitably positioned for C3 proton abstraction is the carbamoylated Lys201, and hence the carbamate group has been proposed as the base [4, 30, 33]. However, as such a role for a carbamate is unprecedented in biology, it is not clear that the Mg-coordinated carbamate would be sufficiently basic. Examination of the X-ray structures suggests Lys175 as a potential acid, as it is well positioned to protonate O2 of the enolate. This suggestion is consistent with mutagenesis studies that show specific impairment of enolization for the Lys175 mutant [34]. Consequently, the aims of the fragment calculations were to evaluate the energetic feasibility of carbamate acting as a base, and to explore possible roles for Lys175, in the enolization step.

Fragment calculations, with and without an ammonium ion representing Lys175, were performed at SCF, DFT, and MP2 levels for the states I–VI shown in Fig. 14.4. The DFT and MP2 results were qualitatively similar and, in comparison with the SCF results, indicated that inclusion of correlation was important, with all states being stabilized relative to state I, especially the transition states [25]. Relative energies for the highest-level (MP2/6-31G(d,p)//RHF/6-31G(d,p)) calculations are shown in Fig. 14.5. The relative energies of states I and II with a neutral ammonia molecule H-bonded to O2 show little effect (0.8 kcal mol$^{-1}$) on the activation energy. However, a positively charged ammonium ion similarly placed causes a significant reduction in the activation energy (by 4.1 kcal mol$^{-1}$ to 17.4 kcal mol$^{-1}$). We deem this activation energy of 17.4 kcal mol$^{-1}$ as conservatively high, due to the simplicity of the model and neglect of other features of the enzyme environment. Together, the results suggest that the Mg-coordinated carbamate group is well capable of functioning as the base in the enolization step and explain why Lys175 mutants show such large reductions in the enolization rate [34].

While, as shown in Fig. 14.5, the energies of the keto (state I) and enol (state V) forms of the 29-atom fragment (i.e., without $NH_4$) are approximately equal ($\Delta = 1.8$ kcal mol$^{-1}$), introduction of the ammonium ion stabilizes the enol form (state VI) by more than 20 kcal mol$^{-1}$. Note that this low value, which would lead to unwanted intermediate trapping, likely represents an over-stabilization due to neglect of other features of the enzyme environment. Nevertheless, the

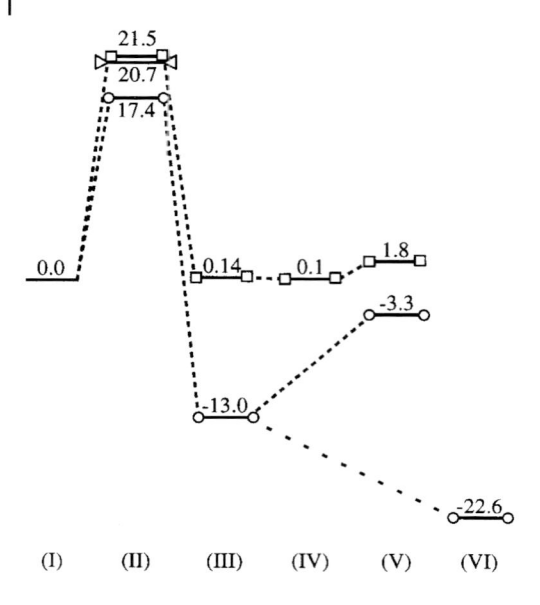

**Fig. 14.5** The relative energies of the states shown in Fig. 14.4 at the single-point MP2/6-31G(d,p)//RHF/6-31G(d,p) level. For the 29-atom fragment, ○; for the 29-atom fragment plus an $NH_4^+$ ion, □; or an $NH_3$ molecule, ▽ (state II only) modeling Lys175. State VI has protonated carbamate, protonated O2, and $NH_3$.

calculations suggest that a protonated Lys175 could play two roles, first in stabilizing the transition state for proton abstraction and second as an acid to protonate O2 of the resulting enediolate.

To investigate the basicity of methylcarbamate, we calculated relative protonation energies for models where methylcarbamate, carbamate, propionate, acetate, and bicarbonate were coordinated to a hydrated magnesium ion or substituted into the fragment (with the ammonium ion). The first set of results (MP2/6-31G(d) level) showed that methylcarbamate and carbamate are significantly more basic (by ~3–10 kcal mol$^{-1}$) than the simple carboxylates. Although this stabilization difference is less for the second set of reaction-pathway states (MP2/6-31G(d)//RHF/6-31G(d) level) with a realistic representation of the active-site Mg coordination, methylcarbamate is still the most basic by 1–2 kcal mol$^{-1}$ [25]. Although apparently small, this is a significant difference in energy that would effectively increase the enolization rate by more than an order of magnitude. Overall, the results suggest an alternative reason for this unique use of a protein carbamate as a general base, namely, that this role could not be duplicated by any other group normally found in proteins. The usual reason advanced for use of the carbamate is in regulating the activity of Rubisco according to the availability of $CO_2$ [35].

As detailed in Section 14.2, fragment-complex calculations for subsequent steps of the carboxylation reaction (Fig. 14.6) were undertaken with a slightly modified fragment model: the structures for the $CO_2$ and water addition steps are shown in Fig. 14.7. As shown, our starting point is the enediolate with O3 deprotonated, whereas the finishing point for the enolization study was the enediol. Deprotonation of O3 is required to direct attack of $CO_2$ to C2; otherwise,

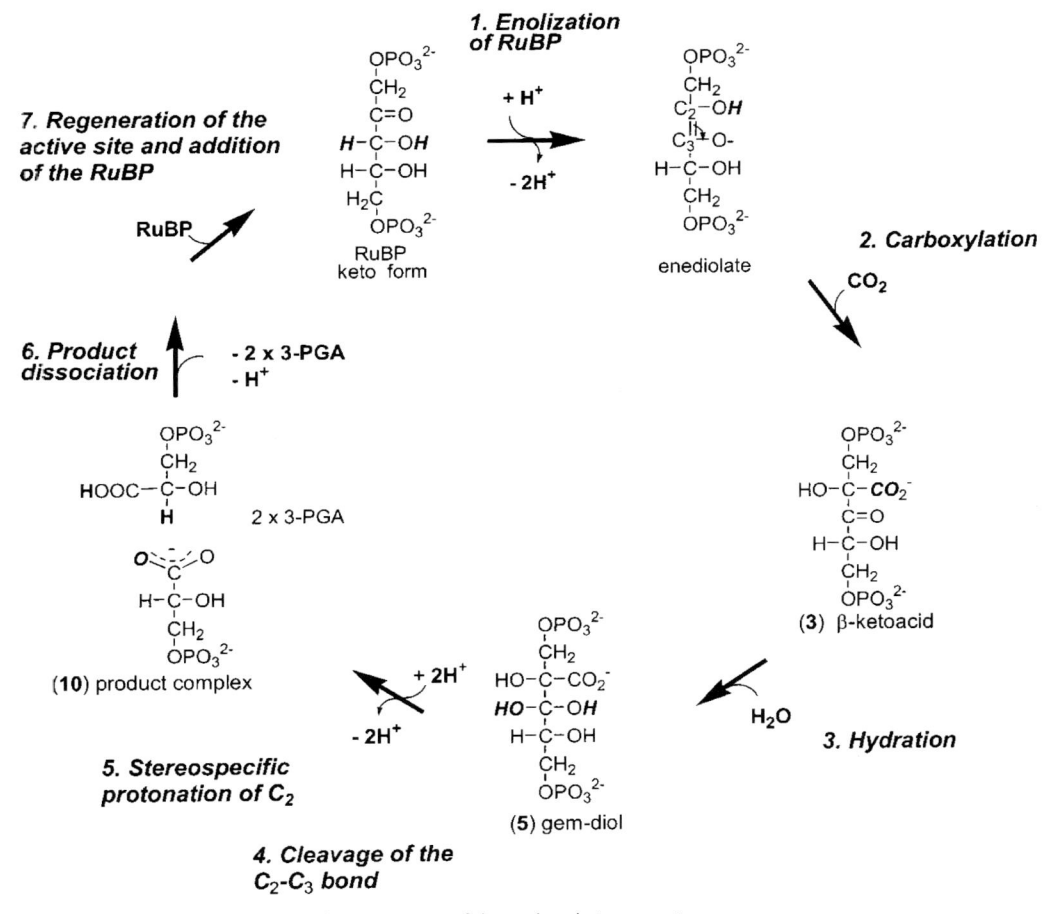

**Fig. 14.6** Mechanism for post-enolization steps of the carboxylation reaction catalyzed by Rubisco. Shown are the key steps of $CO_2$ addition, hydration, and C-C bond cleavage leading from the 2,3-enediolate of RuBP to two molecules of 3-phospho-D-glycerate. (After [26] with permission).

C3 would be equally vulnerable [4, 9, 25, 38]. While His294 or the carbamate might be used for this purpose [25, 38], the later QM/MM results suggest that His294 does not perform this role.

#### 14.3.1.2 Carboxylation Step

The first step after enolization is addition of $CO_2$ to C2 of the enediolate. In the initial calculations, a very weakly bound encounter complex was located in which the $CO_2$ was positioned in the outer coordination sphere with a water molecule still as the sixth Mg ligand (Fig. 14.7) [26]. In subsequent work explor-

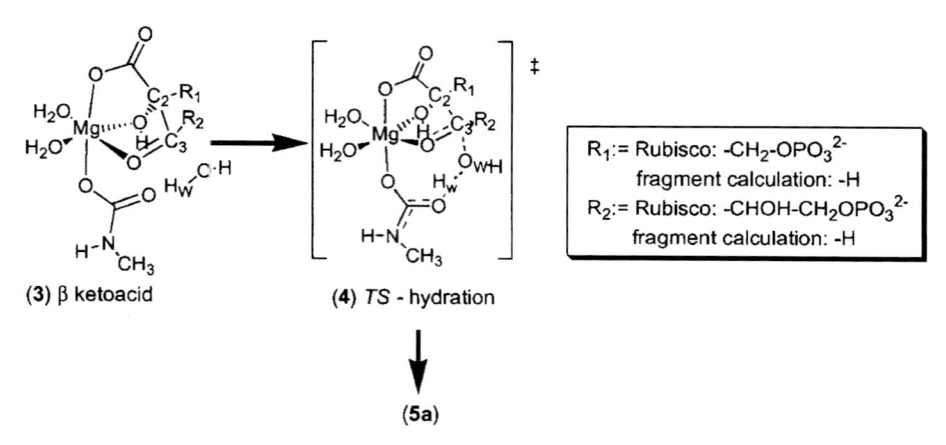

**Fig. 14.7** Carboxylation and hydration reactions starting at the approach of $CO_2$ to the enediolate form of RuBP. Schematic representations for the structures obtained with the active-site fragment model are shown (R1 = R2 = H). "†" indicates a first-order transition state. (After [26] with permission).

ing larger fragments with better potential for complexation of the gas molecule, a more defined structure was located (B. Kannappan et al., unpublished results 2004); however, the earlier structure remains a valid reference for the $CO_2$-addition reaction. These results are consistent with experimental evidence that suggests no Michaelis complex [9]. After overcoming an activation barrier of 9 (7) kcal mol$^{-1}$ (Table 14.1; number1 (number2) are the B3LYP/6-311+G(2d,p)// 6-31G(d) and B3LYP/6-31G(d) results, respectively), the $\beta$-keto acid **3** is formed as a stable intermediate.

**Table 14.1** Energy differences and bond distances for the reaction steps of the Rubisco-catalyzed carboxylation of the 2,3-enediolate of RuBP.[a]

| Complex | B3LYP/6-31(d) (kcal mol$^{-1}$) | B3LYP/6-31+G(2d,p) //6-31(d)s (kcal mol$^{-1}$) | Distance C2-C3 (Å) |
|---|---|---|---|
| 1 | −1.9 | −2.3 | 1.35 |
| 2 | 5.0 | 7.1 | 1.38 |
| 3 | 0.0 | 0.0 | 1.51 |
| 4 | 9.3 | 12.1 | 1.54 |
| 5a | −3.4 | −2.2 | 1.54 |
| 5b | −4.4 | −4.8 | 1.53 |
| 8a | 24.8 | 28.0 | 2.40 |
| 8b | 28.0 | 32.3 | 2.00 |
| 10 | −30.9 | −29.2 | 4.67 |

a) The complex numbering refers to Figs. 14.6 and 14.7.

#### 14.3.1.3 Hydration Step

In the next step, the O atom of a water molecule H-bonded ($H_w$) to the carbamate is added to C3 to form a second TS complex **4**, which is 12 (9) kcal mol$^{-1}$ higher in energy than the β-keto acid. This hydration produces the *gem*-diol **5a**, in which $H_w$ has shifted to protonate O3.

#### 14.3.1.4 Sequential Addition of CO$_2$ and H$_2$O

In contradiction to suggestions in the literature for concerted additions [4, 39], we could not locate a transition structure in which the carboxylation and hydration reactions were concerted. The literature conjecture rests on the observation that the isolated β-keto acid (which is mostly in the unhydrated ketone form), when used as substrate, reacts only slowly [40], suggesting that this process is not on the catalytic pathway. However, the fact that inhibition by its reduced derivative 2′-carboxyarabinitol-P2 is characterized by an equally slow rate of inhibition [41] suggests an alternative explanation, namely, that the rate limitations arise from slow binding of these large molecules.

#### 14.3.1.5 Alternative Conformations of the *Gem*-diol

We discovered that two conformations of the *gem*-diol hydrate, with different metal coordination, are possible and, thus, that there are two possible pathways to continue the reaction sequence. In *gem*-diol **5a**, produced as described above and shown in the general configuration in Fig. 14.7, the relevant Mg ligands are one carboxyl O atom and the O atoms of the C2-OH group and one of the C3-OH groups. However, in the novel configuration **5b**, which is slightly more stable than **5a** (2 (1) kcal mol$^{-1}$; Table 14.1), both C3-OH groups are O-coordinated to Mg, while the C2 atom has rotated away from the Mg; modeling shows that this configuration could be accommodated by the active site. The activation energy for the transition from **5a** to **5b** was estimated to be ca. 10 kcal mol$^{-1}$ (O2 and O3 in *syn* position), which is significantly lower than the TS energy for the cleavage of the C-C bond (see below).

#### 14.3.1.6 C2-C3 Bond Cleavage: Pathway I

Two possible pathways for the C2-C3 bond cleavage were found, depending on the rotamer for the *gem*-diol. The pathway from **5a** requires a number of proton shifts during extension of the C2-C3 bond towards the TS **8a** (Table 14.1). There is an initial proton migration to the carbamate from the C3-OH group coordinated to Mg, followed by a further acid-base shift to protonate the non-Mg-coordinated oxygen of the carboxylate group and to deprotonate the Mg-coordinated C2-OH group. Lys334 and Lys175 are suitably positioned to effect the protonation and deprotonation, respectively, with the latter shift reversing the proton migration proposed for the enolization step. This produces a TS **8a** that is further stabilized by a proton migration from the non-Mg-coordinated C3-OH group back to O2. The activation energy for this pathway is calculated to be 30 (28) kcal mol$^{-1}$.

#### 14.3.1.7 C2-C3 Bond Cleavage: Pathway II

In the alternative pathway starting from *gem*-diol **5b**, a different sequence of proton shifts precedes the TS **8b**. These involve, as for pathway I, a migration of the O3 proton to the carbamate but, in contrast to pathway I, an *intramolecular* protonation of the non-Mg-coordinated O atom of the carboxyl from the second Mg-coordinated C3-OH group. These proton shifts occur when the C2-C3 bond distance is between 1.8 Å and 1.9 Å on the reaction coordinate. This leads to the bond cleavage TS **8b**. The calculated activation energy for this pathway is 37 (32) kcal mol$^{-1}$.

#### 14.3.1.8 Protonation of C2

The final step of the reaction to form two molecules of 3-phospho-D-glycerate (3-PGA) is the stereospecific protonation of C2 of the upper product (Fig. 14.6). For both pathways I and II, Lys175 is well positioned to act as the proton donor: this role is consistent with mutational evidence that side-chain deletion cripples this step [36]. However, while from our modeling Lys175 would be protonated at this stage for pathway I, it would be neutral for pathway II and we need to hypothesize that the proton would come indirectly from Lys175 through the H-bonded network from another proton donor, such as Lys334.

### 14.3.1.9 Dissociation of Products

The catalytic cycle is completed by dissociation of the two 3-PGA molecules. Although the X-ray structure of the product complex shows only one 3-PGA molecule coordinated to Mg [37], we modeled a product complex **10** with two molecules of 3-PGA bound to compare the energies. This structure is (too) strongly stabilized (−31 (29) kcal mol$^{-1}$; Table 14.1).

## 14.3.2
### Summary of Main Findings

1. The rate-limiting step of the enolization step that starts the reaction cycle is removal of the C3 proton by the carbamate group, which has enhanced basicity over simpler carboxylates.
2. Lys175 is suggested to be protonated before enolization commences and to act as the acid that protonates O2 to stabilize the enediol.
3. The carbamate appears to act as a general base at two further steps of the reaction, hydration and C-C cleavage.
4. $CO_2$ is added directly, without assistance of a Michaelis complex, and hydration of the resultant $\beta$-keto acid occurs in a separate subsequent step with a discrete transition state.
5. The finding of two possible conformations of the hydrate (*gem*-diol **5a** and **5b**), with different metal coordination, suggests two alternative pathways for C-C cleavage.
6. The step with the highest activation energy is the C-C bond cleavage. However, the special active-site arrangements of the metal coordination allow bond breaking at remarkably low activation energies (30–37 kcal mol$^{-1}$), which might be reduced further in the complete enzyme environment.

## 14.3.3
### QM/MM+MD Calculations

Overall the results of the fragment-complex calculations were consistent with experimental results, offering some convincing explanations for mutational data and energetic support for suggested roles of groups deduced from their active-site positions in X-ray structures. However, the size restriction of the fragment model did not allow us to discriminate alternative patterns of protonation, which are necessary to test our suggestions of groups outside the fragment acting as acids or bases, or to investigate the feasibility of possible mechanisms for channeling protons produced in the reaction away from the reaction site, e.g., the roles proposed for Lys175 in the enolization step, for His294 or carbamate in preparation of the enediolate for $CO_2$ addition, or for Lys175 and His334 in the C2 deprotonation step. Also, as will be apparent, a mechanism for removal of protons from the carbamate is necessary if it to act several times as a base in the reaction cycle.

Energetically, the results for the $CO_2$ addition and subsequent hydration steps were puzzling, as they would seem to favor the reverse decarboxylation reaction rather than progression to hydration and products (lower activation energy for former than for latter). This is a critical issue given that the fragment calculation results suggest two discrete steps rather than a concerted reaction that would irreversibly trap the $CO_2$, as suggested previously [4]. The experimental evidence suggests no decarboxylation: thus, although the $\beta$-keto acid is quite unstable and decarboxylates readily in solution, when fed to fresh $Mg^{2+}$-activated Rubisco, no decarboxylation was detected and virtually all of it was converted to 3-PGA product [40].

Consequently, we next focused on studying the effects of the enzyme environment on the stationary points (i.e., intermediates and TSs) found from the fragment calculations, using a combined quantum-mechanical/molecular-mechanical (QM/MM) model within MD simulations. In contrast to the fragment calculations that employed ab initio QM calculations, the less computationally expensive semi-empirical PM3 QM model was used for the QM/MM calculations. This makes possible both inclusion in the QM part of a much larger region surrounding the active site and computation of *free* energies by MD simulation averaging, i.e., accounting for entropic effects rather than merely enthalpic energies as given by the fragment calculations (see Fig. 14.2). However, the semi-empirical QM description is less accurate, and the calculated free energy differences with this QM/MM+MD model can be expected to be overestimated, based on our experience with other enzyme problems. This other work, however, has shown that the model is sufficiently reliable to distinguish possible reaction pathways [20, 21].

The QM/MM description allowed consideration of enzyme groups within the second coordination shell around Mg at the QM level, therefore, allowing proton shifts to or from these groups and the inner shell used previously for the fragment calculations. Thus, in addition to carbamoylated Lys201, Asp203, and Glu204, residues Lys175, Lys177, Lys334, and His294 were included. Also, the complete substrate molecule (i.e., five carbons with two terminal phosphates), rather than just a two-carbon fragment, was included in the QM region, allowing assessment of the protonation states of the phosphate groups in the active site and during the reaction. This expanded QM model allows us to compare the stabilities of possible tautomeric arrangements that can be reached via proton exchanges of the reaction center with neighboring side chains during the catalytic sequence.

We present some key initial results from these simulations (H. Mauser, L. Andrees, T.J. Andrews and J.E. Gready, unpublished results), with emphasis on the carboxylation (non-)reversibility puzzle.

### 14.3.3.1 $CO_2$ Addition: Early vs. Late Protonation of the Carboxylate

Previous experimental studies have shown that the absence of Lys334 specifically prevents the reaction steps after enolization and before processing the enediolate intermediate [42, 43], i.e., $CO_2$ and hydration steps. Hence, we investigated the possible role of this residue in proton transfer reactions in the active site. In our fragment calculations, we had found that protonation of the carboxylate group was necessary for a low-energy cleavage of the C-C bond and suggested that Lys334 was suitably positioned to act as a proton donor. X-ray structures (Fig. 14.3) show Lys334 in the direct neighborhood of the $\beta$-keto acid **3**, forming an H bond to the non-Mg-coordinated carboxylate oxygen. Now in the QM/MM calculations, where Lys334 is explicitly incorporated into the QM part of the model, we investigated the effect of this proton transfer at this late stage or at an early stage during $CO_2$ addition. Free energies for stationary points, numbered the same as for the fragment calculations, are given in Table 14.2: those for **2-H** (TS for $CO_2$ addition), **3-H** (intermediate $\beta$-keto acid), **4-H** (TS for hydration), and **5-H** (intermediate *gem*-diolate) correspond to the species with the carboxyl group already protonated (proton has moved from Lys334).

Consideration of these energies shows a dramatic effect on the likelihood of the reversal of $CO_2$ addition compared with forward progression of the reaction, depending on the presence of Lys334 and the carboxyl protonation state. Thus, the calculated free activation energies for the backwards reaction from **3-H** and **3** via unprotonated TS **2** are ca. 32 kcal mol$^{-1}$ and 17 kcal mol$^{-1}$, respectively. In contrast, hydration from **3-H** via TS structure **4-H** requires only ca. 13 kcal mol$^{-1}$, with only a small barrier (6 kcal mol$^{-1}$) for hydration of **3** via **4**. The differences of ca. –19 kcal mol$^{-1}$ and –11 kcal mol$^{-1}$ demonstrate that in the presence of the polarization by protonated Lys334, the hydration is favored over reversal of the $CO_2$ addition, allowing the reaction to progress. In the case of a full proton transfer from Lys334 (i.e., at **3-H** stage), the large difference in the activation energies suggests an almost irreversible $CO_2$ addition with the protonated $\beta$-keto acid **3-H** as favored intermediate. As may be seen from Table 14.2, proton transfer at the TS stage of the $CO_2$ addition to form **2-H**, although possible, leads to a much higher activation energy than **2** and is unlikely.

**Table 14.2** Relative free energies and C2-C3 bond distances for the Rubisco-catalyzed carboxylation of RuBP.

| Complex | Free energy[a] (kcal mol$^{-1}$) | Distance C2-C3[b] (Å) |
|---|---|---|
| 1 | –46±5 | 1.49 |
| 2 | 17±1 | 1.45 |
| 2-H | 39±0.4 | 1.57 |
| 3 | 0±2 | 1.54 |
| 3-H | –15±1 | 1.62 |
| 4 | 6±2 | 1.55 |
| 4-H | –2±1 | 1.71 |
| 5-H | –22±0.4 | 1.66 |

a) Energies are simulated over 240 ps (including 48-ps calibration time) and are equilibrated over 192 ps. All energies are relative to $\beta$-keto acid **3**.
b) The distance is averaged over the last 20 ps.

#### 14.3.3.2 Hydration of the β-Keto Acid

Table 14.2 shows that the free energy for the early-protonated TS (**4-H**) is decreased by 8 kcal mol$^{-1}$ compared with that for the late protonation, facilitating hydration to the *gem*-diol (**5-H**); however, both transition states are energetically accessible. After hydration the stable intermediate *gem*-diol (**5-H**) is formed: an alternative structure corresponding to *gem*-diol **5** with deprotonated carboxylate and protonated Lys334 was not stable during the MD simulations. However, it is known from mutation studies that catalysis can progress from the β-keto acid even when loop 6 containing Lys334 is deleted [43]. This indicates that protonation of the carboxylate by Lys334 is not required obligately for hydration, although the relatively low turnover rates suggest a higher barrier for hydration, as we find (**4** vs. **4-H**). Hence, we assume that in the absence of Lys334, the reaction pathway corresponds to the late protonation pathway via transition state **4**.

#### 14.3.3.3 His294 Protects Intermediates from Decarboxylation

While the early protonation route provides one mechanism to prevent decarboxylation, we were also interested in examining the role of His294, as Harpel et al. [38] had found that a His294 mutant preferentially catalyzes decarboxylation of carboxylated reaction intermediate instead of forward processing. Starting from the crystal structures, we found the most favorable orientation and protonation pattern to be when protonated at the ε-N, allowing it to be H-bonded to the backbone NH of Ala 296 and to O3 of RuBP, at the beginning of the cycle. We found that this arrangement persists throughout the simulations of the reaction cycle, with the strong H bond protecting O3 from protonation except very transiently before the C-C cleavage. Prevention of protonation of O3 effectively suppresses decarboxylation.

#### 14.3.3.4 The Tightly Coupled Active-site Environment

The Rubisco active site contains a complex arrangement of H-bond donors and acceptors, integrated into a tightly bound H-bond network involving as many as seven charged residues. Compared against the results for the fragment calculations, the QM/MM results demonstrate the necessity of considering this wider active-site environment: we have identified several residues that, although not directly involved in the reaction chemistry, assist it by directing functional groups or by directly or indirectly providing or channeling protons. Thus, intermediary protonation/deprotonation lowers transition states (**4-H** and also the C-C cleavage TS **8**) or assists in guiding the catalysis by quenching products to prevent the reverse reaction (**3-H**). In addition to these residue roles, we also found in the first steps of the catalysis (complexes **1**, **2**, **3-H**) a strong H-bond coupling between the carbamylated Lys201 and the P2-phosphate via the substrate water and RuBP-O4. This interaction could be responsible for recycling of the carbamate that was intermediately protonated during enolization. Additionally the low-energy barrier for proton transfer between Lys177 and Lys175 suggests that this interaction is a part of an extensive proton-transport network.

## 14.4
## Conclusions

Deciphering the Rubisco mechanism provides a huge challenge for both experiment and simulation. Definition of not one but several reactions catalyzed within the same active site, and which use and reuse residue groups in apparently different protonation states, is required. Many of these details are not accessible by experiment, while computing power and methodological issues still limit the size and quality of computer simulation that can be done. Nonetheless, using a staged multi-method approach, we have been successful in resolving some key questions raised by experiment and in making some experimentally testable predictions. Our proposals for roles of Lys175 in the carboxylation reaction have already been tested by others and are supported by two lines of evidence [44]. Perhaps more importantly, the simulations have generated some new ideas on the mechanism, such as alternative C-C bond cleavage pathways, and have imposed an energetic perspective that allows us to discriminate what is important in the mechanism in the Rubisco active site from other chemically plausible proposals. Following these initial studies for enthalpic and free-energy mapping of the reaction profile, we are continuing with further calculations and simulations to refine our models, including models with computationally mutated residues.

## Acknowledgments

The calculations summarized in this chapter were undertaken by postdoctoral fellows Dr. Bill King and Dr. Harald Mauser, with funding from an ANU Strategic Development Grant and a German DAAD Fellowship, respectively, and in collaboration with Prof. John Andrews, Research School of Biological Sciences, ANU. Large grants of supercomputing time from ANUSF/APAC are gratefully acknowledged.

## References

1 Andrews, T.J.; Lorimer, G.H. In *Photosynthesis*; Hatch, M.D.; Boardman, N.K.: Eds; The Biochemistry of Plants: A Comprehensive Treatise; Academic Press, New York, 1987, Vol. 10, pp. 131–218.
2 Hartman, F.C.; Harpel, M.R. *Annu. Rev. Biochem.* 1994, *63*, 197–234.
3 Gutteridge, S.; Gatenby, A.A. *Plant Cell* 1995, *7*, 809–819.
4 Cleland, W.W.; Andrews, T.J.; Gutteridge, S.; Hartman, F.C.; Lorimer, G.H. *Chem. Rev.* 1998, *98*, 549–561.
5 Spreitzer, R.J.; Salvucci, M.E. *Annu. Rev. Plant Biol.* 2002, *53*, 449–475.
6 Parry, M.A.; Andralojc, P.J.; Mitchell, R.A.; Madgwick, P.J.; Keys, A.J. *J. Exp. Bot.* 2003, *54*, 1321–1333.
7 Spreitzer, R.J. *Photosynth. Res.* 1999, *60*, 29–42.
8 Tabita, F.R. *Photosynth. Res.* 1999, *60*, 1–28.
9 Roy, H.; Andrews, T.J. (2000) in *Photosynthesis: Physiology and Metabolism*; Leegood, R.C.; Sharkey, T.D.; von

10 Andersson, I.; Taylor, T. C. *Arch. Biochem. Biophys.* 2003, *414*, 130–140.
11 Portis, A. R., Jr. *Annu. Rev. Plant Physiol. Plant Mol. Biol.* 1992, *43*, 415–437.
12 Portis, A. R., Jr. In *Protein-Protein Interactions in Plant Biology*, Vol. 7. McManus, M. T.; Laing, W. A.; Allen, A. C.; Eds; Sheffield Academic Press, Sheffield, 2001, pp. 30–52.
13 Blow D. *Structure Fold. Des.* 2000, *8*, R77–81.
14 Warshel, A.; Florian, J. *Proc. Natl. Acad. Sci. USA* 1998, *95*, 5950–5595; Villa, J.; Warshel, A. *J. Phys. Chem. B* 2001, *105*, 7887–7907.
15 Ma, B.; Kumar, S.; Tsai, C. J.; Hu, Z.; Nussinov, R. *J. Theor. Biol.* 2000, *203*, 383–397.
16 Zhuang, X.; Rief, M. *Curr. Opin. Struct. Biol.* 2003, *13*, 88–97.
17 Benkovic, S. J.; Hammes-Schiffer, S. *Science* 2003, *301*, 1196–1202.
18 Wang, W.; Donini, O.; Reyes, C. M.; Kollman, P. A. *Annu. Rev. Biophys. Biomolec. Struct.* 2001, *30*, 211–243.
19 Stewart, J. J. P. *Int. J. Quant. Chem.* 1996, *58*, 133–146.
20 Cummins, P. L.; Gready, J. E. *J. Comput. Chem.* 1998, *19*, 977–988.
21 Cummins, P. L.; Greatbanks, S. P.; Rendell, A. P.; Gready, J. E. *J. Phys. Chem. B* 2002, *106*, 9934–9944.
22 Cummins, P. L.; Gready, J. E. *THEOCHEM* 2003, *632*, 245–255.
23 Frisch, M. J. et al. (29 authors) Gaussian 94, 1994, Gaussian Inc., Pittsburgh PA.
24 Frisch, M. J. et al. (56 authors) Gaussian 98, 1998, Gaussian Inc., Pittsburgh, PA.
25 King, W. A.; Gready, J. E.; Andrews, T. J. *Biochemistry* 1998, *37*, 15414–15422.
26 Mauser, H.; King, W. A.; Gready, J. E.; Andrews, T. J. *J. Am. Chem. Soc.* 2001, *123*, 10821–10829.
27 Stewart, J. J. P. *J. Comput. Chem.* 1991, *12*, 320–341.
28 Cornell, W. D.; Cieplak, P.; Bayly, C. I.; Gould, I. R.; Merz, K. M.; Ferguson, D. M.; Spellmeyer, D. C.; Fox, T.; Caldwell, J. W.; Kollman, P. A. *J. Am. Chem. Soc.* 1995, *117*, 5179–5197.
29 Jorgensen, W. L.; Chandrasekhar, J.; Madura, J. D.; Impey, R. W.; Klein, M. L. *J. Chem. Phys.* 1983, *79*, 926–935.
30 Andersson, I. *J. Mol. Biol.* 1996, *259*, 160–174.
31 Siegbahn, P. E. M.; Blomberg, M. R. A. *Annu. Rev. Phys. Chem.* 1999, *50*, 221–249.
32 Aqvist, J.; Medina, C.; Samuelsson, J. E. *Protein Eng.* 1994, *7*, 385–391.
33 Newman, J.; Gutteridge, S. *J. Biol. Chem.* 1993, *268*, 25876–25886.
34 Lorimer, G. H.; Hartman, F. C. *J. Biol. Chem.* 1988, *263*, 6468–6471.
35 Schloss, J. V. In *Enzymatic and Model Carboxylation and Reduction Reactions for Carbon Dioxide Utilization*; Aresta, M.; Schloss, J. V.: Eds; Kluwer Academic Publishers, Dordrecht, 1990, pp. 321–345.
36 Harpel, M. R.; Hartman, F. C. *Biochemistry* 1996, *35*, 13865–13870.
37 Taylor, T. C.; Andersson, I. *Biochemistry* 1997, *36*, 4041–4046.
38 Harpel, M. R.; Larimer, F. W.; Hartman, F. C. *Protein Sci.* 1998, *7*, 730–738.
39 Cleland, W. W. *Biochemistry* 1990, *29*, 3194–3197.
40 Pierce, J.; Andrews, T. J.; Lorimer, G. H. *J. Biol. Chem.* 1986, *261*, 10248–10256.
41 Pierce, J.; Tolbert, N. E.; Barker, R. *Biochemistry* 1980, *19*, 934–942.
42 Lorimer, G. H.; Chen, Y. R.; Hartman, F. C. *Biochemistry* 1993, *32*, 9018–9024.
43 Larson, E. M.; Larimer, F. W.; Hartman, F. C. *Biochemistry* 1995, *34*, 4531–4537.
44 Harpel, M. R.; Larimer, F. W.; Hartman, F. C. *Biochemistry* 2002, *41*, 1390–1397.

# 15
# Carbon-based End Products of Artificial Photosynthesis

*Thomas D. Sharkey*

## 15.1
## Introduction

As we contemplate artificial photosynthesis, we also need to contemplate what products should be made. Photosynthesis can be defined as simply the production of high-energy intermediates by light-driven electron transport, but in plants photosynthesis is normally thought of as involving production of a somewhat stable end product, primarily in the form of a sugar. Photosynthesis is a way of capturing light energy and storing that light energy on carbon-based end products. While most attention is rightfully focused on the immediate conversion of light energy into chemical energy, the full picture can be represented as carbon inputs, carbon processing, and then end-product synthesis for storage (Scheme 15.1). In

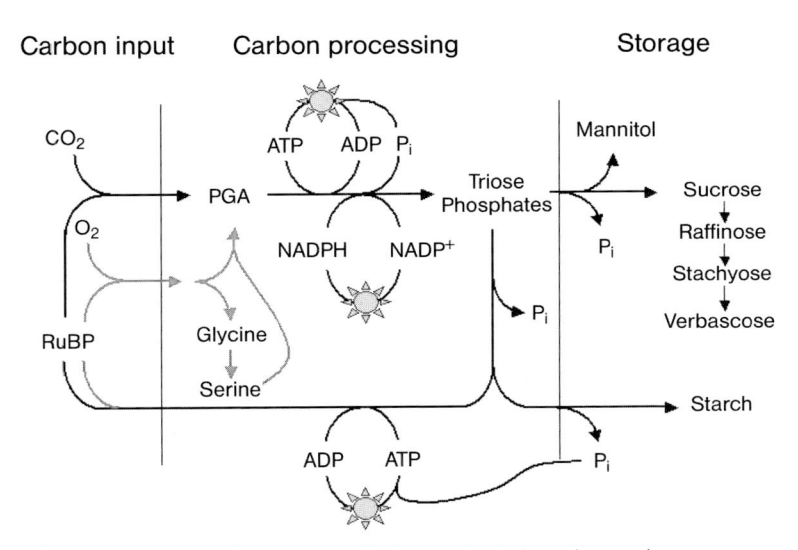

**Scheme 15.1** Three major processes of photosynthesis from the standpoint of storing chemical energy on carbon.

*Artificial Photosynthesis: From Basic Biology to Industrial Application*
Edited by Anthony F. Collings and Christa Critchley
Copyright © 2005 WILEY-VCH Verlag GmbH & Co. KGaA, Weinheim
ISBN: 3-527-31090-8

this chapter I shall describe some of the well-known and some of the lesser-known end products of photosynthesis and then consider whether photosynthesis is ever limited by the production of end products (as opposed to being limited by light or $CO_2$). Finally I will suggest that isoprenoids, relatively abundant carbon-based end products of photosynthesis, have certain advantages in an artificial photosynthesis scenario.

## 15.2
### What Are the End Products of Plant Chloroplast Photosynthesis?

Most textbooks suggest that photosynthesis produces glucose. This is not chemically correct. In plants, photosynthesis takes place in chloroplasts, compartments of the plant cell that are likely endosymbionts and are closely related to bacteria. Bacteria as a group specialize in biochemical flexibility, and photosynthesis is closely associated with bacteria. For this reason one of the first places to look at the end product of photosynthesis is at the product supplied by the symbiont (chloroplast) to the host (the rest of the plant). This product is primarily a three-carbon sugar with phosphate attached (triose phosphate), specifically dihydroxyacetone phosphate. The symbiont exchanges this compound for phosphate from the host. The plant converts the triose phosphate to sucrose, which is then exported around the plant for growth and respiration.

There are a number of other carbon-based end products. Second in terms of amount is starch. Starch is stored inside chloroplasts temporarily and exported at night. All major groups of plants store starch during the day and break it down at night. This provides a mechanism for supplying reduced carbon (sugars) at a constant rate regardless of the input; in other words, products of photosynthesis are exported from chloroplasts continuously despite the fact that the energy to drive photosynthesis is available for only one half of each day. By storing starch, plants can allow photosynthesis to proceed faster during the day than the products of photosynthesis can be used [1]. Since any light not immediately used for photosynthesis is lost forever, this helps plants to take advantage of even very high light levels. This storage can happen over many timescales; for example, the plastoquinone pool in the electron transport chain can provide storage relevant to the millisecond timescale [2]. Starch storage is relevant to storage over 24 hours.

The export of reduced carbon from chloroplasts at night, as a result of starch breakdown, is very different from export during the day. Instead of triose phosphates, chloroplasts export maltose plus some glucose at night [1, 3]. It is not clear why export from chloroplasts is different at night than during the day, but one possibility is the opportunity for additional buffering of the availability of sugar [4]. Regardless of how the carbon is exported from the chloroplast, most is used to make sucrose (Scheme 15.2).

Sugar is the primary product of photosynthesis in terms of amount, but chloroplasts make many other important compounds including several vita-

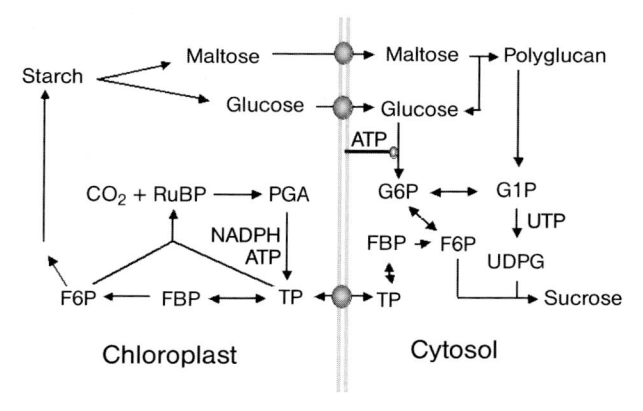

**Scheme 15.2** Carbon dioxide is made into sucrose and starch during the day. Abbreviations: F6P=fructose 6-phosphate; RuBP=ribulose 1,5-bisphosphate; FBP=fructose 1,6-bisphosphate; PGA=3-phosphoglyceris acid; TP=triose phosphate (dihydroxyacetone phosphate plus glyceraldehyde 3-phosphate); G6P=glucose 6-phosphate; ATP=adenosine triphosphate; UTP=uridine triphosphate.

mins, all of the essential amino acids (those amino acids humans cannot make and thus must obtain though diet), and all of the fatty acids in the plants (needed to make fats and oils). In particular, chloroplasts make isoprene (2-methyl-1,3-butadiene) and larger isoprenoids, sometimes in substantial amounts. Global production of isoprene is estimated to exceed 400 Tg yr$^{-1}$ [5].

## 15.3
### Does End-product Synthesis Ever Limit Photosynthesis?

How important is it to consider the production of the end products? Can the production of the end products limit the overall rate of photosynthesis? There is evidence that photosynthesis can be limited by end-product synthesis but not by light or $CO_2$ supply [6]. This occurs most often when photosynthesis is proceeding at high rates. Plants with plenty of nutrients, light, and $CO_2$ can become limited by end-product synthesis, and, because the metabolic pathways for starch and sucrose synthesis are highly temperature sensitive, end-product synthesis limitation is more likely to occur at low temperature than at high temperature. This is exemplified by photosynthesis of celery leaves (Fig. 15.1). At low $CO_2$, photosynthesis is the same at the three temperatures tested, but at about 30 Pa, photosynthesis no longer responds to increasing $CO_2$, indicating that end-product synthesis is now limiting. When light limits photosynthesis, a slight $CO_2$ sensitivity remains because of the oxygen- and $CO_2$-sensitive process of photorespiration. The open symbols for each line are data obtained in low oxygen (2% instead of 21% oxygen). At the two lower temperatures the low-oxygen measurement is below the line, con-

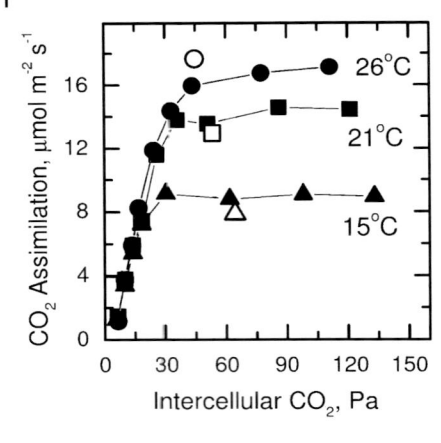

**Fig. 15.1** Rate of $CO_2$ assimilation (photosynthesis) of celery leaves as a function of $CO_2$ partial pressure inside the leaves at three temperatures. Unpublished data of Sharkey and Loescher.

firming that photosynthesis was limited by end-product synthesis. Through many years of evolution, the capacity for photosynthesis and the capacity to form end products are normally well matched, and end-product synthesis limitation occurs rarely under natural conditions (but it does occur). The lesson for designing an artificial photosynthesis system is that the products must be chosen and the capacity to make those products must be developed so that the system does not become limited by the ability to use the products.

## 15.4
### What Would Be a Desirable Carbon-based End Product of Photosynthesis?

One end product envisioned for artificial photosynthesis is hydrogen. However, it is possible that problems associated with storage and transport of hydrogen [7] mean it will be easier to use a carbon-based molecule that can be efficiently reformulated into hydrogen and carbon at the point of use. Since any carbon released in this process would have been taken up in the original manufacture, this system would not be a net source of carbon dioxide. This, in essence, is what is done in natural photosynthesis; oxidation-reduction energy is stored on carbon in the form of sugar, and the energy is converted at the point of use by metabolic reactions. Reformulation of hydrogen is showing great promise, and highly efficient methods have been reported [8].

If it is decided that a carbon-based end product is desirable there are several considerations:

1. The compound should be one that is relatively easy to make from a biochemical perspective.
2. The compound should have value as a fuel.
3. The compound should have development potential; for example, it should have value in its own right so that small-scale production might be profitable and lead to larger-scale production.

4. The compound should allow for continuous production. One of the limitations of natural photosynthesis is that a crop is grown and then harvested and processed, with each step taking substantial energy and thus reducing the net yield.

There are three major groups of compounds that can be considered: sugars, fatty acid-based products, and isoprenoids. The normal carbon-based end products of natural photosynthesis are sugars, or more broadly, carbohydrates. Sugars and other carbohydrates made by plants have value as foods as well as raw materials for fuels, e.g., by hydrogen reformulation. However, as currently processed, sugars cannot be easily adapted to continuous production because it is hard to separate sugars from the enzymes that make them. One exception to this would be the growth of starch granules big enough to be physically removed from solution. To make a liquid fuel, sugars are typically fermented to ethanol. Currently, ethanol production from corn has improved in efficiency and now can provide a net positive energy value of just over 30% of the energy required to produce it [9].

A second major group is compounds based on the malonyl-CoA pathway, especially fatty acids and their derivatives. This includes fats and oils (triglycerides) as well as plant-derived plastics such as polyhydroxybutyrate and, more generally, polyhydroxyalkanoates. Soybean oil can easily be used in place of diesel fuel, although this requires methylation by addition of an alcohol, typically methanol. This is required to remove the glycerol component of the natural vegetable oil. Once esterified, biodiesel can be used in place of petroleum diesel. However, production of the soybean oil requires investment of fossil fuels of about 30% of the fuel value of the soybean oil [10]. Nevertheless, the situation is more favorable than the production of ethanol from carbohydrates as described above. Thus, soybean oil has advantageous properties in that it has a high value as a food product but can also be readily adapted as a fuel. As currently produced, oil production in plants is not a continuous process.

Plants have been engineered to make the fatty acid derivative polyhydroxybutyrate. When this synthesis occurred in the cytosol of cells, it caused the cells to grow slowly. More recently, polyhydroxybutyrate synthesis was targeted to chloroplasts, and polyhydroxybutyrate accumulated to high levels without stunting the growth of the plants [11]. However, polyhydroxybutyrate cannot be handled as a liquid fuel and might be difficult to adapt to a continuous process.

Both oils and plastics (polyhydroxybutyrate) require biochemically complex synthesis of fatty acids. Thus, while these may be suitable end products in plants, they may not be as well suited as end products of artificial photosynthesis.

The third major category has not been developed as an energy source up to now but might have potential. The isoprenoid family of compounds is common hydrocarbons in plants. One branch of this family is the sterols such as cholesterol and many hormones, which likely are not suited as end products of artificial photosynthesis. However, the terpene branch of the family may have some potential. In this branch, hydrocarbons are made from allylic precursors whose lengths are multiples of five carbons. The first compound has just five carbons. Isoprene ($C_5H_8$) is made from dimethylallyl diphosphate. Once thought not to occur naturally [12], isoprene is more common in plants than any other hydrocarbon [13, 14]. Isoprene is made by a recently discovered pathway inside chloroplasts, whose capacity is surprisingly high [5]. Isoprene is a liquid at room temperature but has the disadvantage of boiling at 32 °C. Because of its low solubility in water, isoprene would come into the gas phase above a solution in which it was being synthesized, allowing for continuous production. Currently isoprene is made from petroleum, and more than $10^5$ kg is consumed each year in the manufacture of rubber and pharmaceuticals [15]. Thus, there currently exists a market for high-quality, high-value isoprene. The enzymes that plants use to make isoprene have all been cloned, providing the raw material for designing and making a production system for isoprene.

The terpene family contains larger molecules. The 10-carbon molecule geranyl diphosphate can be converted into any of hundreds of compounds called monoterpenes. The simplest acyclic monoterpenes are ocimene and myrcene. Cyclized monoterpenes are the very familiar, pleasant smells of plants such as lemon and pine scent. Monoterpenes would have many of isoprene's good points, but with less volatility. The next class is the sesquiterpenes, which have 15 carbons and even less volatility.

Isoprene and monoterpenes are made by the recently described methyl erythritol 4-phosphate pathway. All of the steps of this pathway have been worked out recently. The pathway is present in plants and in most bacteria, providing many potential sources of enzymes for an artificial system. Isoprene is made directly from one of the products of the methyl erythritol pathway [16]. The gene for the enzyme that does this, isoprene synthase, has recently been cloned and is available. Making monoterpenes requires an additional step: a prenyl transferase builds the 10-carbon precursor on which monoterpene synthases can act. Monoterpene synthases are very closely related to isoprene synthase.

Terpenes have the advantage of being in high demand for cosmetic products (high value for high-purity products) and rubber production (slightly lower value, but purity can also be slightly lower). Terpenes are flammable, but many of the common terpenes are unsaturated, possibly making hydrogen reformulation more difficult at present. However, little work on hydrogen production from terpenes appears to have been done up to now.

In summary, there are three major types of end products that might be considered. Artificial photosynthesis is not a reasonable alternative to natural photosynthesis for food production, but artificial photosynthesis can be considered for production of industrial raw materials and fuels. Carbohydrates and oils (triglycerides) have been used as industrial end products. In the future, the potential of the terpenes to be used as fuels might profitably be explored.

## Acknowledgments

Work in my laboratory on feedback limitations of photosynthesis was supported by the United States Department of Energy. Studies of isoprene emission from plants were supported by the National Science Foundation and by the United States Department of Agriculture National Research Initiative.

## References

1 Weise, S. E., Weber, A., Sharkey, T. D., *Planta* 2004, *218*, 474–482.
2 Tennessen, D. J., Bula, R. J., Sharkey, T. D., *Photosynth. Res.* 1995, *44*, 261–269.
3 Nittylä, T., Messerli, G., Trevisan, M., Chen, J., Smith, A. M., Zeeman, S. C., *Science* 2004, *303*, 87–89.
4 Lu, Y. and Sharkey, T. D., *Planta* 2004, *218*, 466–473.
5 Sharkey, T. D. and Yeh, S., *Annu. Rev. Plant Physiol. Plant Mol. Biol.* 2001, *52*, 407–436.
6 Sharkey, T. D., *Bot. Mag. Tokyo* 1990, special issue 2, 87–105.
7 Keith, D. W. and Farrell, A. E., *Science* 2003, *301*, 315–316.
8 Deluga, G. A., Salge, J. R., Schmidt, L. D., Verykios, X. E., *Science* 2004, *303*, 993–997.
9 Shapouri, H., Duffield, J. A., Wang, M. 2002. *www.usda.gov/agency/oce/oepnu/aer-814.pdf*, United States Department of Agriculture. Agricultural Economic Report.
10 Sheehan, J., Camobreco, V., Duffield, J. A., Gabroski, M., Shapouri, H. 1998. *www.nrel.gov/docs/legosti/fy98/24772.pdf*, National Renewable Energy Laboratory.
11 Nawrath, C., Poirier, Y., Somerville, C., *Proc. Nat. Acad. Sci.* 1994, *91*, 12760–12764.
12 Kaufman, P. B., Cseke, L. J., Warber, S., Duke, J. A., Brielmann, H. L., Natural Products from Plants (CRC Press, Boca Raton, 1999), pp. 1–343.
13 Guenther, A., Hewitt, C. N., Erickson, D., Fall, R., Geron, C., Graedel, T., Harley, P., Klinger, L., Lerdau, M., McKay, W. A., Pierce, T., Scholes, B., Steinbrecher, R., Tallamraju, R., Taylor, J., Zimmerman, P., *J. Geophys. Res.* 1995, *100*, 8873–8892.
14 Fuentes, J. D., Lerdau, M., Atkinson, R., Baldocchi, D., Bottenehiem, J. W., Ciccioli, P., Lamb, B., Geron, C., Gu, L., Guenther, A., Sharkey, T. D., Stockwell, W., *Bull. Amer. Met. Soc.* 2000, *81*, 1537–1575.
15 Weitz, H. M. and Loser, E., in *Ullman's Encyclopedia of Industrial Chemistry*, Vol. A 14. B. Elvers, S. Hawkins, M. Ravenscroft, G. Schulz, Eds. VCH, Weinheim, 1989.
16 Silver, G. M. and Fall, R., *Plant Physiol.* 1991, *97*, 1588–1591.

# 16
# The Artificial Photosynthesis System: An Engineering Approach

*Dilip K. Desai*

## 16.1
## Introduction

Man-made emission of $CO_2$ into the earth's atmosphere presently exceeds $10^{10}$ tons $yr^{-1}$ [1]. Atmospheric $CO_2$ accumulation and the resultant global warming are now well recognized as global problems. The governments of many nations agreed in 1996 (Kyoto) to take action to curtail $CO_2$ emissions. To this end, scientists all over the world are trying to develop technologies for capture, sequestration, and utilization of $CO_2$ [2–4].

The artificial photosynthesis system (APS) has been proposed as a means of not only curtailing $CO_2$ emissions but also converting $CO_2$ into some useful products [5]. This chapter explores a practical realization of this concept from an engineer's perspective. The potential benefits of such efforts are envisaged to be superior, "intensified" technologies for the production of (1) nutraceuticals (omega-3 acids, carotenoids, xanthins, etc.), (2) chemical feedstock (isoprene), (3) green fuel (methane, ethanol, hydrogen, etc.), (4) specialized aquaculture feed, and (5) high-protein livestock feed.

## 16.2
## Engineering Approach to APS

As observed in plants, the basic photosynthesis reaction is [6a]

$$CO_2 + H_2O + \text{Sunlight} + \text{Nutrients} \rightarrow \text{Carbohydrates} + O_2 \qquad (1)$$

APS refers to any man-made arrangement that carries out this basic reaction in a manner or setting that is different from that of nature. An engineering or industrial realization of APS implies intensification and widespread use (and/or large-scale application) of Reaction (1), so that potentially thousands of tons of $CO_2$ per year could be converted into useful products.

*Artificial Photosynthesis: From Basic Biology to Industrial Application*
Edited by Anthony F. Collings and Christa Critchley
Copyright © 2005 WILEY-VCH Verlag GmbH & Co. KGaA, Weinheim
ISBN: 3-527-31090-8

## 16.3
### Elements of the Engineering Approach

For the evaluation (development) of an APS system, some of the important issues are [7]: (1) economic value; (2) functional specifications; (3) system choice; (4) system development, including (a) modules development, (b) integration of modules, and (c) lab-scale testing; (5) scale-up; and (6) implementation.

### 16.3.1
### Economic Value

The viability of an APS will largely be determined by the economic and environmental benefit it could generate for society and by the safety, health, and environmental acceptability of the system. It is apparent that

1. the value of the APS system will essentially be generated by its potential impact on the issues of greenhouse gas buildup, energy production, sustainable food production and water use faced by the world;
2. energy and food needs are currently met by fossil fuels, forests, and farms; and
3. APS needs to overcome the inherent limitations of natural photosynthesis systems to improve upon the photosynthesis of farms and forests.

### 16.3.2
### Limitations of Natural Photosynthesis Systems (NPS)

#### 16.3.2.1 Speed of NPS
It is evident that natural photosynthesis is a very slow process. The primary reason is that the rate of capture of $CO_2$ by plants is limited by (1) plant physiology and anatomy and (2) the surrounding climatic and soil conditions that are not readily controllable. By one estimate [8], this rate is thought to be $<0.1$ kg m$^{-3}$ day$^{-1}$. To justify investment in its development, an APS should enhance the rate of $CO_2$ capture substantially beyond this figure. It could then be termed an *accelerated* photosynthesis system.

#### 16.3.2.2 Energy Efficiency of NPS
Fig. 16.1 shows intervals of typical values for the efficiency of solar energy utilization by biomass as estimated by Sorenson [9].

It is interesting to note that the global average value is estimated to be less than 0.25%. It is also interesting that in some cases this efficiency could be improved by interventions such as optimization of conditions for alga and nutrient subsidy for corn. Thus, APS development should target a solar energy utilization efficiency of greater than 2% to offer substantial advantage.

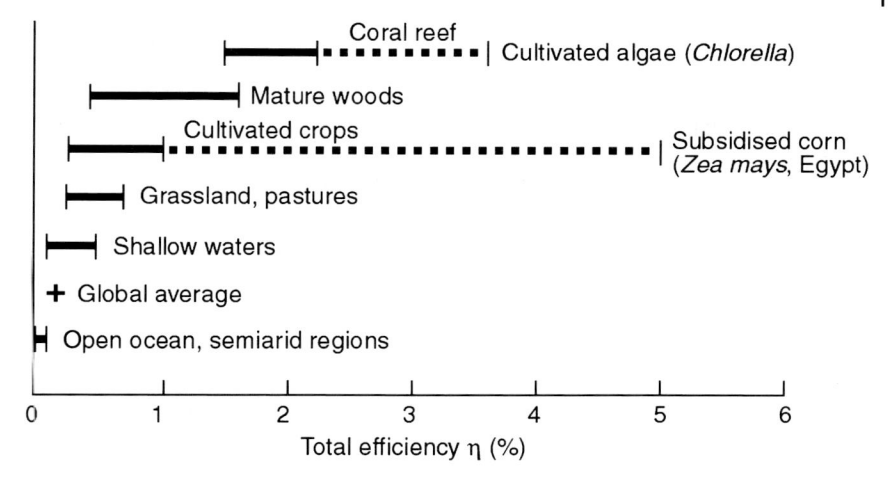

**Fig. 16.1** Typical values of total photosynthetic efficiency for different plants and communities [9].

#### 16.3.2.3 Water Requirements of NPS

Plants in farms and forests use a large amount of water per gram of dry biomass produced. For instance, wheat plants use more than 200 L water per kilogram of dry mass [10a] and cotton uses more than 1000 L water per kilogram of cotton [5]. It is interesting to note that the basic photosynthesis reaction (Reaction (1)) requires only one molecule of water per molecule of $CO_2$. APS should target substantially less water usage.

#### 16.3.2.4 Land Use for NPS

Forests and agriculture require large amounts of arable land to produce one kilogram of biomass. Table 16.1 shows the typical annual productivity of forests and crops per square meter of land [10b]. It is evident that to produce the crops required to feed nations, vast areas of land are needed. With the increase in the earth's population, land available per capita has been in continual decline. An APS would need far less land and would make use of non-agricultural, non-forestry land.

### 16.3.3
### Scale of Operation

The other factor influencing economic viability is the scale of operation, the well-known "economy of scale" effect. The exact scale of the operation will depend upon many factors, including the geographic location, what the primary product is, the size and vicinity of the market, the technological production capability of the plant, etc. As a starting point, the APS system should aim at $10^5$ t yr$^{-1}$ of $CO_2$ capture. This could be achieved by either a large number of small operations or vice versa.

**Table 16.1** Typical productivity of various natural photosynthesis systems [10b].

| Ecosystem | Annual productivity (g dry matter $m^{-2}$ $yr^{-1}$) |
|---|---|
| Tropical | |
|   Perennial crops | 8000 |
|   Rainforest | 3500 |
|   Annual crops | 3000 |
| Temperate | |
|   Perennial crops | 3000 |
|   Annual crops | 2000 |
|   Grassland | 2000 |
|   Evergreen forest | 2000 |
|   Deciduous forest | 1500 |
|   Savanna | 1000 |
| Arctic and arid | |
|   Desert | 100 |

### 16.3.4
### Functional Specification

To demonstrate the economic viability of the APS concept, realistic and achievable targets need to be set for the process intensification. These targets might need to be modified as the project progresses. However, to start with they represent the project team's perception of what is realistically achievable.

The following targets are suggested for the APS system: (1) a $CO_2$ fixation rate of $>5$ kg $m^{-3}$ $day^{-1}$; (2) an energy efficiency of $>5\%$; (3) water usage of $<10–100$ mole $H_2O$ per mole $CO_2$; (4) use of non-agricultural, non-forestry land; and (5) production of non-agricultural, non-forestry products. These functional specifications may be encapsulated in the following vision.

## 16.4
## Elements of Envisaged System

1. The gas pretreatment unit: The $CO_2$ is envisaged to be collected from the stack of an operating factory such as a power plant or a small- to medium-sized boiler. Such a stack gas would normally also have nitrogen, water vapor, solid particles, and other gaseous impurities such as NOx, SOx, etc. This unit may have to be capable of purifying and separating $CO_2$ out of such a gas. Technologies for this are well established [11, 12]. The energy of $CO_2$ separation from the stack gas is 85–125 kJ per mole of $CO_2$, which is about 6–9% of the energy required for the photosynthesis [6b].
2. The reactor: The concept of acceleration and intensification of Reaction (1) will be embodied in the reactor design. This is the subject of this report.

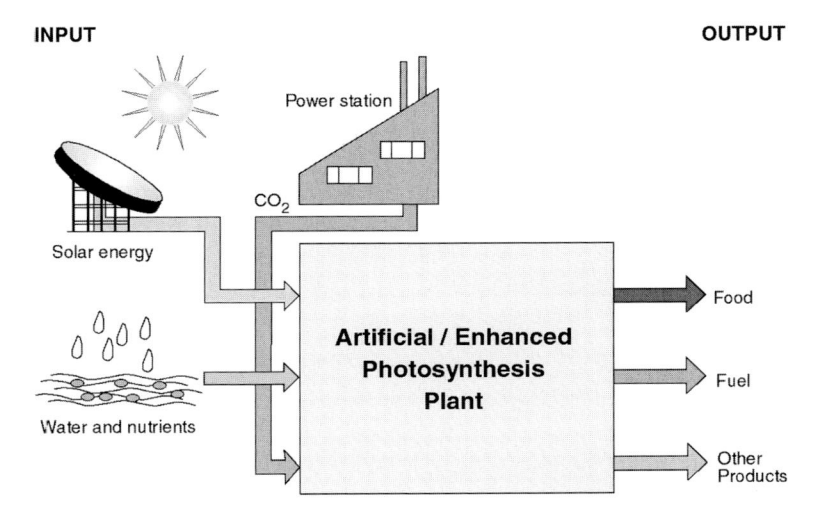

**Fig. 16.2** Vision for APS system.

3. The microorganism: The choice and design of microorganism will depend upon the process and upon the final product sought. So far the author has tried the cyanobacteria *Synechococcus* PCC 7002. The desired product is a high-protein livestock feed and the biomass is to be used in making biofuels.
4. The separators: There will be a need to separate the gas, solid, and liquid phases involved in the process from each other.
5. Sunlight collection and distribution: As shown in Fig. 16.2, sunlight utilization is an essential element of the APS system. The system design will require novel, efficient, and relatively inexpensive ways of sunlight collection and distribution.
6. Product extraction: Depending upon the details of the scheme, a number of different products, mentioned in the introduction, could be produced from the plant.
7. Process integration: This is the integration of the individual elements to produce the desired system output.

## 16.5 Cyanobacteria

Cyanobacteria *Synechococcus* PCC 7002 have been chosen as the microorganisms for the photo-bioreactor. The main reasons for this choice are as follows.

1. They have a role in nature to fix $CO_2$ into the biosphere. In fact, they are estimated to fix roughly one-third of the global $CO_2$ back into the biosphere.

2. They have a feature called the "carbon dioxide concentration mechanism" (CCM) that allows them to build nearly 1000-fold higher $CO_2$ concentrations inside the cell than outside the cell. This leads to one of the best-known $CO_2$ fixation rates available among naturally occurring organisms. The operation of the CCM in cyanobacteria is explained by Price et al. [13].
3. They have a well-understood genome that would facilitate the further improvement of the strain when required.

## 16.6
### Photo-bioreactor

The photo-bioreactor has attracted considerable research interest in the past decade as described by Richmond [14], Pulz [15], and Ogbonna [16]. There are several configurations of closed photo-bioreactors being investigated [15]: tubular, helical, flat plate, and tanks. The author's group has seen the most potential in the flat-plate reactors, with ease of scale-up being an important consideration. Fig. 16.3 shows a photo and a schematic of the photo-bioreactor in the author's lab [17].

## 16.7
### Theory

Experimental efforts have been supplemented by development of a mathematical model. Some of the important factors that affect the performance of such a system are (1) light availability within the reactor, (2) chemical conditions (e.g., pH, concentrations of various species, etc.), (3) temperature, and (4) flow conditions.

To construct the model, the reactor in the Fig. 16.3 is assumed to be a well-mixed, continuously stirred tank reactor (CSTR) operating in a steady-state isothermal mode. The conditions are maintained such that photoinhibition is negligible and cellular growth is limited by the availability of the light. Under such conditions, the main equations describing the operation of the reactor are as follows [17]:

$$\text{Exp}\{-A_1 C\} + A_2 C^2 = A_3 C + 1 \tag{2}$$

$$A_1 = L_1 K \tag{3}$$

$$A_3 = A_2 C_{in} \tag{4}$$

$$A_2 = QKL_1 / [VkI_0 A_i] \tag{5}$$

$$RCO_2 = d[C - C_{in}]Q/V \tag{6}$$

where
- $L_1$ = characteristic length of the reactor (m)
- $k$ = reaction rate constant (g m$^{-3}$ μE$^{-1}$)
- $Q$ = liquid flow rate (m$^3$ s$^{-1}$)
- $V$ = occupied reactor volume (m$^3$)
- $I_0$ = irradiance on the reactor surface (μE m$^{-2}$ s$^{-1}$)
- $A_i$ = area of irradiated surface (m$^2$)
- $C, C_{in}$ = concentrations of the cells in the reactor and in the input stream (g m$^{-3}$)
- $d$ = conversion factor (s d$^{-1}$)
- $K$ = extinction coefficient (m$^2$ g$^{-1}$)
- $RCO_2$ = $CO_2$ uptake rate (g m$^{-3}$ d$^{-1}$)

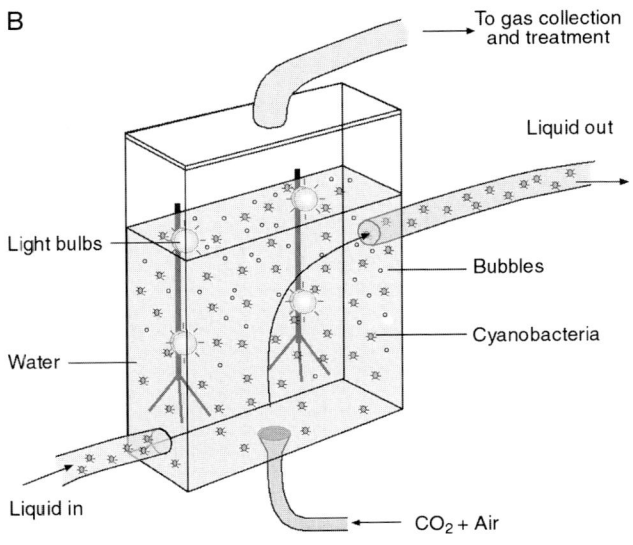

**Fig. 16.3** Photograph (A) and schematic (B) of photo-bioreactor at CSIRO [17].

## 16.8
## Results

The results of the experiments on this configuration look encouraging, as cell densities higher than 4 g (dry mass) $L^{-1}$ and productivities higher than forests plants have been achieved.

The results of the model are shown in Figs. 16.4 and 16.5. The model indicates that by properly adjusting the reactor design and operating parameters (such as length $L_1$ and inlet concentration $C_{in}$), it is possible to enhance the $CO_2$ uptake rate.

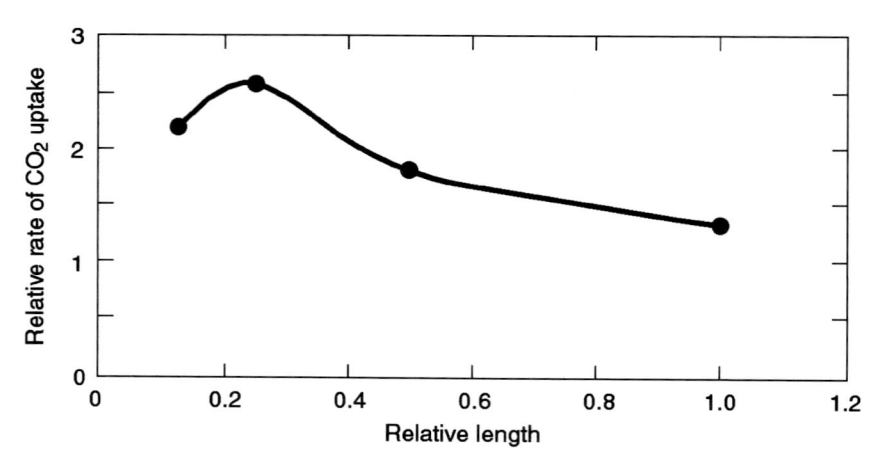

**Fig. 16.4** Model prediction of variation of $CO_2$ uptake rate with length $L_1$ [17].

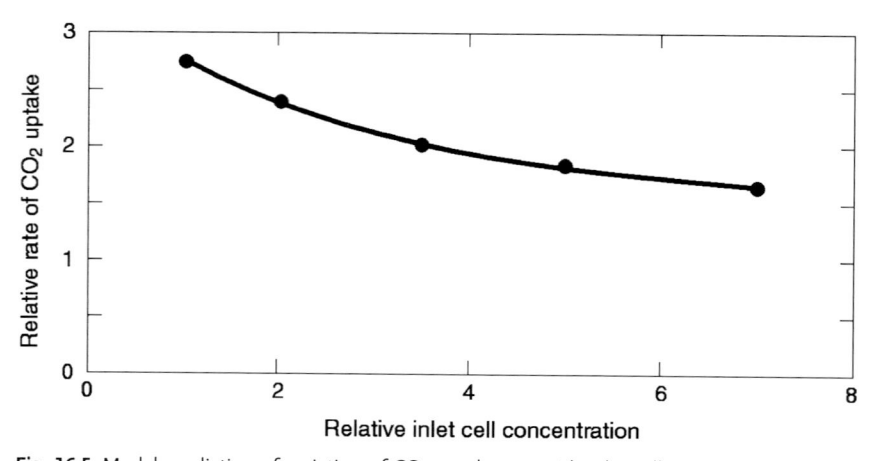

**Fig. 16.5** Model prediction of variation of $CO_2$ uptake rate with inlet cell concentration [17].

## 16.9
## Conclusions

An engineering approach to the concept of an APS has been developed. The conclusions of the initial investigations are as follows:

1. The APS should include production of useful non-agricultural materials with higher than 5% energy efficiency, with higher than 5 kg $CO_2$ per cubic meter of reactor per day, and with less than 10–100 mole $H_2O$ per mole $CO_2$.
2. The APS should be targeted at large-scale application with $CO_2$ fixation capability in excess of $10^5$ t yr$^{-1}$.
3. The cyanobacterial photo-bioreactor system (PRS) could provide a good starting point.
4. A cyanobacterial PRS project is under progress at CSIRO and involves experiments and a mathematical model.
5. The energy for $CO_2$ separation from stack gas is about 6–9% of photosynthesis energy.

## Acknowledgments

The author would like to thank Dr. Dean Price, ANU, for providing the cyanobacteria and for numerous discussions on cyanobacteria and photo-bioreactors. The author would also like to thank Ian Shepherd for discussions and for editing the manuscript. Thanks are also owed to Dina Nolasco, Andrew Martin, Greg Threlfall, Robin Clarke, Martina Schimanski, and Anna Hardie (CSIRO-MIT) for scientific and technical support.

## References

1 Van Ree, R. Technologies for usage and disposal of $CO_2$, *Energy Conversion and Management*, 1995, *36*, 935–938.
2 Proceedings of the 2nd International Conference on Carbon Dioxide Removal, *Energy Conversion & Management*, 1995, *36 (6–9)*, 375–944.
3 Proceedings of the 3rd International Conference on Carbon Dioxide Removal, *Energy Conversion & Management*, 1997, *38 (Special edition)*, S1–S689.
4 5th International conference on Carbon Dioxide Utilisation, (a) *The science of total environment*, 2001, *277 (1–3)*, 1–31 (b) Applied *Organometallic Chemistry*, 2001, *15 (2)*, 87–150.
5 Davidson, S. Light Factories, *Ecos (a CSIRO publication)*, Oct–Dec, 2003, pp. 10–12.
6 Hall, D.O. and Rao, K.K., *Photosynthesis*, Cambridge University Press, UK, 1999, a) p. 2, b) p. 67.
7 Martin, J.N., *System Engineering Guidebook*, CRC Press, 2000, p. 13.
8 Michiki, H., Biological $CO_2$ Fixation and Utilization Project, *Energy Convers Mgmt*, 1995, *36 (6–9)*, 701–705.
9 Sorenson, B. *Renewable Energy*, Academic Press, 1979, p. 316.
10 *Plants in Action*, Atwell, B., Kriedemann, P., Turnbull, C., Eds, Macmillan Education Australia Pty Ltd, Australia, 1999, (a) p. 474, (b) p. 212.

11 Astarita, G., Savage, D., Bisio, A. *Gas treating with chemical solvents*, Wiley and Sons, New York, 1982, pp. 7–9, 201–244.
12 Leci, C. L., 1997, *Energy Convers Mgmt*, 1997, *38*, *Suppl*, S45–S50.
13 Price, G. D., Sultemeyer, D., Klughammer, B., Ludwig, M., and Badger, M. The functioning of $CO_2$ concentration mechanism in several cyanobacterial strains, *Can J Botany*, 1998, *76*, 973–1002.
14 Richmond, A. Principles for attaining maximal microalgal productivity in photo-bioreactors – an Overview, *Hydrobiologica*, 2004, *512*, 33–37.
15 Pulz, O. Photo-bioreactors: Production systems for phototrophic micro-organisms, *Applied Microbiology and biotechnology*, 2001, *57*, 287–293.
16 Ogbonna, J. and Tanaka, H. Light requirement and photosynthetic cell cultivation – development of process for efficient light utilization in photo-bioreactors, *J of Applied Phycology*, 2000, *12*, 207–218.
17 Desai, D. K. and Price, G. D. $CO_2$ uptake in a cyanobacterial photosynthesis reactor, *Proceedings 9th APCChE Congress and Chemeca*, 2002, *Christchurch, NZ*, Poster paper no 204.

# 17
# Greenhouse Gas Technologies: A Pathway to Decreasing Carbon Intensity

*Peter J. Cook*

## 17.1
## Introduction

The concentration of carbon dioxide in the atmosphere has risen from 280 parts per million to 370 parts per million over the last 150 years, mainly from increased use of fossil fuels, particularly for electricity generation and transport. However, a rapid move to meeting all energy needs through alternative/renewable energy sources would be very costly to consumers and damaging to the economy and at the present time is technically impractical. What then are the options for maintaining the benefits of access to low-cost energy derived from fossil fuels, while at the same time significantly decreasing $CO_2$ emissions to the atmosphere? In particular, what are the prospects for the use of greenhouse gas technologies to decrease $CO_2$ emissions?

The most promising technologies for significantly decreasing emissions from large-scale stationary sources of $CO_2$ (coal-fired power stations, cement plants, gas-processing facilities, etc.) involve separating and capturing the $CO_2$ and compressing and then storing that $CO_2$ in geological or other locations where it will not leak back into the atmosphere. How effective are these technologies likely to be for the production of low-emission energy from fossil fuels, and how will they impact on energy costs?

## 17.2
## $CO_2$ Capture

Technologies for capturing $CO_2$ from emission streams have been used for many years to produce a pure stream of $CO_2$ from natural or industrial $CO_2$ emissions for use in the food-processing and chemical industries. The gas industry routinely separates $CO_2$ from natural gas before natural gas is then transported to market by pipeline. Methods currently used for $CO_2$ separation include:

*Artificial Photosynthesis: From Basic Biology to Industrial Application*
Edited by Anthony F. Collings and Christa Critchley
Copyright © 2005 WILEY-VCH Verlag GmbH & Co. KGaA, Weinheim
ISBN: 3-527-31090-8

- physical and chemical solvents, particularly monoethanolamine (MEA),
- various types of membranes,
- adsorption onto zeolites and other solids, and
- cryogenic separation.

These methods can be applied to a range of industrial processes. However, their use for separating out $CO_2$ from high-volume, low-$CO_2$-concentration, low-pressure flue gases, such as those generated by conventional pulverized coal-fired power stations, is much more problematical. The high capital cost of installing the huge post-combustion separation systems needed to process massive volumes of flue gases is a major impediment at the present time to post-combustion capture of $CO_2$. The second problem is the large amount of extra energy (30–40% extra) used to release the $CO_2$ from solvents or from solid adsorbents after separation. Additionally, membranes and other methods have yet to be adequately scaled up to the level where they can be used for capture of $CO_2$ at the scale required for large power stations.

Therefore, there are some major technical and cost challenges to be addressed before retrofit (or new build) of post-combustion capture systems becomes an effective mitigation option. The present cost of post-combustion capture of $CO_2$ is commonly quoted in the range of US$ 30–60 (US$ 30–50 as a retrofit option according to the U.S. Department of Energy) per tonne of $CO_2$ or higher (Dave et al., 2000). Given that this would probably more than double the wholesale price of electricity, does this mean that post-combustion removal of $CO_2$ is a lost cause? No, it does not! Capture technologies will undoubtedly improve in the coming years, just as technologies have greatly improved over the past 30 years for the removal of SOx and NOx from flue gases, and costs will come down.

One key to achieving lower capture costs lies in the production of a more concentrated stream of $CO_2$ (the average PF power station has only 10–14% $CO_2$ in the flue gases), thereby decreasing capture costs considerably and perhaps eliminating them completely. The options for increasing the $CO_2$ concentration (and pressure) in flue gases are pre-combustion capture of $CO_2$ and oxyfuel combustion.

Pre-combustion capture, together with integrated gasification combined cycle (IGCC) power generation, involves combustion of coal or gas in air (or oxygen) to produce CO plus $H_2$, followed by a shift conversion to produce $CO_2$ plus more $H_2$. The $H_2$ is then used in a gas turbine driving the generator, and the stream of pure $CO_2$ is available for storage or use. While IGCC plants are becoming more common, few currently incorporate $CO_2$ capture. Is "new-build" IGCC to produce a pure stream of $CO_2$ likely to be a cheaper option than retrofit of a power station coupled with post-combustion capture? Figures currently available suggest that the answer is yes, even allowing for the initial capital costs of new-build IGCC.

Oxyfuel combustion relies on the relatively simple principle of burning coal in an atmosphere of oxygen rather than air to produce a pure stream of $CO_2$. While the principle is simple, there are major issues to be overcome, including the very high combustion temperatures and the cost of producing the pure stream of $O_2$. Oxyfuel combustion for power generation is an emerging option, though it has yet to confirm its operational and commercial viability. Much the same, the oxycombustion technique is used in steel making, and consequently there may be no insurmountable technical barriers to $CO_2$ storage linked to oxyfuel power generation in the future. Currently, IGCC is commonly regarded as the cheapest option for producing a pure stream of $CO_2$. Thus, there are technical options available either for producing a pure stream of $CO_2$ or for separating and capturing $CO_2$ from low-$CO_2$ flue gases, but the challenge over the next few years will be to bring down costs. It is reasonable to expect that this will happen, in the way that it has also happened in other areas of emerging technology, including removal of SOx and NOx.

If the main pressure on power companies is to decrease electricity costs, it is unrealistic to expect them to voluntarily retrofit post-combustion $CO_2$ capture or move to IGCC or oxyfuels. However, if the main pressure in the future is to reduce $CO_2$ emissions, then post-combustion retrofit will receive increased attention, given that many countries have made massive investments in conventional thermal power stations, though more advanced power-generation technologies linked to $CO_2$ storage will be preferable in many (perhaps most) cases.

## 17.3
## Storing $CO_2$

Once a stream of high-purity $CO_2$ is captured from a major stationary source such as a power station or an industrial process, is it possible to lock up (store) that $CO_2$ for the long term? The answer is yes, and there are several storage options being considered for $CO_2$ at the present time, including ocean storage, mineral storage, and geological storage.

Ocean storage of $CO_2$ has been investigated as an option for the past 15 years or so, with research by Japan, the U.S., and other countries. There are two main options proposed: (1) dispersal of $CO_2$ as droplets at intermediate water depths of around 500–1000 m and (2) disposal at abyssal depths (5000 m or more) as liquid $CO_2$. Given that the ocean is an enormous sink for $CO_2$ and that the system is quite strongly buffered, the injected $CO_2$ would probably have very little effect on the chemistry of the ocean as a whole, but it would result in a measurable drop in the pH of seawater in the immediate vicinity of the injection site and would therefore impact on marine organisms. The ocean is an open system, and it would be difficult, if not impossible, to monitor the distribution of the stored carbon in order to confirm residence times of $CO_2$. Also, the impact of elevated levels of $CO_2$ on the biota is poorly known and difficult to monitor. The potential application of the London Dumping Convention to ocean storage of $CO_2$ also raises legal uncertainties. For all these reasons, ocean storage is most unlikely to be a preferred $CO_2$ storage option in the foreseeable future.

$CO_2$ is used on a modest scale in a number of industrial processes and by the food industry. Much of that $CO_2$ returns to the atmosphere or biosphere after a fairly short time. This, and the very small quantity of $CO_2$ involved, means that most uses of $CO_2$ are unlikely to have a significant impact on $CO_2$ fluxes. Construction materials may be an exception, and with this in mind, new cements and new types of carbonate bricks are being considered, but we must look to large-scale storage if we are to have a major impact on $CO_2$ emissions and to stabilize, and in the longer term reduce, the levels of $CO_2$ emissions to the atmosphere.

Mineral storage – the locking up of $CO_2$ in stable mineral carbonates that can be safely left in situ at or near the earth's surface – has been suggested as a storage option. In some cases it might also be possible to use mineral storage to ameliorate localized environmental problems such as alkaline groundwater. But overall, reaction rates are slow and the quantities of $CO_2$ stored are likely to be very modest. The Zero Emission Coal Alliance (ZECA) has undertaken a program of research into a concept combining a gasification/power plant to produce hydrogen and electricity plus $CO_2$. The $CO_2$ is then stored via reaction with magnesium silicates (derived from serpentinites and other ultrabasics) to produce stable magnesium carbonates and silica. Given that the large amounts of magnesium silicates required for the process need a major mining and crushing operation, the economic prospects for this storage pathway are somewhat doubtful.

The most promising large-scale storage option is not mineral storage but geological storage. For the past 30 years, the oil industry has been injecting up to 30 million tonnes of $CO_2$ (derived mainly from natural $CO_2$ accumulations in Colorado) per year into the subsurface of West Texas for enhanced oil recovery (EOR). The Weyburn Project in Canada is a recently initiated example of an EOR project that uses $CO_2$ derived as a byproduct of a coal gasification plant in Beulah, North Dakota. Acid gas (including $CO_2$) re-injection in the Alberta Basin has been underway for a number of years, and 31 injection sites are currently in operation that also involve associated storage of significant quantities of $CO_2$ in the subsurface.

Geological storage will build on and extend some of the principles underlying EOR and acid gas storage, but it will need to be on a much larger scale and in a wider range of geological settings than has been the case to date.

Because of the added benefit of a valuable byproduct, the use (and storage) of $CO_2$ for enhanced petroleum recovery (EPR) has received a high level of attention. However, most oils are not suitable for EOR, and in fact, much of the $CO_2$ used for EOR is not "stored" but reused. Also, the global opportunities for $CO_2$ storage with EOR are limited. Enhanced gas recovery (EGR) is also an option, but the potential problem of "fingering" of the $CO_2$ into the natural gas has limited its use to date. Enhanced coal bed methane (ECBM) recovery is seen as an opportunity for sequestering $CO_2$ in unmineable coal seams and obtaining improved production of coal bed methane as a valuable byproduct. A demonstration ECBM project in the San Juan Basin produced a positive though modest enhancement of the rate of methane recovery. An ECBM project (RECOPOL) has recently been undertaken by the IEA Greenhouse Gas Program in Poland, and the results of that program are awaited with interest.

The true potential of $CO_2$-ECBM is difficult to gauge at this time. Coal bed methane is an important energy source in the U.S., but it has not been significant in most other countries to date. One reason for this is that many countries lack existing infrastructure and expertise. Also, compared to U.S. coals, those in Australia, Europe, and many other areas often have low permeabilities; this in turn will drive up capital and operational costs (because of the need to drill a large number of $CO_2$ injection wells), potentially making it uneconomic. An additional concern is that any increase in the leakage of methane (a very powerful greenhouse gas) as a result of ECBM could counter the potential benefits of $CO_2$ storage. $CO_2$ storage linked to EPR is an intriguing option, but for the present the extent to which that potential will be realizable is uncertain.

By far the greatest potential for geological storage of $CO_2$ involves injection of compressed $CO_2$ into the subsurface, down to a depth of 6–800 m. The pure stream of $CO_2$ is compressed to a super-critical state; it is then very much denser than gaseous $CO_2$, thereby minimizing problems posed by the large amounts of $CO_2$ that will need to be stored if this method is to have a major impact on current emission levels. At a depth of 6–800 m, much of the $CO_2$ will initially remain in a super-critical state, while some of it may react with the bedrock to form carbonate minerals and some will go into solution. Over time, more will go into solution, but provided that the injection site is carefully chosen, the $CO_2$ will remain stored for very long periods of time – geological time.

An obvious site for geological storage is depleted oil and particularly gas reservoirs. In the U.S., it is estimated (by the U.S. Department of Energy) that the storage capacity of depleted gas reservoirs is on the order of 80–100 Gt of $CO_2$, or enough to store all of the U.S. emissions of $CO_2$ from major stationary sources for 50 years or more. The use of deep unmineable coal seams for geological storage (without ECBM) is also an option for storage, but the low permeability of many deep coals is likely to be an issue. Storage of $CO_2$ in deep natural caverns is possible, but the total potential is likely to be very limited.

By far the greatest potential for storing very large amounts of $CO_2$ exists in deep saline water-saturated reservoir rocks, particularly sandstones, with the $CO_2$ stored as a result of hydrodynamic trapping. One major project to store $CO_2$ in deep saline aquifers is already being conducted by the Norwegian company Statoil, in the sediments of the North Sea Basin. There, a million tonnes of $CO_2$ per year are being injected into the Utsira Formation at a depth of around 1000 m below the sea floor. Much of the initial impetus for this storage project was provided by the potential imposition of a carbon tax by the Norwegian government, and in the future, a carbon tax undoubtedly will be a driver for other projects. But many companies now wish to do (and to be seen doing) "the right thing." For example, BP has recently announced a major $CO_2$ injection project in Algeria, an Annex B country that is not affected by the Kyoto Protocol and that has no plans to impose a carbon tax.

A comprehensive regional analysis of the storage potential of saline reservoirs has been undertaken in Australia as part of the GEODISC project. This study has indicated a $CO_2$ storage potential for Australia that is adequate to store $CO_2$

emissions for many hundreds of years at the current rates of emission. It is hoped that a major demonstration project will be underway in Australia in 2005–2006 to start realizing some of that potential. The GESTCO study of Western Europe also suggests a large $CO_2$ storage potential for the North Sea Basin. The storage potential of areas such as the Alberta Basin in western Canada and parts of the mid-continent U.S. also appears to be very large. Because of its complex geology, Japan represents something of a challenge for geological storage of $CO_2$, but Japanese organizations hope to embark on a pilot project there in the near future. In most other parts of the world, there has been no systematic study of geological storage potential, but it is likely to be very large indeed in many sedimentary basins.

What is the cost of geological storage of $CO_2$? A number of projects are presently underway to better establish costs, but they are certainly significantly less than the costs of $CO_2$ capture. Commonly quoted storage costs are on the order of US$10 or less per tonne of $CO_2$, suggesting that, excluding the costs of $CO_2$ capture, the costs of geological storage are likely to be cost-competitive with other sequestration options. However, storage costs are highly dependant on the distance between a major stationary source of $CO_2$ and the geological storage site: the longer the $CO_2$ pipeline, the greater the cost of storage, because of both the higher capital costs of the pipeline and the higher recurrent costs of compression. In other words, storage costs are project-dependent. $CO_2$ source-sink matching in Australia indicates a considerable range in project costs (Allinson and Nguyen 2002; Allinson et al. 2003 a, b).

When the cost of capture also has to be factored into project costs, such as in the case of post-combustion capture linked to a conventional coal-fired power station, the total costs rise considerably. However, when an existing power station is close to a good geological storage option, the prospects for $CO_2$ storage linked to retrofit of capture systems to an existing power station cannot be dismissed.

## 17.4
**Australian Initiatives: Capture and Storage Technologies**

For the past four years the Australian Petroleum Cooperative Research Centre (APCRC) has been undertaking research into geological storage of $CO_2$ through the GEODISC program (Cook et al. 2000; Rigg et al. 2001). This program has been highly successful in determining the overall potential of the Australian continent for geological storage of $CO_2$.

At a more fundamental level, the program has developed technologies for the assessment and risking of potential storage sites, including modeling the subsurface behavior of $CO_2$ (Bradshaw et al. 2002). This preliminary work has demonstrated that geological storage is a technology that could have a major impact on Australia's $CO_2$ emission profile in the future, but that more work must to be done to convert that technical potential into economic and environmental benefit.

In December 2002 the Australian government announced federal funding for a new Cooperative Research Centre for Greenhouse Gas Technologies (CO2CRC). Over the next seven years the Centre will undertake a major program of research into $CO_2$ capture and storage.

Together with collaborating overseas organizations, CO2CRC has the aim of significantly reducing capture costs over the next seven years. On the storage side, it intends to establish the scientific basis for an acceptable monitoring and verification regime for Australian geological storage. Additionally, CO2CRC intends to carry out one or more major demonstration projects in Australia.

A very important part of the approach of CO2CRC will be to work in close collaboration with major international programs, recognizing the scale of the technological challenges to be overcome. CO2CRC believes that by working with a range of major research organizations and industries, $CO_2$ capture and storage will become an acceptable and cost-effective technology with the potential to significantly decrease the rate of $CO_2$ emissions to the atmosphere from major stationary sources.

## 17.5 Conclusions

There are obviously many issues still to be resolved regarding capture and storage of $CO_2$ in terms of costs, preferred technologies, monitoring, etc., but the technologies offer great promise. Post-combustion capture costs are high but will decrease. Advanced power-generation technologies such as IGCC and oxyfuel combustion offer a potentially cheaper path to pure $CO_2$ streams, thereby avoiding capture costs. $CO_2$ storage technologies are now reasonably well known. Ocean storage appears to be an unlikely option, while mineral storage may be useful but will make only a modest contribution. Enhanced petroleum recovery has some attractive features, but it is likely to have only a modest impact on $CO_2$ storage. Storage in depleted gas fields may be useful in the short to medium term in some areas. Storage in saline reservoir rocks is by far the most widely occurring storage option with the greatest potential – probably enough to store a significant proportion of the world's stationary $CO_2$ emissions for very many years to come.

There are obviously uncertainties about what the future carbon-trading regime will be or what steps will be taken to limit $CO_2$ emissions post-Kyoto. Some power companies may argue that it is best to do nothing until these uncertainties are clarified. But it would be rash, if not financially irresponsible, to plan for any new thermal power stations without making some provision for installation of future capture technologies. It would also be foolish to site a new power station without doing a comprehensive survey of geological storage options prior to deciding on the siting. In the future there will be community and governmental expectations of decreased carbon intensity, and these will impact on the fossil fuel and power industries. Capture and storage technologies will help to decrease the future financial risk to these industries.

Capture and storage of $CO_2$ is not a "silver bullet" that will overcome all our greenhouse gas problems, but it is one of the most promising options that we now have for decreasing $CO_2$ emissions. Greater fuel efficiency, increased use of low carbon intensity fuels, enhancement of natural carbon sinks, and greater use of renewable energy will be important to the total range of measures necessary to decrease our total carbon intensity. But among all the options, only $CO_2$ capture and storage technologies have the potential for making deep cuts in $CO_2$ emissions in the short to medium term while enabling us to continue benefiting from access to widely available, low-cost fossil fuels.

## References

Allinson, W. G., and Nguyen, D. (2002 a) $CO_2$ Geological Storage Economics, International Conference on Greenhouse Gas Control Technologies, 6, 2002, Kyoto, Japan. GHGT-6, 3.

Allinson, W. G. and Nguyen, D. (2002 b) The economics of $CO_2$ Capture and Geological Storage. SPE Asia Pacific Oil and Gas Conference and Exhibition held in Melbourne, Australia, 8–10 October 2002. SPE 77810, 5.

Allinson, W. G., Nguyen, D. N. and Bradshaw, J. (2003) The Economics of Geological Storage of $CO_2$ in Australia. APPEA Journal, 43.

Bradshaw, J., Bradshaw, B. E., Alinson, G., Rigg, A. J., Nguyen, V, and Spencer, L. J. (2002) The potential for Geological Sequestration of $CO_2$ in Australia – Preliminary technical and commercial findings. APPEA Journal, 42 (1), 25–46.

Cook, P. J., Rigg, A., and Bradshaw, J. (2000) Putting it back where it came from: Is geological disposal of carbon dioxide an option for Australia. APPEA Journal, 40, 634–666.

Dave, N. C., Duffy, G. J., Edwards, J. H., and Lowe, A. (2000) Economic Evaluation of Capture and Sequestration of $CO_2$ from Australian Black Coal-fired Power Stations. International Conference on Greenhouse Gas Control Technologies, 5, 2000, Cairns, Qld., GHGT-5, 68/69, 173–178.

Rigg, A., Allinson, J., Bradshaw, J., Ennis-King, J., Gibson-Poole, C., Hillis, R. R., Lang, S. C., and Streit, J. E. (2001) The search for sites for geological sequestration of $CO_2$ in Australia: a progress report on GEODISC APPEA Journal, 41, 711–736.

# Subject Index

## a

acetyl-CoA 216
agriculture 17
– conventional 17
– dry 17
– waterless 17
Alberta Basin 306
alcohol 32
alcohol dehydrogenase 217
algae, biotechnology 229, 233
anaerobiosis 215
antenna-reaction center complexes 199ff.
ATP
– synthase 29
– synthesis 204ff.
– – mechanism 189, 204
ATPase 16
Australian Petroleum Cooperative Research Centre 306

## b

bacterial reaction centre 78, 131
biodiesel 287
bioenergetic converter 29
bioenergetics 189
bioluminescence assay 205
bio-mimetic 22
biotechnology, green algae 229
BRC see bacterial reaction centre
British Petroleum (BP) 305

## c

calcium, transport 206ff.
Calvin cycle 14
carbohydrate 287
carbon
– end products 283ff.
– fixation 32
– fixation rate 294
– sequestration 13
– tax 305
carbon dioxide 5, 7, 14, 302
– assimilation 286
– – rate 255, 286
– capture 301, 306
– emission 301
– mitigation 302
– ocean 303
– – geological 304ff.
– – mineral 304
– separation 294
– storage 303ff.
carboxylation 245
– cycle 265
carotenoid 192
CD see circular dichroism
cell 9
– culture 229
chaperone 253
charge recombination 196
charge separation 109ff.
chemical energy 187, 283
chemistry, computational 266
*Chlamydomonas reinhardtii* 214, 221, 230, 235
*Chlorella fusca* 221
chlorophyll 22, 230ff., 234
– antenna 13
– – size 230ff., 234
chlorophyll a oxygenase 233
chlorophyll b-less mutant 233
chloroplast 17
– genome 243
– metabolism 216
– plant 255, 284
– transcription 258
chromophore 21
– linkage 147ff.
– scaffolds 24

## Subject Index

- spacing   147
- zinc porphyrin   190, 198
circular dichroism   94
Clark electrode   219
climate change   5
- global   10
*Clostridium pasteurianum*   222
CO$_2$ *see* carbon dioxide
coal   3 ff.
combustion, oxyfuel   302
computational programs   267 ff.
Cooperative Research Centre   8
- Australien Petroleum   306
- Greenhouse Gas Technologies   306
cosmetic products   288
current, biolelectric   14
cyanobacteria, carbon concentrating mechanism   296
cytochrome b/c1 complex   189
cytochrome b6/f complex   15, 215

### d

dark reactions   13
detergent   204
Dexter mechanism   192
DNA shuffling
- insertional mutagenesis   230, 235
- shuffling   249
driving force   195
DSSC *see* dye-sensitized solar cell
dyad, porphyrin fullerene   196

### e

ecology, riverine   10
economic growth   9
efficiency, photosynthetic   293
electricity   10
- generation   304
electrochemical energy   187, 194
electrolysis   13
electron
- acceptor   134 ff.
- donor   136 ff.
- transport   220
electronic coupling   195
emission   10
energy
- chemical   187, 283
- efficiency   170, 292, 294
- - thermodynamic   6, 8, 13
- electrochemical   187, 194
- needs   9
- nuclear   7

- quenching   79
- solar   6, 8, 13
- thermodynamic   13
- transfer, singlet   192
- use   3 ff.
engineering   292
- molecular   221
environment, problems   9
enzyme bed reactor   28
enzyme reactor   31
*Escherichia coli*   32
ethanol   217
evolution, artificial   249

### f

FCCP   205
feedstock   17
Fe-S cluster   135 ff.
- structure   135
fiber   17
*Flaveria bidentis*   250
fluorescence, excitation spectrum   192
fluorescent dye   203
Foerster mechanism   190
Foerster theory   67 ff.
- conventional   69
- generalized   69 ff.
food   13, 17, 287
- production   13
Franck-Condon factor   195
fuel   4, 187
- coal   3 f., 7
- diesel   287
- energy   3, 5
- fossil   5
- green   287, 291
- liquid   17
- natural gas   3
- oil   3
- wood   4

### g

*Galdieria sulfuraria*   255
gas turbines   172
gene deletion   231
genetic control   232
GEODISC project   305 f.
geo-sequestration   7
GESTCO study   305
global warming   8, 291
Graetzel cell   38
greenhouse gas   5
- technologies   301 ff.

## h

hexad 200
hydrogen
– economy 25
– production 19, 214, 229, 304
– photoproduction 213 ff.
hydrogenase
– algal 213 ff., 229
– anaerobic induction 224
– DNA sequences 223
– oxygen inhibition 224
– oxygen tolerance 224
– protein 215
– Western blot 223
– X-ray structure 225

## i

industry
– algal biotechnology 233
– gas 301 ff.
isoprene 32, 285, 288
isoprenoid, synthesis 284, 287

## k

Kyoto protocol 1 ff.

## l

land, use 293
leaves 14
light
– penetration 231
– reactions 13
– utilization efficiency 229
light harvesting
– complex 231
– regulation 79 ff.
lipid bilayer membrane 201
liposome vesicle 201

## m

Mehler reaction 218
melanin 50 ff., 61 ff.
membrane 13
– technology 20
metabolism, aerobic 215
microorganism 229
mitochondrion, metabolism 216
Mn cluster 20
molecular genetics 229
molecular rotary motor 204
monoterpene 288

## n

NADPH 13, 29 ff.
nanomorphology 42 ff.
nanotechnology 20
– natural 187
NMR see nuclear magnetic resonance
North Sea Basin 305 f.
nuclear magnetic resonance 133
nucleus, genome 243, 250

## o

operon 255
oxidative phosphorylation 216
oxygenase 17
oxygen evolution, rate 214
oxygen-evolving complex 15
oxygenation 245, 263
– cycle 265

## p

peer review 6
pentad 197
peptide see protein 138
*Phaeodactylum tricornutum* 255
phenome 244
photo-bioreactor
– design 219, 294 ff.
photoelectric generator 22
photoinhibition 232
photon 9
– dissipation 232
– harvesting 61 ff.
photoprotection 78
photorespiration 245, 263
photosynthesis
– artificial 5, 6, 17 ff., 187, 190 ff.
– – antenna systems 190 ff.
– – reaction center 194 ff.
– bacterial 109 ff.
– – apparatus 188
– light saturation curve 234
– natural 13 ff., 188 ff.
– – limitations 292
– – productivity 294, 236
– – reaction center 127 ff.
– oxygenic 13
photosystem I 14, 75 ff.
– capacity 218
photosystem II 14, 87 ff.
– capacity 218
– complex 16
– P680 14 f.
– – special pair 109 ff.

– photochemical activity 220
– reaction centre
– – cyanobacteria 94 ff.
– – plant 94 ff.
– X-ray crystal structure 88
photovoltaics
– in silico 169
– organic 21
– thin film 170
phthalocyanine 192 f.
physiology
– algal 213
– plant 244 f.
pigment
– antenna
– – bacteriochlorophyl 50 ff.
– – chlorophyll 230
– – clorophyll b 234
– – euromelanin 50 ff.
– – melanin 50 ff.
– mutations 22, 230, 234
– photosynthetic 22
– reaction centre
– – artificial 194 ff.
– – PSI 131
– – PSII 131
– – purple bacterial 131
– – synthetic 130 ff.
plant 17
– carbon fixation 32
plastics 287
plastocyanin, structure 135
plastome 255
plastoquinone 16, 220
political issues 3 ff.
polyhydroxyalkanoate 287
polyhydroxybutyrate 287
polymer 17
polynorbonane
– dyads 151 ff.
– scaffolds 147 ff.
porphyrin 22
protein 15
– design 140
– natural 132, 214, 263
– synthetic 132
proton
– channel 266
– pumping 201
– – transmembrane 202
proton motive force 201
PSI see Photosystem I
PSII see Photosystem II

## q
qE see energy quenching
quantum mechanics 266
quenching, feedback re-excitation 79 ff.
quinine 134 ff.
– shuttle 202

## r
reaction center, antenna-reaction center complexes 199 ff.
Research Council 7
Resource Assessment Commission 8
respiration 214
*Rhodobacter sphaeroides* 110
*Rhodopseudomonas viridis* 133
*Rhodospirillum rubrum* 248, 254
rubber 288
rubisco
– activase 243, 247, 255
– active site 268, 280
– activity regulation 244, 247
– assembly 244
– chemistry 264
– – carboxylation 273 ff., 280
– – enolization 271 ff.
– – hydration 275 ff., 280
– directed molecular evolution 249, 259
– Form-I 248
– Form-II 248
– gene expression 258
– genetic manipulation 243 ff., 250 ff., 257
– hypothetical 246
– kinetic efficiency 247
– mechanism 281
– – catalytic 244, 265
– – chemical 263 ff.
– – $CO_2$ addition 278 ff.
– mutagenesis 256, 265
– "perfect" 246 f.
– protein 20, 215
– red algal 246
– spinach 268
– – X-ray structure 269
– subunits 243, 255
– – assembly 258 ff.
– – hybrids 256
– – mutation 257
– synthesis 244
– tobacco 243 ff.

## s

*Scenedesmus obliquus* 221
semiconductor 21
sesquiterpene 288
silicon
– thin film 173 ff.
– third generation 178 ff.
– wafer 169, 171 ff.
simulations
– computational 263 ff.
– spectral 116 ff.
social issues 3 ff.
solar cell 173 ff.
special pair 109 ff.
spectroscopy
– absorbance-difference kinetic 229
– circular dichroism 94
– optical 90 ff.
– vibrational infrared 110
starch 214, 284
Stark effect 125
Statoil 305
sucrose 284
sulfur, deprivation 214
sustainability science 9, 10

## t

technology 6
– military 8
thylakoids 13
translation 253
transpiration 263
tricarboxylic acid cycle 216
triose phosphate 284

## u

U. S. Department of Energy 302

## v

valinomycin 204
water
– requirement 293
– sources 13
– usage 263

## w

Western blot 234
Weyburn Project 304
wind generators 172

## z

Zero Emission Coal Alliance 304
Z-scheme 14